BIOCHEMISTRY

BIOCHEMISTRY

Noer Kasanah
Eli Bridge
Rachel Powers

Kruger Brentt
Publishers
2025

Kruger Brentt Publishers UK. LTD.
Company Number 9728962

Regd. Office: 68 St Margarets Road, Edgware, Middlesex HA8 9UU

ISBN: 978-1-78715-448-3

Disclaimer:

Every possible effort has been made to ensure that the information contained in this book is accurate at the time of going to press, and the publisher and author cannot accept responsibility for any errors or omissions, however caused. No responsibility for loss or damage occasioned to any person acting, or refraining from action, as a result of the material in this publication can be accepted by the editor, the publisher or the author. The Publisher is not associated with any product or vendor mentioned in the book. The contents of this work are intended to further general scientific research, understanding and discussion only. Readers should consult with a specialist where appropriate.

Every effort has been made to trace the owners of copyright material used in this book, if any. The author and the publisher will be grateful for any omission brought to their notice for acknowledgement in the future editions of the book.

For information on all our publications visit our website at http://krugerbrentt.com/

PREFACE

Biochemistry is a pivotal field of science that explores the intricate chemical processes and substances that occur within living organisms. It serves as a bridge between biology and chemistry, focusing on the molecular mechanisms that underlie various biological functions.

At its core, biochemistry investigates the structure, function, and interactions of biomolecules, including proteins, nucleic acids, carbohydrates, and lipids. These biomolecules play essential roles in cellular processes, serving as enzymes, structural components, signaling molecules, and genetic material. For instance, proteins are critical for catalyzing biochemical reactions as enzymes, while nucleic acids, such as DNA and RNA, carry genetic information necessary for the synthesis of proteins and the regulation of cellular activities. The study of metabolism is also a crucial aspect of biochemistry, as it encompasses the chemical reactions that convert nutrients into energy and the building blocks needed for growth and maintenance.

Metabolic pathways, which involve a series of enzymatic reactions, enable cells to efficiently utilize energy sources and synthesize vital biomolecules. Biochemistry has far-reaching implications in various fields, including medicine, agriculture, and biotechnology. Understanding the biochemical basis of diseases has led to significant advancements in medical research and therapeutic interventions. For example, insights into the biochemical pathways involved in cancer, diabetes, and cardiovascular diseases have paved the way for the development of targeted treatments and personalized medicine approaches.

Moreover, biochemistry plays a critical role in the development of diagnostic tools, such as enzyme-linked immunosorbent assays (ELISAs) and polymerase chain reactions (PCR), which enable the detection of specific biomolecules and pathogens. In agriculture, biochemistry contributes to the development of genetically modified organisms (GMOs) that exhibit enhanced traits, such as resistance to pests and improved nutritional content. Additionally, biochemistry underpins advancements in biotechnology, leading to the production of recombinant proteins, vaccines, and biofuels. As the field continues to evolve, techniques such as genomics, proteomics, and metabolomics are increasingly used to explore complex biological systems and understand the molecular basis of life.

The interdisciplinary nature of biochemistry, combining principles from chemistry, biology, and physics, makes it an essential area of study for addressing some of the most pressing challenges in health, sustainability, and technology in the 21st century.

Noer Kasanah

Eli Bridge

Rachel Powers

CONTENTS

Preface v

1. Biochemistry in Space and Time 1

1.1 The Spatial Dimension: Biochemical Gradients and Microenvironments 1

1.2 Temporal Biochemistry: Circadian Rhythms and Cellular Timers 1

1.3 Biochemical Adaptations Across Evolutionary Time Scales 2

1.4 Biochemistry in Extreme Environments: Life Beyond Earth 2

1.5 The Role of Technology in Mapping Biochemistry in Space and Time 3

1.6 Biochemistry and the Future: Emerging Frontiers in Space-Time Research 3

1.7 Conclusion 4

2. Protein Composition and Structure 5

2.1 The Building Blocks: Amino Acids and Peptide Bonds 5

2.2 Primary Structure: The Sequence of Amino Acids 5

2.3 Secondary Structure: Alpha Helices and Beta Sheets 6

2.4 Tertiary Structure: The Three-Dimensional Folding 6

2.5 Quaternary Structure: Multi-Subunit Complexes 6

2.6 The Relationship between Structure and Function 7

2.7 Protein Misfolding and Disease 7

2.8 Conclusion 8

3. Binding and Molecular Recognition 9

3.1 Types of Molecular Interactions: The Basis of Binding 9

3.2 Binding Affinity and Specificity: Determining Strength and Selectivity 10

3.3 Enzymes and Molecular Recognition: Catalysis and Substrate Specificity 10

3.4 Receptors and Ligand Binding: Signal Transduction Pathways 11

3.5 Applications in Therapeutic Development: Targeting Molecular Recognition 12

3.6 Conclusion 12

4. **Protein Methods** **13**

4.1 Protein Purification: Isolating Proteins for Analysis 13

4.2 Electrophoresis: Analyzing Protein Size and Charge 14

4.3 Spectroscopic Methods: Assessing Protein Structure and Dynamics 14

4.4 X-ray Crystallography: Determining Atomic-Resolution Structures 15

4.5 Nuclear Magnetic Resonance (NMR) Spectroscopy: Studying Proteins in Solution 15

4.6 Cryo-Electron Microscopy (Cryo-EM): Imaging Large Protein Complexes 15

4.7 Mass Spectrometry: Analyzing Protein Composition and Modifications 16

4.8 Conclusion 16

5. **Enzymes: Core Concepts and Kinetics** **17**

5.1 Enzyme Structure and Function: The Basis of Catalysis 17

5.2 Mechanisms of Enzyme Catalysis: Lowering Activation Energy 18

5.3 Enzyme Kinetics: Understanding Reaction Rates 18

5.4 The Michaelis-Menten Model: Describing Enzyme Behavior 19

5.5 Factors Affecting Enzyme Activity 19

5.6 Enzyme Inhibition and Regulation: Controlling Biological Pathways 20

5.7 Conclusion 21

6. **Enzyme Catalytic Strategies** **23**

6.1 Proximity and Orientation Effects: Bringing Substrates Together 23

6.2 Acid-Base Catalysis: Donating and Accepting Protons 23

6.3 Covalent Catalysis: Forming Transient Bonds with Substrates 24

6.4 Metal Ion Catalysis: Utilizing Metal Ions as Catalytic Agents 25

6.5 Transition State Stabilization: Lowering Activation Energy 25

6.6 Conclusion 26

7. **Enzyme Regulatory Strategies** **27**

7.1 Allosteric Regulation: Fine-Tuning Enzyme Activity 27

7.2 Covalent Modification: Reversible Chemical Changes 27

7.3 Proteolytic Activation: Irreversible Enzyme Activation 28

7.4 Feedback Inhibition: Regulating Pathways through End-Product Control 29

7.5 Enzyme Compartmentalization: Spatial Regulation within Cells 29

7.6 Isozymes: Variants for Specialized Regulation 30

7.7 Conclusion 30

8. **DNA, RNA and the Flow of Genetic Information** **31**

8.1 Structure and Function of DNA 31

8.2 The Role of RNA in Genetic Information Transfer 31

8.3 Transcription: The First Step in Gene Expression 32
8.4 Translation: The Synthesis of Proteins 32
8.5 Regulation of Gene Expression 33
8.6 The Importance of Genetic Information Flow 34
8.7 Conclusion 34
9. Nucleic Acid Methods 35
9.1 DNA Extraction: Isolating Genetic Material 35
9.2 Polymerase Chain Reaction (PCR): Amplifying DNA 35
9.3 DNA Sequencing: Deciphering Genetic Information 36
9.4 Hybridization Techniques: Probing Nucleic Acids 36
9.5 CRISPR-Cas9: A Revolutionary Tool for Genome Editing 37
9.6 Quantitative PCR: Measuring Gene Expression 38
9.7 Conclusion 38
10. Exploring Evolution and Bioinformatics 39
10.1 Understanding Evolutionary Biology 39
10.2 The Role of Bioinformatics in Evolutionary Studies 39
10.3 Genomic Sequencing and Comparative Genomics 40
10.4 Phylogenetics: Unraveling Evolutionary Relationships 40
10.5 Evolutionary Genomics: Exploring Adaptation and Speciation 41
10.6 The Intersection of Evolutionary Biology and Conservation 41
10.7 Future Directions in Evolution and Bioinformatics 42
10.8 Conclusion 42
11. Carbohydrates and Glycoproteins 45
11.1 Carbohydrates 45
11.2 Glycoproteins 56
12. Lipids and Biological Embrances 65
12.1 Lipids 65
12.2 Biological Membrances 77
13. Membrane-Channels and Pumps 83
13.1 The Structure of Cell Membranes 83
13.2 Membrane Channels: Facilitating Passive Transport 83
13.3 Membrane Pumps: Active Transport Mechanisms 84
13.4 Regulation of Membrane Channels and Pumps 85
13.5 Physiological Implications of Membrane Transport 86
13.6 Conclusion 86

14. Signal-Transduction Pathways ... 89

14.1 An Introduction 89
14.2 History 90
14.3 Stimuli 92
14.4 Receptors 94
14.5 Second Messengers 113
14.6 Cellular Responses 114
14.7 Major Pathways 115

15. Metabolism: Basic Concepts and Themes ... 117

15.1 An Introduction 117
15.2 History 118
15.3 Key Biochemicals 119
15.4 Catabolism 122
15.5 Energy Transformations 125
15.6 Anabolism 126
15.7 Xenobiotics and Redox Metabolism 130
15.8 Thermodynamics of Living Organisms 131
15.9 Regulation and Control 131
15.10 Evolution 133
15.11 Investigation and Manipulation 134

16. Glycolysis and Gluconeogenesis ... 137

16.1 Glycolysis 137
16.2 Gluconeogenesis 157

17. Pyruvate Dehydrogenase and the Citric Acid Cycle ... 165

17.1 Pyruvate Dehydrogenase 165
17.2 The Citric Acid Cycle 168

18. Oxidative Phosphorylation ... 181

18.1 An Introduction 181
18.2 History 182
18.3 Chemiosmosis 182
18.4 Electron and Proton Transfer Molecules 183
18.5 Eukaryotic Electron Transport Chains 184
18.6 Prokaryotic Electron Transport Chains 190
18.7 ATP Synthase (Complex V) 192

18.8 Oxidative Phosphorylation—Energetics 194
18.9 Reactive Oxygen Species 194
18.10 Oxidative Phosphorylation in Hypoxic/Anoxic Conditions 195
18.11 Inhibitors 195
19. Phototrophy and the Light Reactions of Photosynthesis 197
19.1 Understanding Phototrophy and Photosynthesis 197
19.2 The Light Reactions: Harnessing Light Energy 198
19.3 Importance of the Light Reactions in Photosynthesis 199
19.4 Evolutionary Significance of Phototrophy 200
19.5 Future Directions in Phototrophy Research 200
19.6 Conclusion 200
20. The Calvin-Benson Cycle and the Pentose Phosphate Pathway 201
20.1 The Calvin-Benson Cycle 201
20.2 Light-Dependent Regulation 207
20.3 The Pentose Phosphate Pathway 208
21. Glycogen Metabolism 213
21.1 An Introduction 213
21.2 History 214
21.3 Structure 214
21.4 Functions 216
21.5 Structure Type 218
21.6 Metabolism 218
21.7 Clinical Relevance 219
22. Fatty Acid and Triacylglycerol Metabolism 221
22.1 Fatty Acid 221
22.2 Nomenclature 223
22.3 Triacylglycerol Metabolism 232
23. Protein Turnover and Amino Acid Catabolism 237
23.1 Protein Turnover 237
23.2 Amino Acid Catabolism 238
24. Integration of Energy Metabolism 243
24.1 Overview of Energy Metabolism Pathways 243
24.2 Key Organs in Energy Metabolism 244
24.3 Hormonal Regulation of Energy Metabolism 244
24.4 Metabolic States: Fed, Fasting, and Starvation 245

24.5 Integration of Carbohydrate, Fat, and Protein Metabolism 246
24.6 Pathophysiological Aspects of Energy Metabolism Imbalance 246
24.7 Conclusion 247

25. Biosynthesis of Amino Acids..........249

25.1 Overview of Amino Acid Biosynthesis 249
25.2 Families of Amino Acids Based on Precursors 249
25.3 Regulation of Amino Acid Biosynthesis 252
25.4 Physiological Significance of Amino Acid Biosynthesis 252
25.5. Metabolic Disorders Associated with Amino Acid Biosynthesis 253
25.6 Conclusion 253

26. Nucleotide Biosynthesis..........255

26.1 Understanding the Nucleotide 255
26.2 Overview of Nucleotide Biosynthesis 262
26.3 Purine Biosynthesis 263
26.4 Pyrimidine Biosynthesis 263
26.5 Conversion to Deoxyribonucleotides 264
26.6 Nucleotide Degradation and Recycling 264
26.7 Physiological Roles of Nucleotides 264
26.8 Metabolic Disorders Associated with Nucleotide Biosynthesis 265
26.9 Therapeutic Targeting of Nucleotide Biosynthesis 265
26.10 Conclusion 266

27. Biosynthesis of Membrane Lipids and Steroids..........267

27.1 Overview of Membrane Lipids 267
27.2 Overview of Steroids 270
27.3 Biosynthesis of Phospholipids 284
27.4 Biosynthesis of Glycolipids 285
27.5 Biosynthesis of Cholesterol 286
27.6 Biosynthesis of Steroid Hormones 287
27.7 Physiological Roles of Membrane Lipids and Steroids 288

28. DNA Replication, Repair and Recombination..........289

28.1 DNA Replication 289
28.2 DNA Repair 305
28.3 DNA Recombination 329
28.4 Interconnections between Replication, Repair, and Recombination 335
28.5 Conclusion 335

29. RNA Functions, Biosynthesis and Processing....337

29.1 Functions of RNA 337

29.2 RNA Biosynthesis 338

29.3 RNA Processing 338

29.4 RNA Stability and Degradation 339

29.5 Conclusion 339

30. Protein Biosynthesis....341

30.1 Overview of Protein Biosynthesis 341

30.2 Transcription: From DNA to Mrna 341

30.3 RNA Processing 342

30.4 Translation: The Synthesis of Proteins 342

30.5 Post-Translational Modifications 343

30.6 Regulation of Protein Biosynthesis 343

30.7 Conclusion 344

31. Control of Gene Expression....345

31.1 Importance of Gene Expression Control 345

31.2 Transcriptional Regulation 345

31.3 Post-Transcriptional Regulation 346

31.4 Translational Regulation 347

31.5 Post-Translational Modifications 348

31.6 Conclusion 348

32. Principles of Drug Discovery and Development....349

32.1 Target Identification and Validation 349

32.2 Lead Discovery 349

32.3 Preclinical Development 350

32.4 Clinical Development 350

32.5 Regulatory Considerations 351

32.6 Post-Marketing Surveillance 351

32.7 Commercialization of Drugs 351

32.8 Conclusion 352

Bibliography....353

Index....357

CHAPTER 1

BIOCHEMISTRY IN SPACE AND TIME

Biochemistry is central to understanding life's processes, a science intricately connected to the molecules and reactions that underpin biological functions. When examined across both space and time, biochemistry transforms from a study of static cellular activities to a dynamic exploration of how biochemical processes operate across spatial environments and temporal scales. In fields as diverse as evolutionary biology, medicine, and even space exploration, understanding the biochemical changes that occur over time and across different environments provides profound insights into how life functions and adapts.

1.1 THE SPATIAL DIMENSION: BIOCHEMICAL GRADIENTS AND MICROENVIRONMENTS

Within living cells, biochemistry does not operate uniformly. Instead, biochemical reactions are highly localized, creating gradients and microenvironments essential for cellular function. Enzymes, proteins, and other molecules are arranged in such a way that particular cellular regions foster specific reactions, optimizing the energy efficiency and speed of metabolic processes. For example, mitochondria concentrate enzymes related to ATP synthesis, while the nucleus contains DNA and associated repair enzymes. This spatial organization ensures that reactions occur exactly where they are needed, preserving cellular energy and function.

Spatial biochemistry also includes cell-to-cell interactions, where signaling molecules released by one cell influence the behavior of neighboring cells. In multicellular organisms, these signals orchestrate tissue development and response to environmental changes. For example, in wound healing, cells release biochemicals that guide immune cells to the injury site, stimulating clot formation and tissue repair. Understanding spatial organization in biochemistry is crucial not only for cell biology but also for developing medical interventions that target specific cells or tissues, such as localized drug delivery.

1.2 TEMPORAL BIOCHEMISTRY: CIRCADIAN RHYTHMS AND CELLULAR TIMERS

Biochemistry does not occur in isolation from time; indeed, many biochemical processes are regulated by internal timers known as circadian rhythms. Circadian rhythms are roughly 24-hour cycles that influence physiological processes, such as sleep, hormone release, and digestion. Controlled by "clock genes," these rhythms adjust the timing of enzymatic reactions and cellular metabolism to the day-night cycle, enhancing the organism's adaptation to environmental changes.

For example, melatonin production increases during nighttime, signaling the body to prepare for sleep, while cortisol levels peak in the morning to promote alertness. Disruptions to circadian rhythms, such as those caused by jet lag or shift work, can lead to biochemical imbalances that impact overall health, increasing the risk of metabolic disorders and cognitive decline. Temporal biochemistry also influences seasonal adaptations in animals, including changes in metabolic rates and hormone levels that prepare organisms for winter or mating periods. By studying the temporal aspects of biochemistry, researchers gain insights into both normal physiological functioning and the underlying causes of various disorders.

1.3 BIOCHEMICAL ADAPTATIONS ACROSS EVOLUTIONARY TIME SCALES

Examining biochemical processes across evolutionary time scales provides a window into how life has adapted to changing environments. The evolution of metabolic pathways, for instance, reveals how organisms have optimized energy production and resource utilization. Anaerobic respiration, an ancient metabolic pathway, allowed early organisms to produce energy in oxygen-poor environments. As Earth's atmosphere accumulated oxygen, organisms evolved aerobic respiration, a more efficient form of energy production that fueled the development of complex multicellular organisms.

Genetic mutations and selective pressures also drive biochemical adaptations. The enzyme lactase, for instance, originally only functioned during infancy in mammals, aiding in milk digestion. However, in certain human populations, a genetic mutation enabled lactase production into adulthood, allowing the consumption of dairy products as a calorie-rich food source. These adaptations illustrate how biochemical changes provide evolutionary advantages, enabling organisms to survive in diverse environments.

1.4 BIOCHEMISTRY IN EXTREME ENVIRONMENTS: LIFE BEYOND EARTH

One of the most intriguing applications of biochemistry is studying how life might exist in environments vastly different from those on Earth, such as in space. Biochemical research conducted in microgravity aboard the International Space Station (ISS) has revealed that cells and organisms respond to space conditions in surprising ways. For example, studies show that microgravity affects cell growth, gene expression, and even microbial virulence. Some bacteria become more resistant to antibiotics in space, presenting challenges for treating infections in astronauts.

The discovery of extremophiles—organisms that thrive in extreme conditions such as high radiation, acidity, or salinity—suggests that biochemistry may be flexible enough to support life beyond Earth. Understanding how these organisms biochemically adapt to harsh conditions helps researchers speculate about potential life forms on planets like Mars or moons like Europa, where extreme conditions are common. Biochemical research in extreme environments also informs the design of technologies for long-duration space travel, ensuring that astronauts can remain healthy in space for extended periods.

1.5 THE ROLE OF TECHNOLOGY IN MAPPING BIOCHEMISTRY IN SPACE AND TIME

Advancements in technology have dramatically enhanced our ability to study biochemistry in spatial and temporal dimensions. Techniques such as fluorescence microscopy and mass spectrometry allow scientists to visualize and analyze biochemical processes within cells and tissues at an unprecedented level of detail. With fluorescence microscopy, researchers can observe real-time biochemical reactions, such as enzyme activities, within living cells, providing a dynamic view of cellular biochemistry.

Mass spectrometry enables the identification of thousands of biomolecules within a sample, offering insights into how biochemical composition varies across different tissues or time points. Genomic sequencing has also expanded our understanding of temporal biochemistry, revealing how gene expression changes in response to environmental factors. Together, these tools have transformed biochemistry from a static field to a dynamic science that captures the intricate interplay of molecular processes over space and time.

1.6 BIOCHEMISTRY AND THE FUTURE: EMERGING FRONTIERS IN SPACE-TIME RESEARCH

The study of biochemistry across space and time has profound implications for the future of science and medicine. In personalized medicine, for example, researchers are working to map individual biochemical profiles that reflect how genetic, environmental, and lifestyle factors interact over time. Such data could allow for treatments tailored to an individual's specific biochemistry, improving the effectiveness of therapies and minimizing side effects.

In space exploration, understanding biochemical adaptations to extreme environments is critical for future missions to Mars or other distant planets. Long-duration space missions expose astronauts to unique biochemical stressors, including prolonged radiation exposure and microgravity. By studying how biochemistry adapts in space, researchers can develop interventions to maintain astronaut health and potentially prepare for establishing human settlements beyond Earth.

1.7 CONCLUSION

Biochemistry in space and time opens up new dimensions in understanding life's complexity, from the cellular level to the scale of evolution and interplanetary adaptation. This dynamic field reveals that biochemical processes are not confined to isolated reactions but are shaped by their spatial context within cells, tissues, and ecosystems and by temporal factors such as circadian rhythms and evolutionary changes. As technology advances, our ability to explore biochemistry in increasingly intricate detail across space and time will undoubtedly lead to groundbreaking discoveries. Whether examining cellular gradients, evolutionary shifts, or adaptations to extraterrestrial environments, biochemistry in space and time offers a richer understanding of life and its boundless potential to adapt and thrive.

CHAPTER 2
PROTEIN COMPOSITION AND STRUCTURE

Proteins are essential molecules that perform a vast array of functions within living organisms, acting as enzymes, structural components, transporters, and signaling molecules. Comprising amino acids linked by peptide bonds, proteins are integral to the biochemistry of life and the functioning of nearly every cellular process. The structure of a protein is intricately tied to its function, with even small changes in composition or structure having significant effects on biological activity.

2.1 THE BUILDING BLOCKS: AMINO ACIDS AND PEPTIDE BONDS

Proteins are made up of smaller units known as amino acids, which are organic molecules consisting of a central carbon atom (α-carbon) attached to an amino group, a carboxyl group, a hydrogen atom, and a unique side chain (R-group). The R-group varies among the 20 common amino acids and determines the chemical properties of each amino acid, such as whether it is hydrophobic, hydrophilic, acidic, or basic. This diversity of amino acids is what allows proteins to adopt complex structures and fulfill a wide range of functions.

Amino acids link together through peptide bonds, forming a chain known as a polypeptide. Peptide bonds are formed by a dehydration reaction between the amino group of one amino acid and the carboxyl group of another, releasing a molecule of water. The resulting polypeptide chain has an N-terminus (amino end) and a C-terminus (carboxyl end), establishing a directionality that is crucial for protein synthesis and function. As these polypeptide chains grow and fold, they form the functional three-dimensional structures known as proteins.

2.2 PRIMARY STRUCTURE: THE SEQUENCE OF AMINO ACIDS

The primary structure of a protein is its unique sequence of amino acids. This sequence is determined by the genetic code in an organism's DNA, which dictates the order in which amino acids are linked during protein synthesis. The primary structure is critical because it determines the way a protein will fold and ultimately its function within the cell. For instance, even a single amino acid substitution can have profound effects on protein activity, as seen in the case of hemoglobin in sickle cell anemia, where a single mutation causes red blood cells to take on an abnormal, sickle shape.

Each protein has a specific amino acid sequence that confers its unique properties. Changes or mutations in the primary structure can lead to protein malfunctions or diseases. Thus, the primary structure serves as the foundation upon which all higher levels of protein structure are built, and it is fundamental to the protein's stability and function.

2.3 SECONDARY STRUCTURE: ALPHA HELICES AND BETA SHEETS

The secondary structure of a protein refers to the local folding of the polypeptide chain into specific shapes, primarily alpha helices and beta sheets, stabilized by hydrogen bonds. In an alpha helix, the polypeptide chain coils into a spiral shape, with hydrogen bonds forming between the carbonyl oxygen of one amino acid and the amide hydrogen of another four residues down the chain. This structure is common in many proteins and contributes to their flexibility and elasticity, as seen in keratin, the protein in hair and nails.

Beta sheets, on the other hand, consist of polypeptide chains lying side-by-side, forming a sheet-like structure stabilized by hydrogen bonds between carbonyl and amide groups on adjacent chains. Beta sheets can be parallel or antiparallel and are often found in proteins that require strength and stability, such as silk fibroin. The arrangement of alpha helices and beta sheets is crucial for the formation of a protein's three-dimensional shape and, subsequently, its function.

2.4 TERTIARY STRUCTURE: THE THREE-DIMENSIONAL FOLDING

The tertiary structure of a protein is its fully folded, three-dimensional shape, which is formed when the polypeptide chain bends and twists in response to interactions among its side chains. These interactions include hydrogen bonds, ionic bonds, hydrophobic interactions, and disulfide bridges, which stabilize the protein's structure. The tertiary structure is essential for the protein's functionality, as it determines the spatial arrangement of active sites or binding sites necessary for the protein's specific activity.

For example, enzymes, which are proteins that catalyze biochemical reactions, rely on their tertiary structure to position their active sites in a specific orientation, allowing them to interact with their substrates precisely. Similarly, antibodies have a specific tertiary structure that enables them to recognize and bind to foreign molecules. The tertiary structure is dynamic and can change slightly in response to environmental factors, such as pH or temperature, allowing proteins to adapt to various cellular conditions.

2.5 QUATERNARY STRUCTURE: MULTI-SUBUNIT COMPLEXES

Some proteins consist of more than one polypeptide chain, each with its own primary, secondary, and tertiary structure. The quaternary structure refers to the way these subunits are arranged relative to each other. Quaternary structures are stabilized by the same interactions that stabilize tertiary structures, such as hydrogen bonds, ionic bonds, and hydrophobic interactions. Hemoglobin, the oxygen-carrying protein in blood, is a classic example of a protein with quaternary structure, as it comprises four polypeptide chains that interact to form a functional protein complex.

Quaternary structures allow proteins to undertake functions that individual subunits could not accomplish alone. For example, in hemoglobin, the quaternary structure allows cooperative binding of oxygen, meaning the binding of oxygen to one subunit increases the affinity of the remaining subunits for oxygen. This cooperative interaction is vital for efficiently transporting oxygen in the blood. The quaternary structure also facilitates regulation, as some proteins change their activity when specific subunits interact or dissociate.

2.6 THE RELATIONSHIP BETWEEN STRUCTURE AND FUNCTION

The structure of a protein is intimately connected to its function, as each level of organization—from primary to quaternary—contributes to the overall shape and properties of the molecule. Even slight alterations in protein structure, whether due to genetic mutations or environmental conditions, can impair protein function. This relationship is exemplified in diseases such as cystic fibrosis, where a mutation causes misfolding of the CFTR protein, leading to dysfunctional ion transport in cells and resulting in severe respiratory and digestive issues.

This structural-function relationship underscores the importance of protein folding, a complex process in which the polypeptide chain achieves its native, functional conformation. In cells, specialized proteins known as chaperones assist in the folding process, preventing aggregation and misfolding. When proteins fold incorrectly, they may form aggregates, which are implicated in neurodegenerative diseases like Alzheimer's and Parkinson's disease.

2.7 PROTEIN MISFOLDING AND DISEASE

Protein misfolding can have serious consequences, as improperly folded proteins often lose their functionality or become toxic to cells. Misfolded proteins can aggregate, forming insoluble fibrils that disrupt cellular functions. These aggregates are the hallmark of many neurodegenerative diseases, including Alzheimer's, where amyloid-beta peptides aggregate, forming plaques in the brain. Similarly, in Parkinson's disease, the accumulation of misfolded alpha-synuclein proteins leads to the formation of Lewy bodies, which impair neuronal function.

The study of protein misfolding is essential for understanding the molecular basis of these diseases and developing treatments. Researchers are exploring ways to enhance chaperone activity or design molecules that prevent misfolding, which could offer therapeutic strategies for these disorders. Understanding how proteins fold and misfold provides insights into the stability and longevity of cellular proteins and their ability to adapt to stressors.

2.8 CONCLUSION

Protein composition and structure are central to understanding how these molecules fulfill a vast range of biological functions. From the linear sequence of amino acids to the complex folding and assembly of multi-subunit complexes, each level of protein structure is meticulously organized to enable specific activities within the cell. The relationship between protein structure and function highlights the precision of biochemical processes, with even minor changes in structure leading to significant functional consequences. As research into protein misfolding and aggregation continues, we gain valuable insights into the underlying mechanisms of diseases and the potential for novel therapeutic approaches. The study of protein composition and structure not only advances our understanding of cellular biology but also holds promise for improving human health and treating diseases rooted in molecular dysfunction.

CHAPTER 3

BINDING AND MOLECULAR RECOGNITION

Binding and molecular recognition are fundamental concepts in biochemistry and molecular biology, describing how molecules interact with remarkable specificity to regulate nearly all cellular processes. Molecular recognition refers to the ability of biomolecules to selectively identify and bind to specific targets, enabling complex biological functions such as enzyme catalysis, immune response, signal transduction, and DNA replication. Understanding the mechanisms behind binding and molecular recognition is essential not only for grasping cellular physiology but also for advancing fields like drug design and molecular engineering.

3.1 TYPES OF MOLECULAR INTERACTIONS: THE BASIS OF BINDING

Binding interactions between molecules depend on a variety of forces that operate at the atomic and molecular levels, including hydrogen bonds, ionic interactions, van der Waals forces, and hydrophobic interactions. While each individual interaction may be weak, collectively, they create a binding affinity strong enough to enable specific molecular recognition.

- **Hydrogen Bonds:** These occur when a hydrogen atom is shared between a donor atom, such as oxygen or nitrogen, and an acceptor atom, which also has a partial negative charge. Hydrogen bonds are crucial in stabilizing the secondary structures of proteins, such as alpha helices and beta sheets, and contribute significantly to the binding specificity of enzymes and antibodies.
- **Ionic Interactions:** Also known as electrostatic interactions, these occur between oppositely charged ions or polar molecules. They are particularly important in enzyme-substrate binding, where the charges help align molecules correctly for catalysis. Ionic interactions are often shielded in aqueous environments but can play a vital role in creating binding specificity within the microenvironments of protein structures.
- **Van der Waals Forces:** These weak interactions arise from temporary dipoles created by the movement of electrons around atoms. Though

individually weak, van der Waals forces become significant in large biomolecules and contribute to the overall stability of complexes, such as protein-protein interactions.

- **Hydrophobic Interactions:** Nonpolar molecules or molecular regions tend to avoid water, leading them to cluster together in aqueous environments. These interactions play a major role in protein folding and in the formation of cell membranes, where hydrophobic tails of lipid molecules are sequestered away from the aqueous environment, driving membrane formation.

3.2 BINDING AFFINITY AND SPECIFICITY: DETERMINING STRENGTH AND SELECTIVITY

Binding affinity refers to the strength of the interaction between two molecules, typically quantified by a binding constant (Kd_dd), which measures how tightly a molecule binds to its target. High-affinity int eractions, characterized by low Kd_dd values, involve stable complexes where the interacting partners remain associ ated for extended periods. For instance, t he binding affinity of oxygen to hem oglobin is finely tuned to allow efficient oxygen transport and release throughout the body.

Specificity, on the other hand, describes how selective a molecule is in binding to its target, avoiding interactions with non-target molecules. Specificity is essential in molecular recognition, as it ensures that each molecule interacts with its appropriate partner, preventing unintended interactions that could disrupt cellular processes. For example, antibodies have high specificity for their antigens, recognizing and binding only to specific epitopes on foreign molecules. This specificity is crucial for the immune response, allowing the body to target and neutralize pathogens effectively.

Affinity and specificity are often balanced to suit biological needs; some systems require very high specificity, while others, like certain enzymes, benefit from moderate specificity to accommodate multiple substrates. This balance is a defining characteristic of molecular recognition, dictating how and when biological interactions occur.

3.3 ENZYMES AND MOLECULAR RECOGNITION: CATALYSIS AND SUBSTRATE SPECIFICITY

Enzymes are proteins that act as biological catalysts, accelerating chemical reactions by lowering activation energy. A key component of enzyme function is their ability to recognize and bind to specific substrates with high affinity and specificity, a feature that is central to their catalytic efficiency. The active site of an enzyme is designed to complement the substrate's shape, charge distribution, and chemical properties, enabling selective binding and subsequent reaction.

The "lock-and-key" model and the "induced-fit" model are two common theories explaining enzyme-substrate recognition. The lock-and-key model proposes that the active site of an enzyme has a rigid structure complementary to the substrate, much like a key fitting into a lock. However, the induced-fit model suggests that enzymes are more flexible, and the binding of the substrate induces a conformational change in the enzyme that enhances binding affinity and catalytic activity. The induced-fit model provides a more dynamic view of molecular recognition, illustrating how proteins can adjust their shapes to accommodate specific substrates.

The specificity of enzymes is further enhanced by coenzymes and cofactors, non-protein molecules that assist in enzyme catalysis. For example, NAD+^++ (nicotinamide adenine dinucleotide) acts as a coenzyme in redox reactions, facilitating electron transfer. These molecules extend the range of chemical reactions enzymes can perform, enhancing their versatility in molecular recognition.

3.4 RECEPTORS AND LIGAND BINDING: SIGNAL TRANSDUCTION PATHWAYS

Receptors are proteins that bind to signaling molecules (ligands) and play a critical role in cellular communication by transmitting signals from the outside to the inside of a cell. This process, known as signal transduction, begins when a ligand binds to its specific receptor, triggering a conformational change that activates intracellular signaling cascades.

Ligand-receptor interactions are highly specific, as each receptor is structured to recognize particular ligands, such as hormones, neurotransmitters, or cytokines. This specificity ensures that cells respond only to relevant signals, maintaining the precision of cellular responses. For instance, adrenaline binds to beta-adrenergic receptors, initiating a signaling pathway that increases heart rate and energy mobilization during the "fight-or-flight" response. By contrast, insulin binds to its receptor to regulate glucose uptake and metabolism.

Once a ligand binds to its receptor, the signal is often amplified through second messengers, such as cyclic AMP (cAMP) or calcium ions, which activate downstream proteins and produce a large cellular response. This cascade system underscores the importance of specificity in molecular recognition, as small initial signals can generate substantial physiological effects.

3.5 APPLICATIONS IN THERAPEUTIC DEVELOPMENT: TARGETING MOLECULAR RECOGNITION

Understanding the principles of binding and molecular recognition has transformative implications for drug design and therapeutic development. Drugs are often designed to mimic or inhibit the natural ligands of receptors or substrates of enzymes, binding with high affinity and specificity to their target sites to alter physiological pathways.

One approach in drug design is the development of small-molecule inhibitors that bind to active sites on enzymes or receptors, blocking the natural ligand from binding and thereby disrupting the pathway. For example, ACE inhibitors target the enzyme angiotensin-converting enzyme to treat high blood pressure by blocking the formation of angiotensin II, a molecule that narrows blood vessels.

Another approach involves designing antibodies that specifically recognize and neutralize harmful molecules, such as in cancer immunotherapy. Monoclonal antibodies are engineered to bind specifically to proteins on the surface of cancer cells, marking them for destruction by the immune system. This strategy relies on precise molecular recognition to ensure that the antibody binds only to cancer cells, minimizing damage to healthy tissue.

Advances in computational chemistry and structural biology have further enhanced our ability to design molecules with optimal binding properties, enabling the development of highly specific therapies for diseases previously deemed intractable.

3.6 CONCLUSION

Binding and molecular recognition lie at the heart of biological specificity, driving the interactions that govern cellular functions and physiological responses. Through various types of molecular interactions, proteins and other biomolecules recognize their specific partners with remarkable precision, enabling essential processes such as enzyme catalysis, signal transduction, and immune defense. By understanding the mechanisms that underlie binding affinity and specificity, scientists can manipulate these interactions to develop new therapeutics that target specific proteins or pathways, improving treatment precision and efficacy.

As we continue to unravel the intricacies of molecular recognition, applications in personalized medicine, drug design, and biotechnology expand, opening new frontiers for combating diseases and harnessing molecular recognition for industrial and clinical advancements. In this way, the study of binding and molecular recognition not only enhances our fundamental understanding of biology but also propels innovations that promise to reshape healthcare and improve quality of life.

CHAPTER 4

PROTEIN METHODS

Proteins are essential macromolecules that perform a variety of roles in biological systems, from catalyzing reactions to supporting cellular structures and signaling processes. To understand their roles fully, scientists have developed a broad range of methods to study proteins' structure, function, and interactions. These techniques enable researchers to determine amino acid sequences, observe three-dimensional structures, investigate protein-protein interactions, and assess activity under various conditions.

4.1 PROTEIN PURIFICATION: ISOLATING PROTEINS FOR ANALYSIS

Protein purification is the first step in many protein studies, isolating the target protein from a complex mixture of cellular components. This process is critical to ensuring that analyses focus solely on the protein of interest without interference from other molecules.

Purification usually involves multiple steps to increase purity and yield, starting with cell lysis to release proteins from cells, followed by a series of separation techniques. Chromatography is one of the primary methods, with different types tailored to exploit specific properties of proteins:

- **Ion Exchange Chromatography:** Separates proteins based on charge differences. Proteins bind to a charged resin, and the elution is controlled by varying the pH or salt concentration, allowing proteins with different charges to be collected sequentially.
- **Size-Exclusion Chromatography:** Also known as gel filtration, this method separates proteins based on size. Smaller molecules enter the porous beads in the column, slowing their passage, while larger molecules pass around the beads and elute more quickly.
- **Affinity Chromatography:** This method relies on specific interactions between a protein and a ligand attached to the column. For example, histidine-tagged proteins can be purified using a nickel-affinity column, as histidine residues bind strongly to nickel ions.

Protein purification ensures that the sample used in further analysis is as pure as possible, making it an essential foundation for most protein-based experiments.

4.2 ELECTROPHORESIS: ANALYZING PROTEIN SIZE AND CHARGE

Electrophoresis is a technique used to separate proteins based on size and charge by applying an electric field across a gel matrix. SDS-PAGE (sodium dodecyl sulfate-polyacrylamide gel electrophoresis) is one of the most common electrophoresis methods, often used to estimate the molecular weight of proteins. SDS-PAGE uses the detergent SDS to denature proteins and impart a negative charge, allowing them to migrate toward the positive electrode in proportion to their size. Smaller proteins move faster through the gel, while larger ones are impeded by the gel matrix. The bands in the gel are stained, usually with Coomassie Brilliant Blue or silver stain, to visualize the protein distribution. Two-Dimensional Gel Electrophoresis (2D-PAGE) is an extension of SDS-PAGE that separates proteins based on two properties: isoelectric point (pI) and molecular weight. In the first dimension, proteins are separated by pI using isoelectric focusing, and in the second dimension, they are separated by size through SDS-PAGE. This method allows for high-resolution separation and is widely used in proteomics to analyze complex protein mixtures.

4.3 SPECTROSCOPIC METHODS: ASSESSING PROTEIN STRUCTURE AND DYNAMICS

Spectroscopy is crucial for studying protein structure, dynamics, and interactions in solution. Several spectroscopic methods, including UV-visible, fluorescence, and circular dichroism (CD) spectroscopy, are commonly used.

- **UV-Visible Spectroscopy:** This method relies on the absorbance of light in the ultraviolet and visible range by aromatic amino acids like tryptophan and tyrosine. It provides information on protein concentration and can detect changes in the environment around these amino acids, useful for monitoring folding and conformational changes.
- **Fluorescence Spectroscopy:** Certain amino acids, particularly tryptophan, exhibit fluorescence when excited by specific wavelengths of light. Fluorescence spectroscopy is sensitive to changes in protein structure, as it monitors alterations in the microenvironment around tryptophan residues. It is also valuable for studying protein-ligand interactions and conformational changes.
- **Circular Dichroism (CD) Spectroscopy:** CD measures the difference in absorption of left- and right-circularly polarized light, providing information on protein secondary structure. Alpha helices, beta sheets, and random coils each have distinct CD spectra, allowing researchers to estimate the secondary structure content and monitor structural changes, such as folding or denaturation.

These spectroscopic techniques are essential for analyzing protein stability, folding, and interactions with other molecules, helping to elucidate protein behavior under different conditions.

4.4 X-RAY CRYSTALLOGRAPHY: DETERMINING ATOMIC-RESOLUTION STRUCTURES

X-ray crystallography is a powerful method for determining the three-dimensional structures of proteins at atomic resolution. This technique involves crystallizing the protein and exposing it to an X-ray beam. The X-rays scatter upon hitting the electrons in the protein, creating a diffraction pattern that is captured on a detector. By analyzing this pattern, researchers can calculate the electron density map, which provides information about the positions of atoms in the protein. X-ray crystallography has provided detailed structures for thousands of proteins, revealing the intricate details of active sites, binding pockets, and protein-ligand interactions. However, the technique requires high-quality crystals, which can be difficult to obtain, particularly for large or flexible proteins. Despite this limitation, crystallography remains one of the primary methods for understanding protein structure and guiding drug design by identifying key molecular interactions within binding sites.

4.5 NUCLEAR MAGNETIC RESONANCE (NMR) SPECTROSCOPY: STUDYING PROTEINS IN SOLUTION

NMR spectroscopy is a technique that provides structural and dynamic information about proteins in solution, preserving their natural state. Unlike X-ray crystallography, which requires crystallization, NMR is particularly useful for studying small to medium-sized proteins and their interactions in aqueous environments. NMR detects the magnetic properties of atomic nuclei, typically hydrogen (1^11H) and carbon (13^{13}13C) atoms, in the protein. By applying a magnetic field, NMR produces a spectrum that reveals information about the positions of these atoms and the distances between them. This data can be used to model the protein's three-dimensional structure, although resolution decreases with increasing protein size, limiting NMR's application to proteins typically under 50 kDa. In addition to structure determination, NMR is valuable for studying protein dynamics, conformational flexibility, and protein-ligand interactions, making it ideal for understanding proteins in more physiologically relevant conditions.

4.6 CRYO-ELECTRON MICROSCOPY (CRYO-EM): IMAGING LARGE PROTEIN COMPLEXES

Cryo-electron microscopy (cryo-EM) has emerged as a revolutionary tool in structural biology, particularly suited for studying large protein complexes, membrane proteins, and viruses. Cryo-EM involves flash-freezing proteins in a thin layer of vitreous ice, preserving them in a near-native state. An electron beam is then used to capture images of these proteins at different orientations, and computer algorithms compile these images into a three-dimensional structure. Cryo-EM does not require crystallization and is capable of analyzing large, flexible, or heterogeneous complexes that are challenging for X-ray crystallography or NMR. Recent advancements in detector technology and image processing have improved cryo-EM resolution to near-atomic levels, making it one of the leading methods for studying complex protein structures.

4.7 MASS SPECTROMETRY: ANALYZING PROTEIN COMPOSITION AND MODIFICATIONS

Mass spectrometry (MS) is a powerful technique for identifying and quantifying proteins, studying post-translational modifications, and analyzing protein complexes. In MS, proteins are ionized and fragmented, and the resulting ions are detected based on their mass-to-charge ratio. This information enables the determination of the protein's molecular weight and amino acid sequence. One common application is tandem mass spectrometry (MS/MS), which provides detailed information on protein sequence by analyzing peptide fragments. Protein modification analysis is also possible, allowing researchers to study phosphorylation, glycosylation, and other modifications that affect protein function. In proteomics, MS is indispensable for identifying proteins within complex mixtures and assessing their relative abundances, making it essential for large-scale studies of protein expression, interactions, and modifications.

4.8 CONCLUSION

The diverse array of protein methods available allows researchers to examine proteins from multiple angles, from primary structure and composition to complex, three-dimensional structures and interactions. Techniques such as protein purification, electrophoresis, and spectroscopy form the foundation for understanding protein properties, while advanced methods like X-ray crystallography, NMR, cryo-EM, and mass spectrometry reveal intricate details of protein structure, dynamics, and modifications. Each of these methods has its strengths and limitations, and the choice of technique often depends on the specific protein being studied and the information sought. Together, these protein methods form a powerful toolkit for investigating the molecular machinery of life, enhancing our understanding of cellular processes, and opening pathways for medical and biotechnological advancements. The ability to analyze proteins at such depth is instrumental in drug discovery, diagnostics, and the development of novel therapies, underscoring the importance of protein methods in both fundamental research and applied sciences.

CHAPTER 5

ENZYMES: CORE CONCEPTS AND KINETICS

Enzymes are biological catalysts that accelerate chemical reactions within living organisms, playing critical roles in nearly every biochemical process, from digestion and energy production to DNA replication and cell signaling. As proteins (and sometimes RNA molecules called ribozymes), enzymes enable cellular reactions to occur at speeds compatible with life.

They achieve this by lowering the activation energy required for reactions to proceed, allowing reactants to convert into products more efficiently. Understanding enzyme core concepts and kinetics is essential for biochemistry, as it provides insight into how enzymes function, how they are regulated, and how they can be manipulated for industrial and therapeutic purposes.

5.1 ENZYME STRUCTURE AND FUNCTION: THE BASIS OF CATALYSIS

Enzymes are composed of one or more polypeptide chains that fold into a unique three-dimensional structure. The active site of an enzyme is the region where substrate molecules bind and undergo a chemical reaction. This site is typically a small pocket or groove within the enzyme that provides a specific environment for catalysis. The structure and amino acid composition of the active site determine the enzyme's specificity for its substrate, as well as its catalytic activity.

The binding between an enzyme and its substrate often follows the "induced fit" model. In this model, the enzyme undergoes a conformational change upon substrate binding, optimizing the fit between the enzyme and substrate. This flexibility ensures that the enzyme can facilitate the reaction more effectively by stabilizing the transition state, thereby lowering the activation energy.

In addition to the active site, some enzymes have additional binding sites called allosteric sites, where molecules can bind and modulate enzyme activity. Allosteric regulation is a crucial mechanism for controlling enzyme activity in response to cellular needs, often allowing enzymes to be "turned on" or "off" by binding small molecules, such as metabolic intermediates, which adjust their catalytic activity accordingly.

5.2 MECHANISMS OF ENZYME CATALYSIS: LOWERING ACTIVATION ENERGY

Enzymes accelerate reactions by lowering the activation energy, the initial energy required to start a chemical reaction. Activation energy acts as a barrier that reactants must overcome to transition into products. By providing an alternative pathway with a lower activation energy, enzymes enable reactions to occur more readily.

Enzymes achieve this catalysis through several mechanisms:

- **Proximity and Orientation Effects:** Enzymes bring substrates close together in the correct orientation for the reaction to occur. By aligning reactive groups of the substrate molecules precisely, enzymes increase the probability of effective collisions, promoting the formation of the transition state.
- **Strain and Distortion:** Some enzymes bind substrates in a way that strains particular bonds, making them more susceptible to cleavage. This "induced strain" lowers the energy required to break bonds, facilitating reactions such as hydrolysis, where water is used to break a molecule into smaller parts.
- **Covalent Catalysis:** In some reactions, the enzyme forms a transient covalent bond with the substrate, creating an enzyme-substrate intermediate. This intermediate stabilizes the transition state, which lowers activation energy. After the reaction occurs, the enzyme releases the product and regenerates its original state, ready to catalyze another reaction cycle.
- **Acid-Base Catalysis:** Enzymes may use amino acid residues in their active sites to donate or accept protons during the reaction. This process can stabilize charged intermediates and facilitate the breaking and forming of bonds. Histidine is a common amino acid involved in acid-base catalysis due to its ability to both donate and accept protons at physiological pH.

By utilizing these mechanisms, enzymes are capable of accelerating reactions by factors of millions or more, enabling cells to carry out complex biochemical processes efficiently and selectively.

5.3 ENZYME KINETICS: UNDERSTANDING REACTION RATES

The study of enzyme kinetics provides insights into how quickly enzymes catalyze reactions and how factors such as substrate concentration, temperature, and pH affect these rates. Enzyme kinetics is essential in biochemistry and pharmacology, as it helps to understand enzyme behavior and to predict how enzymes will respond under different conditions.

The rate of an enzyme-catalyzed reaction is typically measured by monitoring the concentration of the substrate or product over time. Initially, the rate increases with substrate concentration, as more substrate molecules are available to bind with the enzyme. However, at high substrate concentrations, the enzyme becomes saturated, and the rate reaches a maximum known as the Vmax_\text{max}max.

5.4 THE MICHAELIS-MENTEN MODEL: DESCRIBING ENZYME BEHAVIOR

The Michaelis-Menten equation is a mathematical model that describes the rate of enzymatic reactions by relating reaction rate to substrate concentration. This model, developed by Leonor Michaelis and Maud Menten, provides a foundation for understanding how enzymes function under different substrate conditions. The equation is as follows:

$$V = \frac{V_{max} [S]}{Km + [S]}$$

Where:

- V is the reaction velocity (rate),
- V_{max} is the maximum reaction rate,
- [S] is the substrate concentration, and
- K_m is the Michaelis constant, representing the substrate concentration at which the reaction rate is half of V_{max}.

The Michaelis constant (K_m) is an important parameter in enzyme kinetics, as it provides information about the enzyme's affinity for its substrate. A low K_m indicates high affinity, meaning that the enzyme requires only a small amount of substrate to reach half of its maximum rate. Conversely, a high K_m suggests a lower affinity.

The Michaelis-Menten model assumes that the formation of the enzyme-substrate complex reaches a steady state, where its concentration remains constant over time. This simplification allows for easier analysis of reaction rates and helps researchers understand how enzymes respond to varying substrate concentrations.

5.5 FACTORS AFFECTING ENZYME ACTIVITY

Enzyme activity is influenced by several factors, including substrate concentration, temperature, pH, and the presence of inhibitors or activators.

- **Substrate Concentration:** Increasing substrate concentration raises the reaction rate until the enzyme reaches saturation. Beyond this point, additional substrate has no effect on the rate, as all enzyme molecules are already bound to substrate.
- **Temperature:** Enzymes have an optimal temperature at which they function most efficiently. Higher temperatures increase reaction rates by providing reactants with more kinetic energy. However, excessive heat can denature enzymes, altering their structure and rendering them inactive.
- **pH:** Each enzyme has an optimal pH range, typically reflecting the environment in which it operates. Deviations from this optimal pH can alter the ionization of amino acids within the active site, affecting enzyme stability and function. For example, pepsin, a digestive enzyme in the stomach, functions best in acidic conditions, while enzymes in the intestine prefer a neutral to slightly basic pH.
- **Inhibitors:** Enzyme inhibitors are molecules that decrease enzyme activity, often by binding to the enzyme and interfering with substrate binding or catalysis. Competitive inhibitors bind to the active site, preventing substrate binding, whereas non-competitive inhibitors bind to other regions, reducing the enzyme's catalytic efficiency. Inhibition is a critical regulatory mechanism in cellular pathways and is also used in drug design to modulate enzyme activity therapeutically.
- **Activators:** Conversely, enzyme activators increase enzyme activity by binding to allosteric sites and promoting conformational changes that enhance substrate binding or catalytic efficiency.

5.6 ENZYME INHIBITION AND REGULATION: CONTROLLING BIOLOGICAL PATHWAYS

Enzyme inhibition is a natural regulatory mechanism in metabolic pathways. Feedback inhibition is one such mechanism where the end product of a pathway inhibits an upstream enzyme, thereby controlling the flow of reactants through the pathway. This self-regulation prevents the overaccumulation of products and conserves cellular resources. Pharmaceutical agents often use enzyme inhibition as a strategy to control disease processes. For example, ACE inhibitors (used to treat hypertension) block the activity of the angiotensin-converting enzyme, reducing the production of angiotensin II and thereby lowering blood pressure. Similarly, statins inhibit HMG-CoA reductase, an enzyme involved in cholesterol biosynthesis, to manage high cholesterol levels.

5.7 CONCLUSION

Enzymes are indispensable for life, driving the biochemical reactions that sustain cellular and physiological functions. Their ability to catalyze reactions by lowering activation energy and their specificity for substrates set them apart as sophisticated biological catalysts. Understanding enzyme kinetics through models like Michaelis-Menten reveals how enzymes function under varying conditions, and studying the factors that affect enzyme activity helps elucidate the regulatory mechanisms that control biochemical pathways.

Enzyme inhibitors and activators allow for the modulation of enzyme activity, providing both natural regulation within cells and therapeutic intervention in disease management. As our understanding of enzymes continues to deepen, we gain new opportunities to harness their catalytic power in industrial applications and medical treatments, making enzymes not only central to biochemistry but also valuable tools in advancing human health and technology.

CHAPTER 6

ENZYME CATALYTIC STRATEGIES

Enzymes are remarkable biological catalysts that facilitate chemical reactions in living organisms with high efficiency, specificity, and speed. To carry out their roles, enzymes employ a range of catalytic strategies, which enable them to lower the activation energy of reactions and stabilize transition states. These strategies, grounded in the principles of chemistry and physics, allow enzymes to perform highly specific tasks under physiological conditions, making them vital to metabolic pathways, cellular regulation, and many other biological functions.

6.1 PROXIMITY AND ORIENTATION EFFECTS: BRINGING SUBSTRATES TOGETHER

One of the fundamental strategies used by enzymes to enhance reaction rates is proximity and orientation. Enzymes work by binding substrates within their active site, positioning them close together in a precise alignment that favors chemical interaction. This effect increases the effective concentration of substrates within the active site and reduces the entropy of the system, making reactions more likely to occur.

In a non-enzymatic reaction, molecules in solution move randomly and may encounter each other only occasionally. In contrast, the binding of substrates in an enzyme's active site brings reactants into close proximity and aligns reactive groups in the correct orientation, reducing the energy required for the reaction. This strategy can increase the reaction rate by several orders of magnitude, as it facilitates the formation of intermediate complexes and reduces the need for random molecular collisions to drive the reaction forward.

6.2 ACID-BASE CATALYSIS: DONATING AND ACCEPTING PROTONS

Acid-base catalysis is another essential strategy in enzymatic reactions, in which enzymes facilitate the transfer of protons (H+^++) to stabilize charged intermediates or promote the formation of bonds. In biochemical reactions, amino acid residues in the active site, such as histidine, glutamate, and aspartate, can act as proton donors or acceptors depending on the reaction requirements. Acid-base catalysis plays a critical role in reactions that involve the cleavage or formation of covalent bonds.

There are two main types of acid-base catalysis:

- **General Acid-Base Catalysis:** In this mechanism, amino acid residues in the enzyme's active site function as general acids or bases by donating or accepting protons to facilitate the reaction. For example, histidine, with its side chain that can easily gain or lose a proton, is commonly involved in acid-base catalysis.
- **Specific Acid-Base Catalysis:** This mechanism relies on protons (from water or another source in the environment) rather than specific amino acid residues to drive the reaction. Specific acid-base catalysis is often dependent on the pH of the surrounding environment.

In enzymes like carbonic anhydrase and chymotrypsin, acid-base catalysis is critical. In the case of carbonic anhydrase, a zinc ion and a histidine residue work together to facilitate proton transfer, which accelerates the conversion of carbon dioxide and water into bicarbonate and hydrogen ions. Acid-base catalysis not only stabilizes the transition state but also helps drive the reaction to completion by altering the protonation state of substrates and intermediates.

6.3 COVALENT CATALYSIS: FORMING TRANSIENT BONDS WITH SUBSTRATES

Covalent catalysis involves the formation of a transient covalent bond between the enzyme and the substrate during the reaction, creating an enzyme-substrate intermediate. This strategy allows enzymes to facilitate bond rearrangements and electron transfer, ultimately making the reaction more favorable. In covalent catalysis, the enzyme temporarily participates in the reaction as a reactant, creating a modified intermediate that eventually breaks down to release the final product and regenerate the enzyme.

Covalent catalysis is often observed in enzymes such as serine proteases (e.g., trypsin, chymotrypsin), where a nucleophilic residue, like serine, attacks the substrate to form a covalent bond. In the case of chymotrypsin, serine acts as a nucleophile and forms a covalent bond with the carbonyl carbon of the peptide bond that is being cleaved. This intermediate is then stabilized by other residues in the active site, enabling the reaction to proceed.

This type of catalysis typically involves three key steps:

- **Nucleophilic Attack:** The enzyme's active site provides a nucleophile (often an amino acid side chain) that attacks an electrophilic center on the substrate, forming a covalent intermediate.
- **Intermediate Stabilization:** The intermediate is stabilized by the enzyme, lowering the activation energy.

- **Release of Product:** The covalent bond between the enzyme and the intermediate breaks, releasing the final product and restoring the enzyme's original state.

Covalent catalysis provides an efficient way for enzymes to temporarily alter their chemical structure to assist the reaction, a tactic that enhances their versatility in catalyzing diverse biochemical transformations.

6.4 METAL ION CATALYSIS: UTILIZING METAL IONS AS CATALYTIC AGENTS

Metal ion catalysis is a strategy in which enzymes utilize metal ions (such as zinc, magnesium, or iron) as cofactors to facilitate reactions. Metal ions can stabilize charged intermediates, activate water molecules, and participate in electron transfer, making them essential in a variety of enzymatic reactions.

Metal ions are particularly important in enzymes like alcohol dehydrogenase, where a zinc ion stabilizes the transition state by interacting with the negatively charged oxygen atom in the reaction. In carbonic anhydrase, a zinc ion coordinates with water molecules, making it easier for the enzyme to deprotonate water and generate hydroxide ions, which are crucial for catalyzing the conversion of CO2_22 to bicarbonate.

There are several roles metal ions play in enzymatic reactions:

- **Electrostatic Stabilization:** Metal ions can stabilize negative charges that develop during the reaction, thereby lowering activation energy.
- **Activation of Water Molecules:** Metal ions, particularly zinc, can increase the nucleophilicity of water, making it more reactive in hydrolytic reactions.
- **Electron Transfer Facilitation:** In redox reactions, metal ions like iron and copper can mediate electron transfer between the enzyme and substrate, enabling oxidation and reduction processes.

Metal ion catalysis is a versatile and highly effective mechanism, particularly in enzymes that deal with redox reactions and hydration or dehydration reactions, where the involvement of a metal ion is critical for reaction efficiency.

6.5 TRANSITION STATE STABILIZATION: LOWERING ACTIVATION ENERGY

One of the most important catalytic strategies employed by enzymes is transition state stabilization. The transition state is the highest energy point along the reaction pathway, and enzymes facilitate catalysis by stabilizing this state, thereby lowering the activation energy required for the reaction.

Transition state stabilization often involves the formation of multiple weak interactions, such as hydrogen bonds and van der Waals forces, between the enzyme and the substrate in the transition state. By binding tightly to the transition state, enzymes reduce its energy and make the formation of products more likely. This effect can be likened to holding the substrate in a strained or "activated" state that is closer in energy to the final product.

A classic example of transition state stabilization is seen in lysozyme, an enzyme that breaks down bacterial cell walls. Lysozyme binds its substrate in a way that distorts certain bonds, bringing the substrate closer to the transition state. By stabilizing this distorted intermediate, lysozyme reduces the activation energy, accelerating the reaction and enabling it to occur under mild cellular conditions.

Enzymes often achieve maximum catalytic efficiency by evolving active sites that are complementary in shape, charge, and hydrophobicity to the transition state rather than to the substrate. This specificity for the transition state allows enzymes to be highly selective, as they favor the transformation of substrates into products over the reverse process.

6.6 CONCLUSION

Enzyme catalytic strategies are finely tuned mechanisms that have evolved to maximize reaction rates and specificity under the constraints of cellular conditions. Through proximity and orientation effects, acid-base catalysis, covalent catalysis, metal ion catalysis, and transition state stabilization, enzymes facilitate a wide range of biochemical reactions, each with its unique challenges and requirements.

These catalytic strategies highlight the sophistication of enzymes as molecular machines, using principles of physics and chemistry to overcome energy barriers and stabilize transition states. Understanding these strategies not only reveals the fundamental workings of life at a molecular level but also provides valuable insights for applications in biotechnology and medicine, where enzymes can be harnessed and engineered for specific tasks, from drug synthesis to biofuel production.

CHAPTER 7

ENZYME REGULATORY STRATEGIES

Enzymes play a vital role in regulating nearly all biochemical reactions within cells, ensuring that processes occur at the right time and location, under optimal conditions, and in coordination with other cellular activities. To maintain cellular homeostasis and respond to environmental changes, enzymes require sophisticated regulatory mechanisms. Through enzyme regulation, cells can control metabolic pathways, respond to signaling molecules, and adapt to fluctuating internal and external conditions.

7.1 ALLOSTERIC REGULATION: FINE-TUNING ENZYME ACTIVITY

Allosteric regulation is a strategy in which enzymes are modulated by the binding of regulatory molecules, known as effectors, at sites other than the active site. These binding sites, called allosteric sites, are separate from the active site but induce conformational changes that impact enzyme activity. Effectors can be either activators (increasing enzyme activity) or inhibitors (decreasing enzyme activity), allowing for fine-tuning of enzyme function based on cellular needs.

Allosteric enzymes, often found at critical control points in metabolic pathways, typically exhibit sigmoidal kinetics instead of the hyperbolic kinetics observed in Michaelis-Menten enzymes. This sigmoidal behavior allows allosteric enzymes to act as molecular switches, where small changes in effector concentration can lead to significant alterations in enzyme activity. Hemoglobin is a classic example of allosteric regulation, as it changes shape in response to oxygen binding and thereby adjusts its affinity for additional oxygen molecules.

An essential feature of allosteric regulation is cooperativity, where binding of an effector molecule to one subunit of a multi-subunit enzyme enhances the activity of the other subunits. This cooperative behavior allows enzymes to respond sensitively to fluctuations in effector concentrations, making allosteric regulation a highly effective means of controlling enzyme activity in metabolic networks.

7.2 COVALENT MODIFICATION: REVERSIBLE CHEMICAL CHANGES

Covalent modification involves the addition or removal of chemical groups, such as phosphate, methyl, or acetyl groups, on specific amino acid residues within an enzyme. This modification can change the enzyme's structure, stability, and activity, often serving as a rapid on/off switch. The most common type of covalent modification is phosphorylation, the addition of a phosphate group, which plays a major role in signal transduction and cellular regulation.

Phosphorylation is typically catalyzed by enzymes known as kinases, while phosphatases catalyze the removal of phosphate groups. Phosphorylation can alter enzyme activity by affecting the active site's shape, thus enabling or disabling substrate binding or by altering enzyme interaction with other proteins. The phosphorylation of enzymes involved in glucose metabolism, such as glycogen phosphorylase and glycogen synthase, is essential for regulating energy availability in cells.

Other types of covalent modifications include acetylation, methylation, and ubiquitination, each with distinct roles in modulating enzyme function and stability. For example, acetylation of histone proteins plays a central role in gene expression regulation by altering chromatin structure and enabling or restricting access to DNA. Covalent modifications provide enzymes with a mechanism for reversible activation and deactivation, allowing cells to rapidly adapt enzyme activity to changing metabolic demands and environmental signals.

7.3 PROTEOLYTIC ACTIVATION: IRREVERSIBLE ENZYME ACTIVATION

Proteolytic activation is a regulatory strategy in which enzymes are synthesized in an inactive form, known as a zymogen or proenzyme, and are activated through the cleavage of specific peptide bonds. This activation mechanism is irreversible, as the inactive form must be cleaved to release the active enzyme, which then can no longer revert to the zymogen state. Proteolytic activation is essential in processes that require rapid and localized enzyme activation, such as digestion, blood clotting, and immune response.

For example, pepsinogen, a zymogen produced in the stomach, is activated to pepsin when exposed to the acidic environment of the stomach, which initiates its role in protein digestion. Similarly, caspases, proteolytic enzymes involved in programmed cell death (apoptosis), are synthesized as inactive pro-caspases and activated in response to apoptotic signals, enabling the controlled degradation of cellular components.

Proteolytic activation offers a means of storing enzymes in an inactive form, ensuring that they are only activated under specific conditions. This regulatory strategy prevents premature or unwanted enzyme activity, which is particularly important in processes where unregulated activity could be harmful, such as in proteolytic enzymes that could damage cellular structures if activated prematurely.

7.4 FEEDBACK INHIBITION: REGULATING PATHWAYS THROUGH END-PRODUCT CONTROL

Feedback inhibition, also known as end-product inhibition, is a mechanism in which the end product of a metabolic pathway inhibits an enzyme involved early in the pathway. This form of regulation prevents the excessive accumulation of the final product and conserves cellular resources by limiting the production of intermediates when they are not needed.

In feedback inhibition, the end product binds to an allosteric site on the enzyme, altering its conformation and reducing its activity. This inhibition is often reversible, allowing the pathway to restart when the end product's concentration decreases. A classic example of feedback inhibition is seen in the regulation of threonine deaminase, an enzyme in the biosynthesis pathway for isoleucine. When isoleucine accumulates, it binds to threonine deaminase, inhibiting its activity and thereby controlling the production of isoleucine based on cellular requirements.

Feedback inhibition is essential for maintaining metabolic balance within cells, as it enables pathways to self-regulate without the need for external signals. This type of regulation also provides specificity, as each pathway can be independently modulated based on the needs of the cell or organism.

7.5 ENZYME COMPARTMENTALIZATION: SPATIAL REGULATION WITHIN CELLS

Enzyme compartmentalization refers to the localization of enzymes within specific organelles or cellular structures, effectively regulating their access to substrates and their involvement in metabolic processes. By confining enzymes to particular compartments, cells can control when and where enzymatic reactions occur, reducing unwanted interactions and increasing reaction efficiency.

For example, enzymes involved in fatty acid oxidation are located in the mitochondria, where they have direct access to substrates generated from carbohydrate metabolism. Similarly, enzymes responsible for DNA replication and repair are confined to the nucleus, preventing DNA damage and ensuring fidelity in genetic material duplication. This spatial regulation prevents competition for substrates and allows the cell to create microenvironments optimized for specific reactions, such as the low-pH conditions in lysosomes necessary for enzyme function in degradation pathways.

Compartmentalization not only contributes to the organization and efficiency of metabolic pathways but also provides an additional layer of regulatory control by restricting enzyme activity to certain cellular locations. This localization allows cells to isolate pathways that might otherwise interfere with one another, a crucial strategy for maintaining cellular homeostasis.

7.6 ISOZYMES: VARIANTS FOR SPECIALIZED REGULATION

Isozymes, or isoenzymes, are different molecular forms of an enzyme that catalyze the same reaction but differ in their amino acid sequence, kinetic properties, and regulatory mechanisms. Isozymes allow organisms to fine-tune enzyme activity for different tissues, developmental stages, or environmental conditions, providing tailored responses to specific physiological requirements.

A well-known example of isozymes is lactate dehydrogenase (LDH), an enzyme involved in anaerobic metabolism. LDH has multiple isoforms, each adapted to different tissues. In the heart, an isozyme of LDH is optimized to convert lactate back to pyruvate for use in aerobic metabolism, while in skeletal muscle, a different isozyme favors the conversion of pyruvate to lactate, supporting anaerobic energy production during intense exercise.

By producing multiple forms of the same enzyme, organisms gain the flexibility to regulate metabolic pathways in a tissue-specific manner. Isozymes thus serve as a versatile regulatory tool, adapting enzyme function to the specific demands of different cellular environments.

7.7 CONCLUSION

Enzyme regulatory strategies provide essential mechanisms for controlling the complex biochemical processes within living organisms. Through methods such as allosteric regulation, covalent modification, proteolytic activation, feedback inhibition, compartmentalization, and the use of isozymes, cells can modulate enzyme activity in response to metabolic needs and environmental changes. Each of these regulatory strategies offers unique advantages, allowing enzymes to be activated or inhibited in a highly controlled manner.

The understanding of enzyme regulatory mechanisms not only offers insights into cellular biochemistry but also has practical applications in medicine, biotechnology, and pharmacology. Drugs targeting specific enzyme regulators can treat diseases by restoring or blocking particular pathways. Additionally, engineered enzymes with customized regulatory features are being developed for industrial and therapeutic uses, showcasing the relevance of enzyme regulation in both natural and artificial systems. As our knowledge of enzyme regulation expands, so does our potential to harness these powerful biological catalysts to address a wide range of challenges in science and medicine.

CHAPTER 8

DNA, RNA AND THE FLOW OF GENETIC INFORMATION

The flow of genetic information from DNA to RNA to proteins is a fundamental process that underpins all life on Earth. This intricate pathway, often referred to as the central dogma of molecular biology, describes how genetic information is encoded, transcribed, and translated to produce the proteins that carry out various functions in living organisms.

8.1 STRUCTURE AND FUNCTION OF DNA

Deoxyribonucleic acid (DNA) is the hereditary material in nearly all living organisms, serving as the blueprint for biological functions. Structurally, DNA consists of two long strands of nucleotides that form a double helix. Each nucleotide comprises a phosphate group, a sugar (deoxyribose), and one of four nitrogenous bases: adenine (A), thymine (T), cytosine (C), or guanine (G). The sequence of these bases encodes genetic information, with specific sequences corresponding to genes that dictate the production of proteins. DNA is organized into structures called chromosomes, which are located in the nucleus of eukaryotic cells. In prokaryotic cells, DNA exists as a single circular chromosome in the cytoplasm. The double-helical structure of DNA allows for efficient packaging and protection of genetic information, as well as accurate replication during cell division. The antiparallel strands of DNA facilitate complementary base pairing, where A pairs with T, and C pairs with G, ensuring fidelity in replication and transcription processes.

8.2 THE ROLE OF RNA IN GENETIC INFORMATION TRANSFER

Ribonucleic acid (RNA) plays a critical role in the expression of genetic information. Unlike DNA, RNA is typically single-stranded and contains ribose sugar and the nitrogenous base uracil (U) instead of thymine. RNA serves several functions in the cell, primarily as a messenger, transfer, and ribosomal molecule.

- **Messenger RNA (mRNA):** mRNA is synthesized during the transcription of DNA and carries the genetic code from the nucleus to the ribosomes, where protein synthesis occurs. The sequence of nucleotides in mRNA is complementary to the DNA template strand, with uracil replacing thymine.

- **Transfer RNA (tRNA):** tRNA is responsible for transporting amino acids to the ribosome during translation. Each tRNA molecule has an anticodon that is complementary to a specific codon on the mRNA, ensuring the correct amino acid is added to the growing polypeptide chain.
- **Ribosomal RNA (rRNA):** rRNA is a structural and functional component of ribosomes, the cellular machinery responsible for protein synthesis. rRNA catalyzes the formation of peptide bonds between amino acids, facilitating the assembly of polypeptides based on the sequence encoded by mRNA.

8.3 TRANSCRIPTION: THE FIRST STEP IN GENE EXPRESSION

Transcription is the process by which the information encoded in a specific segment of DNA is copied into mRNA. This process occurs in the nucleus of eukaryotic cells and involves several key steps:

- **Initiation:** Transcription begins when RNA polymerase binds to a specific region called the promoter, located upstream of the gene. This binding is facilitated by various transcription factors that help recruit RNA polymerase to the correct site. Once bound, RNA polymerase unwinds a small section of the DNA double helix.
- **Elongation:** As RNA polymerase moves along the DNA template strand, it synthesizes a complementary strand of RNA by adding ribonucleotides in the 5' to 3' direction. The RNA strand grows as RNA polymerase continues to move along the DNA, unwinding the double helix ahead and rewinding it behind.
- **Termination:** Transcription concludes when RNA polymerase reaches a termination signal in the DNA sequence, prompting it to detach from the DNA and release the newly synthesized mRNA. In eukaryotic cells, this mRNA undergoes several modifications before it is exported from the nucleus. These modifications include the addition of a 5' cap and a poly-A tail, as well as splicing to remove non-coding regions (introns) and join coding regions (exons).

8.4 TRANSLATION: THE SYNTHESIS OF PROTEINS

Translation is the process by which the sequence of nucleotides in mRNA is translated into a specific sequence of amino acids, resulting in the formation of a protein. Translation occurs in the cytoplasm at the ribosome and involves several key steps:

- **Initiation:** Translation begins when the small ribosomal subunit binds to the mRNA at the start codon (AUG). The initiator tRNA, carrying methionine, pairs with the start codon, and the large ribosomal subunit assembles with the small subunit to form the complete ribosome.

- **Elongation:** During elongation, tRNA molecules, each carrying a specific amino acid, enter the ribosome at the A site (aminoacyl site) according to the codon sequence on the mRNA. The ribosome catalyzes the formation of a peptide bond between the amino acids at the P site (peptidyl site) and the A site, forming a growing polypeptide chain. The ribosome then shifts, moving the tRNA from the A site to the P site, and the process repeats until a stop codon is reached.
- **Termination:** Translation terminates when the ribosome encounters a stop codon (UAA, UAG, or UGA). Release factors bind to the ribosome, promoting the release of the completed polypeptide chain and the disassembly of the ribosomal subunits. The polypeptide may then undergo folding and post-translational modifications to achieve its final functional form.

8.5 REGULATION OF GENE EXPRESSION

Gene expression is a highly regulated process that allows cells to respond to environmental changes and maintain homeostasis. Various mechanisms control the expression of genes at multiple stages, including transcriptional, post-transcriptional, translational, and post-translational regulation.

- **Transcriptional Regulation:** The activity of transcription factors, which can enhance or repress transcription by binding to specific DNA sequences, is a primary means of regulating gene expression. The presence of enhancers and silencers in the regulatory regions of genes allows for complex interactions that determine whether a gene is expressed or silenced in response to cellular signals.
- **Post-Transcriptional Regulation:** After mRNA is synthesized, its stability, splicing, and translation efficiency can be modulated by various factors. RNA-binding proteins and microRNAs (miRNAs) can influence mRNA degradation or translation, thereby affecting the amount of protein produced from a given mRNA transcript.
- **Translational Regulation:** The initiation of translation can be regulated by factors that bind to the mRNA or ribosome, controlling the efficiency of protein synthesis. For example, the availability of specific tRNAs can influence the rate of translation.
- **Post-Translational Regulation:** Once proteins are synthesized, they can undergo modifications, such as phosphorylation, glycosylation, and ubiquitination, which can affect their activity, stability, and localization.

These modifications play crucial roles in signaling pathways, protein degradation, and the overall regulation of cellular functions.

8.6 THE IMPORTANCE OF GENETIC INFORMATION FLOW

The flow of genetic information from DNA to RNA to protein is a central concept in molecular biology and is essential for understanding how traits are inherited and expressed. This process is fundamental not only for growth and development but also for responding to environmental challenges and maintaining homeostasis in organisms.

The significance of this flow of information extends beyond individual cells. In multicellular organisms, the regulation of gene expression allows for cell specialization, where different cell types express unique sets of genes that enable them to perform specific functions. For instance, muscle cells express genes related to contraction, while neurons express genes essential for neurotransmission.

Additionally, the principles of DNA, RNA, and genetic information flow have important implications in biotechnology, medicine, and agriculture. Advances in genetic engineering, such as CRISPR-Cas9 technology, allow scientists to manipulate the flow of genetic information, leading to applications in gene therapy, crop improvement, and the development of new treatments for diseases.

8.7 CONCLUSION

The flow of genetic information from DNA to RNA to proteins is a complex and highly regulated process that is foundational to all life forms. Understanding the roles of DNA and RNA, the mechanisms of transcription and translation, and the regulatory strategies involved in gene expression provides insight into the molecular basis of biology. As research in genetics and molecular biology continues to advance, the implications of these processes will undoubtedly play a crucial role in fields ranging from medicine to biotechnology, paving the way for innovative solutions to many of today's challenges. The study of DNA, RNA, and the flow of genetic information not only enhances our comprehension of life itself but also empowers us to manipulate biological systems for the betterment of society.

CHAPTER 9
NUCLEIC ACID METHODS

Nucleic acids, comprising deoxyribonucleic acid (DNA) and ribonucleic acid (RNA), are fundamental to the storage, transmission, and expression of genetic information. The study of nucleic acids has been revolutionized by various methods that allow researchers to analyze, manipulate, and visualize these molecules. These techniques play crucial roles in molecular biology, genetics, and biotechnology.

9.1 DNA EXTRACTION: ISOLATING GENETIC MATERIAL

DNA extraction is the foundational technique used to isolate deoxyribonucleic acid from cells or tissues. The process is essential for many molecular biology applications, including cloning, sequencing, and genetic analysis. The basic steps of DNA extraction involve cell lysis, removal of proteins and contaminants, and precipitation of DNA.

The first step, cell lysis, is typically achieved using a lysis buffer that disrupts the cell membrane and releases the DNA into solution. The lysis buffer often contains detergents to break down lipids and enzymes such as proteinase K to digest proteins associated with the DNA. Following lysis, proteins and other cellular debris are removed through centrifugation, leaving a solution containing the DNA.

DNA is then precipitated using alcohol (ethanol or isopropanol), which causes the DNA to aggregate and form visible strands. After washing and drying, the extracted DNA is resuspended in a suitable buffer or water for downstream applications. This method is crucial in various fields, including forensic science, where it enables the analysis of genetic material from crime scenes, and in medical diagnostics, where it allows for genetic testing and disease identification.

9.2 POLYMERASE CHAIN REACTION (PCR): AMPLIFYING DNA

The polymerase chain reaction (PCR) is a powerful technique used to amplify specific DNA sequences, making millions of copies from a small initial sample. PCR has transformed molecular biology by enabling the analysis of minute amounts of DNA, which is critical for applications such as cloning, sequencing, and gene expression analysis.

PCR involves three main steps: denaturation, annealing, and extension. During denaturation, the double-stranded DNA is heated to around 94-98°C, causing the strands to separate. The temperature is then lowered to approximately 50-65°C during the annealing step, allowing short DNA primers to bind to complementary sequences flanking the target region. Finally, during the extension phase, the temperature is raised to around 72°C, where a thermostable DNA polymerase synthesizes new DNA strands by adding nucleotides to the primers.

One of the key innovations in PCR is the use of Taq polymerase, a heat-stable enzyme derived from the bacterium *Thermus aquaticus*. This allows the enzyme to remain active throughout the multiple cycles of heating and cooling necessary for PCR. The ability to amplify specific DNA sequences rapidly and efficiently has made PCR a cornerstone of modern molecular biology, with applications in medical diagnostics, genetic research, and forensic analysis.

9.3 DNA SEQUENCING: DECIPHERING GENETIC INFORMATION

DNA sequencing is the process of determining the exact order of nucleotides in a DNA molecule. This method has become fundamental in genomics, evolutionary biology, and personalized medicine. Various sequencing technologies exist, but two primary methods have dominated: Sanger sequencing and next-generation sequencing (NGS).

Sanger sequencing, developed by Frederick Sanger in the 1970s, is based on the selective incorporation of chain-terminating dideoxynucleotides during DNA replication. In this method, a mixture of normal deoxynucleotides and modified dideoxynucleotides is used, causing the synthesis of fragments of varying lengths that can be separated by capillary electrophoresis. The resulting fragments can be visualized, allowing researchers to deduce the DNA sequence.

On the other hand, next-generation sequencing (NGS) encompasses a range of modern sequencing technologies that enable the simultaneous sequencing of millions of DNA fragments.

NGS platforms, such as Illumina and Ion Torrent, utilize massively parallel sequencing techniques, significantly reducing the time and cost of sequencing whole genomes. NGS has revolutionized fields such as cancer genomics, metagenomics, and population genetics by enabling researchers to analyze large datasets and identify genetic variations on a scale previously unattainable.

9.4 HYBRIDIZATION TECHNIQUES: PROBING NUCLEIC ACIDS

Nucleic acid hybridization techniques are essential for detecting and analyzing specific nucleic acid sequences. These methods rely on the ability of complementary strands of DNA or RNA to anneal to each other, forming stable double-stranded structures. Hybridization techniques are widely used in applications such as gene expression analysis, genotyping, and diagnostic assays.

One of the most common hybridization techniques is Southern blotting, named after its inventor, Edwin Southern. This method allows for the detection of specific DNA sequences within a complex mixture. After DNA is digested with restriction enzymes and separated by gel electrophoresis, it is transferred to a membrane and hybridized with a labeled probe complementary to the target sequence. The bound probe can be detected, revealing the presence and size of the target DNA.

Another important technique is Northern blotting, which is used to study RNA. Similar to Southern blotting, Northern blotting involves the separation of RNA samples by gel electrophoresis, followed by transfer to a membrane and hybridization with a labeled probe. This technique allows researchers to analyze gene expression levels and alternative splicing events.

In addition to blotting techniques, in situ hybridization is a powerful method for localizing specific nucleic acid sequences within tissues or cells. By using labeled probes that hybridize to target RNA or DNA within fixed cells or tissue sections, researchers can visualize the spatial distribution of genes, providing valuable insights into gene expression patterns during development or disease.

9.5 CRISPR-CAS9: A REVOLUTIONARY TOOL FOR GENOME EDITING

CRISPR-Cas9 is a groundbreaking technology that allows for precise editing of genomic DNA. This method, derived from a natural defense mechanism in bacteria, has transformed molecular biology and genetic engineering by enabling targeted modifications of specific genes with unprecedented efficiency and accuracy.

The CRISPR-Cas9 system consists of two key components: the Cas9 endonuclease and a synthetic guide RNA (gRNA).

The gRNA is designed to be complementary to a specific target DNA sequence, directing the Cas9 enzyme to the desired location in the genome. Once the gRNA-Cas9 complex binds to the target DNA, the Cas9 enzyme introduces a double-strand break in the DNA.

Cells have inherent mechanisms to repair double-strand breaks, primarily through two pathways: non-homologous end joining (NHEJ) and homology-directed repair (HDR). NHEJ often results in insertions or deletions (indels) at the break site, leading to gene knockout, while HDR can be exploited to introduce precise changes by providing a donor template with the desired modification.

The CRISPR-Cas9 technology has numerous applications, including gene knockout studies, gene therapy for genetic disorders, and agricultural biotechnology for developing crops with improved traits. Its simplicity, cost-effectiveness, and versatility have made CRISPR-Cas9 one of the most significant advancements in modern genetics.

9.6 QUANTITATIVE PCR: MEASURING GENE EXPRESSION

Quantitative PCR (qPCR), also known as real-time PCR, is a technique used to quantify the amount of specific DNA or RNA in a sample. This method provides a powerful means to assess gene expression levels and analyze genetic variations in various contexts, including cancer research, infectious disease diagnostics, and developmental biology.

In qPCR, the amplification of target DNA is monitored in real-time through the use of fluorescent dyes or probes that emit signals proportional to the amount of DNA generated during each cycle. As the PCR progresses, the increase in fluorescence is measured, allowing for the quantification of the initial amount of target nucleic acid. This data can be used to compare gene expression levels across different samples or conditions.

qPCR can be applied in various experimental designs, such as absolute quantification, where the exact copy number of a target gene is determined, or relative quantification, which compares expression levels between different samples. This technique has become indispensable in molecular biology research, enabling researchers to gain insights into gene regulation, cellular responses, and disease mechanisms.

9.7 CONCLUSION

Nucleic acid methods are integral to the advancement of molecular biology and genetics, enabling researchers to analyze, manipulate, and visualize DNA and RNA with remarkable precision. Techniques such as DNA extraction, PCR, sequencing, hybridization, CRISPR-Cas9, and qPCR have transformed scientific research and have wide-ranging applications in medicine, agriculture, and biotechnology.

As technology continues to evolve, novel nucleic acid methods will likely emerge, providing even greater capabilities for understanding the complexities of genetic information. The ongoing development and refinement of these techniques will further enhance our ability to explore the intricacies of life at the molecular level, leading to significant advancements in health, disease prevention, and biotechnological innovation. The study of nucleic acids remains a cornerstone of modern science, with the potential to unlock new discoveries that will shape our understanding of biology for years to come.

CHAPTER 10

EXPLORING EVOLUTION AND BIOINFORMATICS

Evolution is a fundamental concept in biology that explains the diversity of life on Earth and how species adapt to their environments over time. It is the process by which organisms change through variations in their genetic material, influenced by natural selection, genetic drift, mutations, and gene flow. As researchers seek to understand evolutionary processes, the field of bioinformatics has emerged as a vital tool, integrating biology, computer science, and mathematics to analyze and interpret complex biological data.

10.1 UNDERSTANDING EVOLUTIONARY BIOLOGY

Evolutionary biology examines how species evolve over time and the mechanisms driving these changes. The theory of evolution, first articulated by Charles Darwin in the 19th century, posits that all living organisms share a common ancestor and that species diverge through a process of adaptation to their environments. Central to this theory are concepts such as natural selection, which explains how advantageous traits increase an organism's chances of survival and reproduction.

The evolutionary history of organisms is often represented through phylogenetic trees, which illustrate the relationships between different species based on shared ancestry.

These trees are constructed using various data types, including morphological traits, fossil records, and molecular sequences. By examining these relationships, scientists can infer how species have diverged and adapted over time, providing insight into the mechanisms of evolution.

10.2 THE ROLE OF BIOINFORMATICS IN EVOLUTIONARY STUDIES

Bioinformatics is a multidisciplinary field that combines biology, computer science, and statistics to analyze and interpret biological data. In the context of evolutionary studies, bioinformatics provides tools and methodologies for examining large datasets, such as genomic sequences, to uncover evolutionary patterns and relationships. The integration of bioinformatics into evolutionary biology has revolutionized the way researchers study evolutionary processes, allowing for more precise and comprehensive analyses.

One of the key advancements in bioinformatics is the ability to sequence genomes rapidly and at a lower cost. Next-generation sequencing (NGS) technologies have made it possible to generate vast amounts of genomic data from multiple species, enabling researchers to compare genetic sequences across different organisms. This comparative analysis is crucial for identifying conserved genes, understanding genetic variations, and inferring evolutionary relationships.

10.3 GENOMIC SEQUENCING AND COMPARATIVE GENOMICS

Genomic sequencing involves determining the complete DNA sequence of an organism's genome. This information provides a comprehensive view of an organism's genetic makeup, revealing insights into its evolutionary history. Comparative genomics is the field that focuses on comparing the genomes of different species to identify similarities and differences in their genetic sequences.

By comparing genomic sequences, researchers can identify conserved genes that have remained relatively unchanged throughout evolution, suggesting their importance in essential biological functions. Additionally, comparative genomics can reveal unique genes that may contribute to the adaptations of specific species to their environments. For example, researchers studying the genomes of different vertebrates can identify genetic adaptations related to respiratory systems, sensory organs, and immune responses, shedding light on how these species have evolved to survive in diverse habitats.

Bioinformatics tools, such as sequence alignment algorithms, play a critical role in comparative genomics. These tools enable researchers to align DNA or protein sequences from different organisms, highlighting regions of similarity and divergence. Phylogenetic analysis can then be performed using these aligned sequences to construct evolutionary trees that depict the relationships between species and trace their evolutionary paths.

10.4 PHYLOGENETICS: UNRAVELING EVOLUTIONARY RELATIONSHIPS

Phylogenetics is the study of the evolutionary relationships between organisms, often represented as phylogenetic trees. These trees provide a visual representation of the evolutionary history of species, illustrating how they have diverged from common ancestors over time. Bioinformatics plays a vital role in constructing and analyzing these trees by providing computational methods to process large datasets.

There are two primary approaches to phylogenetic analysis: distance-based methods and character-based methods. Distance-based methods, such as the neighbor-joining algorithm, rely on calculating the genetic distances between pairs of sequences to construct trees. In contrast, character-based methods, such as maximum likelihood and Bayesian inference, use models of evolution to estimate the likelihood of observing the data given different tree topologies.

Advancements in bioinformatics have led to the development of software tools that facilitate phylogenetic analysis, allowing researchers to analyze large datasets and construct phylogenetic trees with greater accuracy. These tools can also incorporate uncertainty into the analysis, providing confidence estimates for the inferred relationships. By utilizing these methods, researchers can gain insights into the timing of evolutionary events, such as speciation and extinction, and explore patterns of diversification across different lineages.

10.5 EVOLUTIONARY GENOMICS: EXPLORING ADAPTATION AND SPECIATION

Evolutionary genomics is an interdisciplinary field that combines genomics and evolutionary biology to study the genetic basis of adaptation and speciation. Through the lens of bioinformatics, researchers can analyze genomic data to identify genes associated with adaptive traits and understand the mechanisms underlying speciation events.

One of the significant areas of interest in evolutionary genomics is genomic adaptation. By comparing the genomes of populations or species that inhabit different environments, researchers can identify specific genetic changes that contribute to adaptive traits. For example, studies on the genomes of cichlid fish in African lakes have revealed genetic variations linked to differences in feeding strategies, coloration, and habitat preferences. Such research not only elucidates the genetic basis of adaptation but also enhances our understanding of the evolutionary processes that drive biodiversity.

Speciation, the process by which new species arise, is another key focus of evolutionary genomics. By examining genomic data from closely related species, researchers can identify genetic changes associated with reproductive isolation and divergence. Bioinformatics tools enable the detection of genomic regions under selection, providing insights into the genes and pathways involved in the speciation process.

10.6 THE INTERSECTION OF EVOLUTIONARY BIOLOGY AND CONSERVATION

Understanding evolution and the genetic diversity of species is crucial for conservation efforts. Bioinformatics provides the tools needed to assess genetic diversity, population structure, and the evolutionary potential of species. By analyzing genomic data, conservation biologists can make informed decisions about the management of endangered species and the preservation of biodiversity.

For example, genomics can reveal the genetic health of a population, informing conservation strategies aimed at maintaining genetic diversity. Genetic bottlenecks, which occur when a population undergoes a significant reduction in size, can lead to decreased genetic diversity and increased inbreeding.

By identifying these risks through genomic analysis, conservationists can implement strategies to enhance genetic diversity, such as translocation or captive breeding programs.

Furthermore, bioinformatics can assist in understanding the evolutionary history of species, allowing for the identification of evolutionary significant units (ESUs) that are critical for conservation. ESUs are populations that possess distinct genetic characteristics and evolutionary histories, warranting protection to maintain biodiversity and evolutionary processes.

10.7 FUTURE DIRECTIONS IN EVOLUTION AND BIOINFORMATICS

The field of bioinformatics continues to evolve, driven by technological advancements and the growing availability of genomic data. As researchers collect more data from diverse organisms, new bioinformatics tools and methods will be developed to analyze and interpret this information effectively. The integration of machine learning and artificial intelligence into bioinformatics is poised to enhance data analysis capabilities, enabling the discovery of novel patterns and relationships in large datasets.

Furthermore, the application of bioinformatics in evolutionary studies will expand to include metagenomics, the study of genetic material recovered directly from environmental samples.

This approach allows researchers to explore the diversity of microorganisms in various habitats, revealing insights into their evolutionary histories and ecological roles. Metagenomic studies have the potential to uncover previously unknown species and expand our understanding of microbial evolution.

As the field of bioinformatics continues to advance, it will play an increasingly vital role in addressing pressing global challenges, such as climate change and habitat loss. By understanding the evolutionary processes that shape biodiversity, researchers can contribute to conservation strategies that promote the resilience of ecosystems in the face of environmental change.

10.8 CONCLUSION

The exploration of evolution and bioinformatics represents a dynamic and rapidly advancing field that has transformed our understanding of biological processes. By integrating computational tools with evolutionary biology, researchers can analyze vast amounts of genomic data, uncovering patterns and relationships that illuminate the history of life on Earth. The intersection of these disciplines not only enhances our knowledge of evolutionary mechanisms but also informs conservation efforts aimed at preserving biodiversity and promoting the resilience of ecosystems.

As we continue to unravel the complexities of evolution through bioinformatics, we gain valuable insights into the fundamental processes that shape life. This knowledge empowers scientists and conservationists alike to address the challenges facing our planet, ensuring that the richness of biodiversity is preserved for future generations. The synergy between evolution and bioinformatics promises to drive innovative research and solutions, paving the way for a deeper understanding of life and its evolutionary journey.

CHAPTER 11

CARBOHYDRATES AND GLYCOPROTEINS

11.1 CARBOHYDRATES

A carbohydrate is a biomolecule consisting of carbon (C), hydrogen (H) and oxygen (O) atoms, usually with a hydrogen–oxygen atom ratio of 2:1 (as in water) and thus with the empirical formula $C_m(H_2O)_n$ (where *m* may or may not be different from *n*), which does not mean the H has covalent bonds with O (for example with CH_2O, H has a covalent bond with C but not with O). However, not all carbohydrates conform to this precise stoichiometric definition (e.g., uronic acids, deoxy-sugars such as fucose), nor are all chemicals that do conform to this definition automatically classified as carbohydrates (e.g., formaldehyde and acetic acid).

The term is most common in biochemistry, where it is a synonym of saccharide a group that includes sugars, starch, and cellulose. The saccharides are divided into four chemical groups: monosaccharides, disaccharides, oligosaccharides, and polysaccharides. Monosaccharides and disaccharides, the smallest (lower molecular weight) carbohydrates, are commonly referred to as sugars. While the scientific nomenclature of carbohydrates is complex, the names of the monosaccharides and disaccharides very often end in the suffix *-ose*, which was originally taken from the word glucose and is used for almost all sugars (e.g., fructose (fruit sugar), sucrose (cane or beet sugar), ribose, lactose (milk sugar)).

Carbohydrates perform numerous roles in living organisms. Polysaccharides serve as an energy store (e.g., starch and glycogen) and as structural components (e.g., cellulose in plants and chitin in arthropods and fungi). The 5-carbon monosaccharide ribose is an important component of coenzymes (e.g., ATP, FAD and NAD) and the backbone of the genetic molecule known as RNA. The related deoxyribose is a component of DNA. Saccharides and their derivatives include many other important biomolecules that play key roles in the immune system, fertilization, preventing pathogenesis, blood clotting, and development.

Carbohydrates are central to nutrition and are found in a wide variety of natural and processed foods. Starch is a polysaccharide and is abundant in cereals (wheat, maize, rice), potatoes, and processed food based on cereal flour, such as bread, pizza or pasta. Sugars appear in human diet mainly as table sugar (sucrose, extracted from sugarcane or sugar beets), lactose (abundant in milk), glucose and fructose, both of which occur naturally in honey, many fruits, and some vegetables. Table sugar, milk, or honey is often added to drinks and many prepared foods such as jam, biscuits and cakes.

Cellulose, a polysaccharide found in the cell walls of all plants, is one of the main components of insoluble dietary fiber. Although it is not digestible by humans, cellulose and insoluble dietary fiber generally help maintain a healthy digestive system by facilitating bowel movements. Other polysaccharides contained in dietary fiber include resistant starch and inulin, which feed some bacteria in the microbiota of the large intestine, and are metabolized by these bacteria to yield short-chain fatty acids.

11.1.1 History

The history of the discovery regarding carbohydrates dates back around 10,000 years ago in Papua New Guinea during the cultivation of sugarcane during the Neolithic agricultural revolution. The term "carbohydrate" was first proposed by German chemist Carl Schmidt (chemist) in 1844. In 1856, glycogen, a form of carbohydrate storage in animal livers, was discovered by French physiologist Claude Bernard.

11.1.2 Structure

Formerly the name "carbohydrate" was used in chemistry for any compound with the formula $C_m (H_2O)_n$. Following this definition, some chemists considered formaldehyde (CH_2O) to be the simplest carbohydrate, while others claimed that title for glycolaldehyde. Today, the term is generally understood in the biochemistry sense, which excludes compounds with only one or two carbons and includes many biological carbohydrates which deviate from this formula. For example, while the above representative formulas would seem to capture the commonly known carbohydrates, ubiquitous and abundant carbohydrates often deviate from this. For example, carbohydrates often display chemical groups such as: *N*-acetyl (e.g., chitin), sulfate (e.g., glycosaminoglycans), carboxylic acid and deoxy modifications (e.g., fucose and sialic acid).

Natural saccharides are generally built of simple carbohydrates called monosaccharides with general formula $(CH_2O)_n$ where n is three or more. A typical monosaccharide has the structure $H–(CHOH)_x(C=O)–(CHOH)_y–H$, that is, an aldehyde or ketone with many hydroxyl groups added, usually one on each carbon atom that is not part of the aldehyde or ketone functional group. Examples of monosaccharides are glucose, fructose, and glyceraldehydes. However, some biological substances commonly called "monosaccharides" do not conform to this formula (e.g., uronic acids and deoxy-sugars such as fucose) and there are many chemicals that do conform to this formula but are not considered to be monosaccharides (e.g., formaldehyde CH_2O and inositol $(CH_2O)_6$). The open-chain form of a monosaccharide often coexists with a closed ring form where the aldehyde/ketone carbonyl group carbon (C=O) and hydroxyl group (–OH) react forming a hemiacetal with a new C–O–C bridge. Monosaccharides can be linked together into what are called polysaccharides (or oligosaccharides) in a large variety of ways. Many carbohydrates contain one or more modified monosaccharide units that have had one or more groups replaced or removed. For example, deoxyribose, a component of DNA, is a modified version of ribose; chitin is composed of repeating units of N-acetyl glucosamine, a nitrogen-containing form of glucose.

11.1.3 Division

Carbohydrates are polyhydroxy aldehydes, ketones, alcohols, acids, their simple derivatives and their polymers having linkages of the acetal type. They may be classified according to their degree of polymerization, and may be divided initially into three principal groups, namely sugars, oligosaccharides and polysaccharides.

Table 1: The Major Dietary Carbohydrates

Class (Degree of Polymerization)	Subgroup	Components
Sugars (1–2)	Monosaccharides	Glucose, galactose, fructose, xylose
	Disaccharides	Sucrose, lactose, maltose, isomaltulose, trehalose
	Polyols	Sorbitol, mannitol
Oligosaccharides (3–9)	Malto-oligosaccharides	Maltodextrins
	Other oligosaccharides	Raffinose, stachyose, fructo-oligosaccharides
Polysaccharides (>9)	Starch	Amylose, amylopectin, modified starches
	Non-starch polysaccharides	Glycogen, Cellulose, Hemicellulose, Pectins, Hydrocolloids

11.1.4 Monosaccharides

Monosaccharides are the simplest carbohydrates in that they cannot be hydrolyzed to smaller carbohydrates. They are aldehydes or ketones with two or more hydroxyl groups. The general chemical formula of an unmodified monosaccharide is $(C\cdot H_2O)_n$, literally a "carbon hydrate". Monosaccharides are important fuel molecules as well as building blocks for nucleic acids. The smallest monosaccharides, for which n=3, are dihydroxyacetone and D- and L-glyceraldehydes.

Classification of Monosaccharides

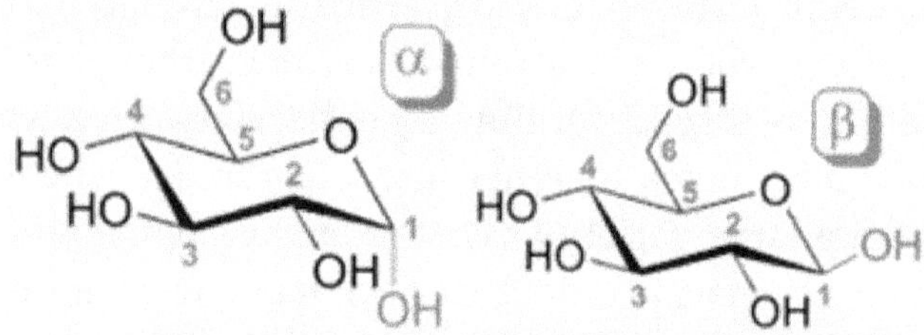

Fig 1: The α and β Anomers of Glucose

The position of the hydroxyl group (red or green) on the anomeric carbon relative to the CH_2OH group bound to carbon 5: they either have identical absolute configurations (R,R or S,S) (α), or opposite absolute configurations (R,S or S,R) (β).

Monosaccharides are classified according to three different characteristics: the placement of its carbonyl group, the number of carbon atoms it contains, and its chiral handedness. If the carbonyl group is an aldehyde, the monosaccharide is an aldose; if the carbonyl group is a ketone, the monosaccharide is a ketose. Monosaccharides with three carbon atoms are called trioses, those with four are called tetroses, five are called pentoses, six are hexoses, and so on. These two systems of classification are often combined. For example, glucose is an aldohexose (a six-carbon aldehyde), ribose is an aldopentose (a five-carbon aldehyde), and fructose is a ketohexose (a six-carbon ketone).

Each carbon atom bearing a hydroxyl group (-OH), with the exception of the first and last carbons, are asymmetric, making them stereo centers with two possible configurations each (R or S). Because of this asymmetry, a number of isomers may exist for any given monosaccharide formula. Using Le Bel-van't Hoff rule, the aldohexose D-glucose, for example, has the formula $(C{\cdot}H_2O)_6$, of which four of its six carbons atoms are stereogenic, making D-glucose one of 2^4=16 possible stereoisomers. In the case of glyceraldehydes, an aldotriose, there is one pair of possible stereoisomers, which are enantiomers and epimers. 1,3-dihydroxyacetone, the ketose corresponding to the aldose glyceraldehydes, is a symmetric molecule with no stereo centers. The assignment of D or L is made according to the orientation of the asymmetric carbon furthest from the carbonyl group: in a standard Fischer projection if the hydroxyl group is on the right the molecule is a D sugar, otherwise it is an L sugar. The "D-" and "L-" prefixes should not be confused with "d-" or "l-", which indicate the direction that the sugar rotates plane polarized light. This usage of "d-" and "l-" is no longer followed in carbohydrate chemistry.

Ring-Straight Chain Isomerism

The aldehyde or ketone group of a straight-chain monosaccharide will react reversibly with a hydroxyl group on a different carbon atom to form a hemiacetal or hemiketal, forming a heterocyclic ring with an oxygen bridge between two carbon atoms. Rings with five and six atoms are called furanose and pyranose forms, respectively, and exist in equilibrium with the straight-chain form.

During the conversion from straight-chain form to the cyclic form, the carbon atom containing the carbonyl oxygen, called the anomeric carbon, becomes a stereogenic center with two possible configurations: The oxygen atom may take a position either above or below the plane of the ring. The resulting possible pair of stereoisomers is called anomers. In the α *anomer*, the -OH substituent on the anomeric carbon rests on the opposite side (trans) of the ring from the CH_2OH side branch. The alternative form, in which the CH_2OH substituent and the anomeric hydroxyl are on the same side (cis) of the plane of the ring, is called the β *anomer*.

Use in Living Organisms

Monosaccharides are the major fuel source for metabolism, being used both as an energy source (glucose being the most important in nature as it is the product of photosynthesis in plants) and in biosynthesis. When monosaccharides are not immediately needed, they are often converted to more space-efficient (i.e., less water-soluble) forms, often polysaccharides. In many animals, including humans, this storage form is glycogen, especially in liver and muscle cells. In plants, starch is used for the same purpose.

The most abundant carbohydrate, cellulose, is a structural component of the cell wall of plants and many forms of algae. Ribose is a component of RNA. Deoxyribose is a component of DNA. Lyxose is a component of lyxoflavin found in the human heart. Ribulose and xylulose occur in the pentose phosphate pathway. Galactose, a component of milk sugar lactose, is found in galactolipids in plant cell membranes and in glycoproteins in many tissues.

Mannose occurs in human metabolism, especially in the glycosylation of certain proteins. Fructose, or fruit sugar, is found in many plants and humans, it is metabolized in the liver, absorbed directly into the intestines during digestion, and found in semen. Trehalose, a major sugar of insects, is rapidly hydrolyzed into two glucose molecules to support continuous flight.

11.1.5 Disaccharides

Two joined monosaccharides are called a disaccharide, the simplest kind of polysaccharide. Examples include sucrose and lactose. They are composed of two monosaccharide units bound together by a covalent bond known as a glycosidic linkage formed via a dehydration reaction, resulting in the loss of a hydrogen atom from one monosaccharide and a hydroxyl group from the other. The formula of unmodified disaccharides is $C_{12}H_{22}O_{11}$. Although there are numerous kinds of disaccharides, a handful of disaccharides are particularly notable.

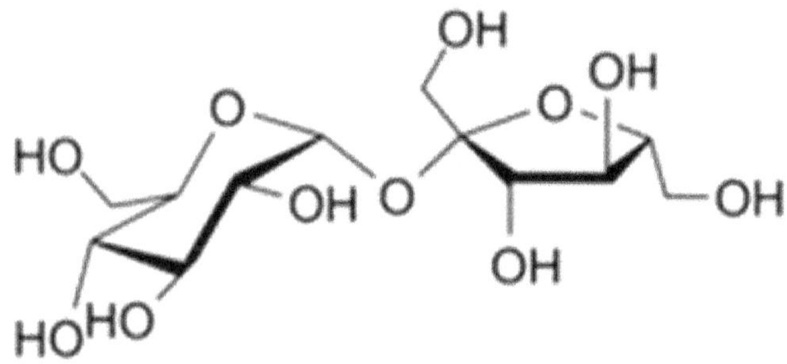

Fig 2: Sucrose, also Known as Table Sugar, is a Common Disaccharide. It is Composed of Two Monosaccharides: D-Glucose (Left) and D-Fructose (Right)

Sucrose, pictured to the right, is the most abundant disaccharide, and the main form in which carbohydrates are transported in plants. It is composed of one D-glucose molecule and one D-fructose molecule. The systematic name for sucrose, *O*-α-D-glucopyranosyl-(1→2)-D-fructofuranoside, indicates four things:

- **Its Monosaccharides:** Glucose and fructose.
- **Their Ring Types:** Glucose is a pyranose and fructose is a furanose.
- **How They are Linked Together:** The oxygen on carbon number 1 (C1) of α-D-glucose is linked to the C2 of D-fructose.
- The *-oside* suffix indicates that the anomeric carbon of both monosaccharides participates in the glycosidic bond.

Lactose, a disaccharide composed of one D-galactose molecule and one D-glucose molecule, occurs naturally in mammalian milk.

The systematic name for lactose is *O*-β-D-galactopyranosyl-(1→4)-D-glucopyranose. Other notable disaccharides include maltose (two D-glucoses linked α-1,4) and cellobiose (two D-glucoses linked β-1,4).

Disaccharides can be classified into two types: reducing and non-reducing disaccharides. If the functional group is present in bonding with another sugar unit, it is called a reducing disaccharide or biose.

11.1.6 Oligosaccharides and Polysaccharides

Oligosaccharides

Oligosaccharides are saccharide polymers composed of three to ten units of monosaccharides, connected via glycosidic linkages, similar to disaccharides. They are usually linked to lipids or amino acids glycosic linkage with oxygen or nitrogen to form glygolipids and glycoproteins, though some, like the raffinose series and the fructooligosaccharides, do not. They have roles in cell recognition and cell adhesion.

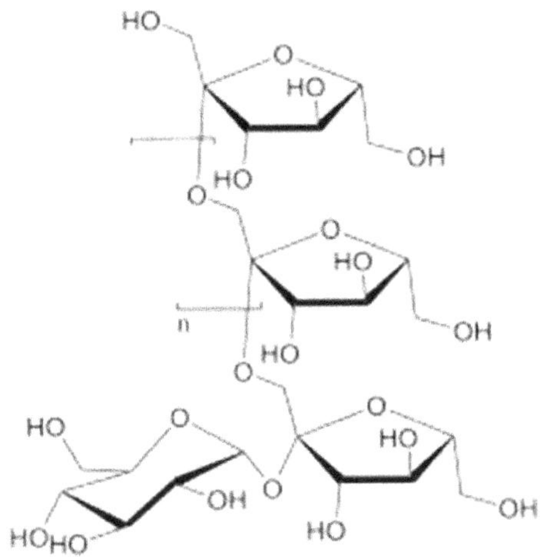

Fig 3: The Structure of Fructooligosaccharide

11.1.7 Nutrition

Carbohydrate consumed in food yields 3.87 kilocalories of energy per gram for simple sugars, and 3.57 to 4.12 kilocalories per gram for complex carbohydrate in most other foods. Relatively high levels of carbohydrate are associated with processed foods or refined foods made from plants, including sweets, cookies and candy, table sugar, honey, soft drinks, breads and crackers, jams and fruit products, pastas and breakfast cereals. Lower amounts of digestible carbohydrate are usually associated with unrefined foods as these foods have more fiber, including beans, tubers, rice, and unrefined fruit. Animal-based foods generally have the lowest carbohydrate levels, although milk does contain a high proportion of lactose.

Organisms typically cannot metabolize all types of carbohydrate to yield energy. Glucose is a nearly universal and accessible source of energy. Many organisms also have the ability to metabolize other monosaccharides and disaccharides but glucose is often metabolized first. In *Escherichia coli*, for example, the lac operon will express enzymes for the digestion of lactose when it is present, but if both lactose and glucose are present the *lac* operon is repressed, resulting in the glucose being used first.

Polysaccharides are also common sources of energy. Many organisms can easily break down starches into glucose; most organisms, however, cannot metabolize cellulose or other polysaccharides like chitin and arabinoxylans. These carbohydrate types can be metabolized by some bacteria and protists. Ruminants and termites, for example, use microorganisms to process cellulose. Even though these complex carbohydrates are not very digestible, they represent an important dietary element for humans, called dietary fiber. Fiber enhances digestion, among other benefits.

The Institute of Medicine recommends that American and Canadian adults get between 45 and 65% of dietary energy from whole-grain carbohydrates. The Food and Agriculture Organization and World Health Organization jointly recommend that national dietary guidelines set a goal of 55–75% of total energy from carbohydrates, but only 10% directly from sugars (their term for simple carbohydrates). A 2017 Cochrane Systematic Review concluded that there was insufficient evidence to support the claim that whole grain diets can affect cardiovascular disease.

Classification

Nutritionists often refer to carbohydrates as either simple or complex. However, the exact distinction between these groups can be ambiguous. The term *complex carbohydrate* was first used in the U.S. Senate Select Committee on Nutrition and Human Needs publication *Dietary Goals for the United States* (1977) where it was intended to distinguish sugars from other carbohydrates (which were perceived to be nutritionally superior). However, the report put "fruit, vegetables and whole-grains" in the complex carbohydrate column, despite the fact that these may contain sugars as well as polysaccharides. This confusion persists as today some nutritionists use the term complex carbohydrate to refer to any sort of digestible saccharide present in a whole food, where fiber, vitamins and minerals are also found (as opposed to processed carbohydrates, which provide energy but few other nutrients). The standard usage, however, is to classify carbohydrates chemically: simple if they are sugars (monosaccharides and disaccharides) and complex if they are polysaccharides (or oligosaccharides).

In any case, the simple vs. complex chemical distinction has little value for determining the nutritional quality of carbohydrates. Some simple carbohydrates (e.g., fructose) raise blood glucose rapidly, while some complex carbohydrates (starches), raise blood sugar slowly. The speed of digestion is determined by a variety of factors including which other nutrients are consumed with the carbohydrate, how the food is prepared, individual differences in metabolism, and the chemistry of the carbohydrate. Carbohydrates are sometimes divided into "available carbohydrates", which are absorbed in the small intestine and "unavailable carbohydrates", which pass to the large intestine, where they are subject to fermentation by the gastrointestinal microbiota.

The USDA's *Dietary Guidelines for Americans 2010* call for moderate- to high-carbohydrate consumption from a balanced diet that includes six one-ounce servings of grain foods each day, at least half from whole grain sources and the rest are from enriched.

The glycemic index (GI) and glycemic load concepts have been developed to characterize food behavior during human digestion. They rank carbohydrate-rich foods based on the rapidity and magnitude of their effect on blood glucose levels. Glycemic index is a measure of how quickly food glucose is absorbed, while glycemic load is a measure of the total absorbable glucose in foods. The insulin index is a similar, more recent classification method that ranks foods based on their effects on blood insulin levels, which are caused by glucose (or starch) and some amino acids in food.

Health Effects of Dietary Carbohydrate Restriction

Low-carbohydrate diets may miss the health advantages – such as increased intake of dietary fiber – afforded by high-quality carbohydrates found in legumes and pulses, whole grains, fruits, and vegetables.

A "meta-analysis, of moderate quality," included as adverse effects of the diet halitosis, headache and constipation.

Carbohydrate-restricted diets can be as effective as low-fat diets in helping achieve weight loss over the short term when overall calorie intake is reduced. An Endocrine Society scientific statement said that "when calorie intake is held constant [...] body-fat accumulation does not appear to be affected by even very pronounced changes in the amount of fat vs carbohydrate in the diet." In the long term, effective weight loss or maintenance depends on calorie restriction, not the ratio of macronutrients in a diet.

The reasoning of diet advocates that carbohydrates cause undue fat accumulation by increasing blood insulin levels, and that low-carbohydrate diets have a "metabolic advantage", is not supported by clinical evidence. Further, it is not clear how low-carbohydrate dieting affects cardiovascular health, although two reviews showed that carbohydrate restriction may improve lipid markers of cardiovascular disease risk.

Carbohydrate-restricted diets are no more effective than a conventional healthy diet in preventing the onset of type 2 diabetes, but for people with type 2 diabetes, they are a viable option for losing weight or helping with glycemic control.

There is limited evidence to support routine use of low-carbohydrate dieting in managing type 1 diabetes. The American Diabetes Association recommends that people with diabetes should adopt a generally healthy diet, rather than a diet focused on carbohydrate or other macronutrients.

An extreme form of low-carbohydrate diet – the ketogenic diet – is established as a medical diet for treating epilepsy.

Through celebrity endorsement during the early 21st century, it became a fad diet as a means of weight loss, but with risks of undesirable side effects, such as low energy levels and increased hunger, insomnia, nausea, and gastrointestinal discomfort. The British Dietetic Association named it one of the "top 5 worst celeb diets to avoid in 2018".

11.1.8 Sources

Most dietary carbohydrates contain glucose, either as their only building block (as in the polysaccharides starch and glycogen), or together with another monosaccharide (as in the hetero-polysaccharides sucrose and lactose).

Table 2: Sugar Content of Selected Common Plant Foods (In Grams Per 100 g)

Food Item	Carbohydrate, Total, Including Dietary Fiber	Total Sugars	Free Fructose	Free Glucose	Sucrose	Ratio of Fructose/ Glucose	Sucrose as Proportion of Total Sugars (%)
Fruits							
Apple	13.8	10.4	5.9	2.4	2.1	2.0	19.9
Apricot	11.1	9.2	0.9	2.4	5.9	0.7	63.5
Banana	22.8	12.2	4.9	5.0	2.4	1.0	20.0
Fig, dried	63.9	47.9	22.9	24.8	0.9	0.93	0.15
Grapes	18.1	15.5	8.1	7.2	0.2	1.1	1
Navel orange	12.5	8.5	2.25	2.0	4.3	1.1	50.4
Peach	9.5	8.4	1.5	2.0	4.8	0.9	56.7
Pear	15.5	9.8	6.2	2.8	0.8	2.1	8.0
Pineapple	13.1	9.9	2.1	1.7	6.0	1.1	60.8
Plum	11.4	9.9	3.1	5.1	1.6	0.66	16.2
Vegetables							
Beet, red	9.6	6.8	0.1	0.1	6.5	1.0	96.2
Carrot	9.6	4.7	0.6	0.6	3.6	1.0	77
Red -pepper, sweet	6.0	4.2	2.3	1.9	0.0	1.2	0.0
Onion, sweet	7.6	5.0	2.0	2.3	0.7	0.9	14.3
Sweet potato	20.1	4.2	0.7	1.0	2.5	0.9	60.3
Yam	27.9	0.5	Traces	Traces	Traces	—	Traces
Sugar cane		13–18	0.2–1.0	0.2–1.0	11–16	1.0	high
Sugar beet		17–18	0.1–0.5	0.1–0.5	16–17	1.0	high
Grains							
Corn, sweet	19.0	6.2	1.9	3.4	0.9	0.61	15.0

Unbound glucose is one of the main ingredients of honey. Glucose is extremely abundant and has been isolated from a variety of natural sources across the world, including male cones of the coniferous tree Wollemia nobilis in Rome, the roots of Ilex asprella plants in China, and straws from rice in California.

11.1.9 Metabolism

Carbohydrate metabolism is the series of biochemical processes responsible for the formation, breakdown and interconversion of carbohydrates in living organisms.

The most important carbohydrate is glucose, a simple sugar (monosaccharide) that is metabolized by nearly all known organisms. Glucose and other carbohydrates are part of a wide variety of metabolic pathways across species: plants synthesize carbohydrates from carbon dioxide and water by photosynthesis storing the absorbed energy internally, often in the form of starch or lipids. Plant components are consumed by animals and fungi, and used as fuel for cellular respiration. Oxidation of one gram of carbohydrate yields approximately 16 kJ (4 kcal) of energy, while the oxidation of one gram of lipids yields about 38 kJ (9 kcal). The human body stores between 300 and 500 g of carbohydrates depending on body weight, with the skeletal muscle contributing to a large portion of the storage. Energy obtained from metabolism (e.g., oxidation of glucose) is usually stored temporarily within cells in the form of ATP. Organisms capable of anaerobic and aerobic respiration metabolize glucose and oxygen (aerobic) to release energy, with carbon dioxide and water as byproducts.

Catabolism

Catabolism is the metabolic reaction which cells undergo to break down larger molecules, extracting energy. There are two major metabolic pathways of monosaccharide catabolism: glycolysis and the citric acid cycle.

In glycolysis, oligo- and polysaccharides are cleaved first to smaller monosaccharides by enzymes called glycoside hydrolases. The monosaccharide units can then enter into monosaccharide catabolism. A 2 ATP investment is required in the early steps of glycolysis to phosphorylate Glucose to Glucose 6-Phosphate (G6P) and Fructose 6-Phosphate (F6P) to Fructose 1,6-biphosphate (FBP), thereby pushing the reaction forward irreversibly. In some cases, as with humans, not all carbohydrate types are usable as the digestive and metabolic enzymes necessary are not present.

11.1.10 Carbohydrate Chemistry

Carbohydrate chemistry is a large and economically important branch of organic chemistry. Some of the main organic reactions that involve carbohydrates are:

- Amadori rearrangement.
- Carbohydrate acetalisation.

- Carbohydrate digestion.
- Cyanohydrin reaction.
- Koenigs–Knorr reaction.
- Lobry de Bruyn–Van Ekenstein transformation.
- Nef reaction.
- Wohl degradation.
- Tipson-Cohen reaction.
- Ferrier rearrangement.
- Ferrier II reaction.

11.1.11 Chemical Synthesis

Carbohydrate synthesis is a sub-field of organic chemistry concerned specifically with the generation of natural and unnatural carbohydrate structures. This can include the synthesis of monosaccharide residues or structures containing more than one monosaccharide, known as oligosaccharides. Selective formation of glycosidic linkages and selective reactions of hydroxyl groups are very important, and the usage of protecting groups is extensive.

Common reactions for glycosidic bond formation are as follows:

- Chemical glycosylation.
- Fischer glycosidation.
- Koenigs-Knorr reaction.
- Crich beta-mannosylation.

While some common protection methods are as below:

- Carbohydrate acetalisation.
- Trimethylsilyl.
- Benzyl ether.
- p-Methoxybenzyl ether.

11.2 GLYCOPROTEINS

Glycoproteins are proteins which contain oligosaccharide (sugar) chains covalently attached to amino acid side-chains. The carbohydrate is attached to the protein in a cotranslational or posttranslational modification. This process is known as glycosylation. Secreted extracellular proteins are often glycosylated.

In proteins that have segments extending extracellularly, the extracellular segments are also often glycosylated. Glycoproteins are also often important integral membrane proteins, where they play a role in cell–cell interactions. It is important to distinguish endoplasmic reticulum-based glycosylation of the secretory system from reversible cytosolic-nuclear glycosylation. Glycoproteins of the cytosol and nucleus can be modified through the reversible addition of a single GlcNAc residue that is considered reciprocal to phosphorylation and the functions of these are likely to be an additional regulatory mechanism that controls phosphorylation-based signalling. In contrast, classical secretory glycosylation can be structurally essential. For example, inhibition of asparagine-linked, i.e. N-linked, glycosylation can prevent proper glycoprotein folding and full inhibition can be toxic to an individual cell. In contrast, perturbation of glycan processing (enzymatic removal/addition of carbohydrate residues to the glycan), which occurs in both the endoplasmic reticulum and Golgi apparatus, is dispensable for isolated cells (as evidenced by survival with glycosides inhibitors) but can lead to human disease (congenital disorders of glycosylation) and can be lethal in animal models. It is therefore likely that the fine processing of glycans is important for endogenous functionality, such as cell trafficking, but that this is likely to have been secondary to its role in host-pathogen interactions. A famous example of this latter effect is the ABO blood group system.

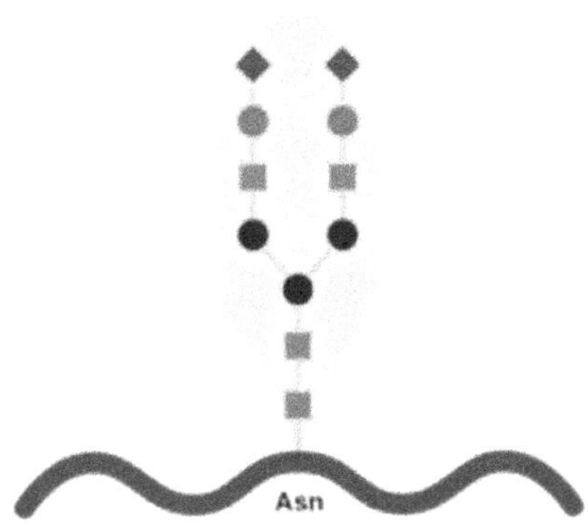

Fig 4: ***N*-Linked Protein Glycosylation (*N*-Glycosylation of *N*-Glycans) at Asn Residues (Asn-x-Ser/Thr Motifs) in Glycoproteins**

Though there are different types of glycoproteins, the most common are *N*-linked and *O*-linked glycoproteins. These two types of glycoproteins are distinguished by structural differences that give them their names. Glycoproteins vary greatly in composition, making many different compounds such as antibodies or hormones. Due to the wide array of functions within the body, interest in glycoprotein synthesis for medical use has increased. There are now several methods to synthesize glycoproteins, including recombination and glycosylation of proteins.

Glycosylation is also known to occur on nucleocytoplasmic proteins in the form of *O*-GlcNAc.

11.2.1 Types of Glycosylation

There are several types of glycosylation, although the first two are the most common.

- In *N*-glycosylation, sugars are attached to nitrogen, typically on the amide side-chain of asparagine.
- In *O*-glycosylation, sugars are attached to oxygen, typically on serine or threonine, but also on tyrosine or non-canonical amino acids such as hydroxylysine and hydroxyproline.
- In *P*-glycosylation, sugars are attached to phosphorus on a phosphoserine.
- In *C*-glycosylation, sugars are attached directly to carbon, such as in the addition of mannose to tryptophan.
- In *S*-glycosylation, a beta-GlcNAc is attached to the sulfur atom of a cysteine residue.
- In glypiation, a GPI glycolipid is attached to the C-terminus of a polypeptide, serving as a membrane anchor.
- In glycation, also known as non-enzymatic glycosylation, sugars are covalently bonded to a protein or lipid molecule, without the controlling action of an enzyme, but through a Maillard reaction.

11.2.2 Monosaccharides

Monosaccharides commonly found in eukaryotic glycoproteins include:

Table 3: The Principal Sugars Found in Human Glycoproteins

Sugar	Type	Abbreviation
β-D-Glucose	Hexose	Glc
β-D-Galactose	Hexose	Gal
β-D-Mannose	Hexose	Man
α-L-Fucose	Deoxyhexose	Fuc
N-Acetylgalactosamine	Aminohexose	GalNAc
N-Acetylglucosamine	Aminohexose	GlcNAc
N-Acetylneuraminic acid	Aminononulosonic acid (Sialic acid)	NeuNAc
Xylose	Pentose	Xyl

The sugar group(s) can assist in protein folding, improve proteins' stability and are involved in cell signalling.

11.2.3 Structure

The critical structural element of all glycoproteins is having oligosaccharides bonded covalently to a protein. There are 10 common monosaccharides in mammalian glycans including: glucose (Glc), fucose (Fuc), xylose (Xyl), mannose (Man), galactose (Gal), N-acetylglucosamine (GlcNAc), glucuronic acid (GlcA), iduronic acid (IdoA), N-acetylgalactosamine (GalNAc), sialic acid, and 5-N-acetylneuraminic acid (Neu5Ac). These glycans link themselves to specific areas of the protein amino acid chain.

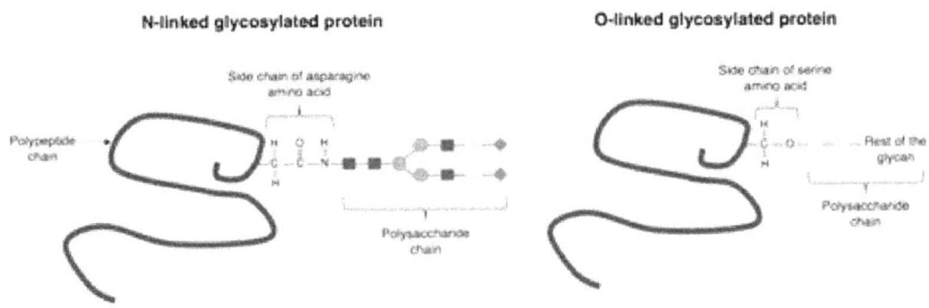

Fig 5: ***N*-Linked and *O*-Linked Glycoproteins**

The two most common linkages in glycoproteins are *N*-linked and *O*-linked glycoproteins. An *N*-linked glycoprotein has glycan bonds to the nitrogen containing an asparagine amino acid within the protein sequence. An *O*-linked glycoprotein has the sugar is bonded to an oxygen atom of a serine or threonine amino acid in the protein.

Glycoprotein size and composition can vary largely, with carbohydrate composition ranges from 1% to 70% of the total mass of the glycoprotein. Within the cell, they appear in the blood, the extracellular matrix, or on the outer surface of the plasma membrane, and make up a large portion of the proteins secreted by eukaryotic cells. They are very broad in their applications and can function as a variety of chemicals from antibodies to hormones.

Glycomics

Glycomics is the study of the carbohydrate components of cells. Though not exclusive to glycoproteins, it can reveal more information about different glycoproteins and their structure. One of the purposes of this field of study is to determine which proteins are glycosylated and where in the amino acid sequence the glycosylation occurs. Historically, mass spectrometry has been used to identify the structure of glycoproteins and characterize the carbohydrate chains attached.

11.2.4 Examples

The unique interaction between the oligosaccharide chains have different applications. First, it aids in quality control by identifying misfolded proteins. The oligosaccharide chains also change the solubility and polarity of the proteins that they are bonded to. For example, if the oligosaccharide chains are negatively charged, with enough density around the protein, they can repulse proteolytic enzymes away from the bonded protein. The diversity in interactions lends itself to different types of glycoproteins with different structures and functions.

One example of glycoproteins found in the body is mucins, which are secreted in the mucus of the respiratory and digestive tracts. The sugars when attached to mucins give them considerable water-holding capacity and also make them resistant to proteolysis by digestive enzymes.

Glycoproteins are important for white blood cell recognition. Examples of glycoproteins in the immune system are:

- Molecules such as antibodies (immunoglobulins), which interact directly with antigens.
- Molecules of the *major histocompatibility complex* (or MHC), which are expressed on the surface of cells and interact with T cells as part of the adaptive immune response.
- Sialyl Lewis X antigen on the surface of leukocytes.

H antigen of the ABO blood compatibility antigens. Other examples of glycoproteins include:

- Gonadotropins (luteinizing hormone and follicle-stimulating hormone).
- Glycoprotein IIb/IIIa, an integrin found on platelets that is required for normal platelet aggregation and adherence to the endothelium.
- Components of the zona pellucida, which surrounds the oocyte, and is important for sperm-egg interaction.
- Structural glycoproteins, which occur in connective tissue. These help bind together the fibers, cells, and ground substance of connective tissue. They may also help components of the tissue bind to inorganic substances, such as calcium in bone.
- Glycoprotein-41 (gp41) and glycoprotein-120 (gp120) are HIV viral coat proteins.

Soluble glycoproteins often show a high viscosity, for example, in egg white and blood plasma.

- Miraculin, is a glycoprotein extracted from *Synsepalum dulcificum* a berry which alters human tongue receptors to recognize sour foods as sweet.

Variable surface glycoproteins allow the sleeping sickness *Trypanosoma* parasite to escape the immune response of the host. The viral spike of the human immunodeficiency virus is heavily glycosylated. Approximately half the mass of the spike is glycosylation and the glycans act to limit antibody recognition as the glycans are assembled by the host cell and so are largely 'self'. Over time, some patients can evolve antibodies to recognise the HIV glycans and almost all so-called 'broadly neutralising antibodies (bnAbs) recognise some glycans. This is possible mainly because the unusually high density of glycans hinders normal glycan maturation and they are therefore trapped in the premature, high-mannose, state. This provides a window for immune recognition. In addition, as these glycans are much less variable than the underlying protein, they have emerged as promising targets for vaccine design.

P-glycoproteins are critical for antitumor research due to its ability block the effects of antitumor drugs. P-glycoprotein, or multidrug transporter (MDR1), is a type of ABC transporter that transports compounds out of cells. This transportation of compounds out of cells includes drugs made to be delivered to the cell, causing a decrease in drug effectiveness. Therefore, being able to inhibit this behavior would decrease P-glycoprotein interference in drug delivery, making this an important topic in drug discovery. For example, P-Glycoprotein causes a decrease in anti-cancer drug accumulation within tumor cells, limiting the effectiveness of chemotherapies used to treat cancer.

11.2.5 Hormones

Hormones that are glycoproteins include:

- Follicle-stimulating hormone.
- Luteinizing hormone.
- Thyroid-stimulating hormone.
- Human chorionic gonadotropin.
- Alpha-fetoprotein.
- Erythropoietin (EPO).

11.2.6 Distinction between Glycoproteins and Proteoglycans

Quoting from recommendations for IUPAC:

- A glycoprotein is a compound containing carbohydrate (or glycan) covalently linked to protein. The carbohydrate may be in the form of a monosaccharide, disaccharide(s). oligosaccharide(s), polysaccharide(s), or their derivatives (e.g. sulfo- or phospho-substituted). One, a few, or many carbohydrate units may be present. Proteoglycans are a subclass of glycoproteins in which the carbohydrate units are polysaccharides that contain amino sugars. Such polysaccharides are also known as glycosaminoglycans.

11.2.7 Functions

Table 4: Some Functions Served by Glycoproteins

Function	Glycoproteins
Structural molecule	Collagens
Lubricant and protective agent	Mucins
Transport molecule	Transferrin, ceruloplasmin
Immunologic molecule	Immunoglobulins, histocompatibility antigens
Hormone	Human chorionic gonadotropin (HCG), thyroid-stimulating hormone (TSH)
Enzyme	Various, e.g., alkaline phosphatase, patatin
Cell attachment-recognition site	Various proteins involved in cell–cell (e.g., sperm–oocyte), virus–cell, bacterium–cell, and hormone–cell interactions
Antifreeze protein	Certain plasma proteins of coldwater fish
Interact with specific carbohydrates	Lectins, selectins (cell adhesion lectins), antibodies
Receptor	Various proteins involved in hormone and drug action
Affect folding of certain proteins	Calnexin, calreticulin
Regulation of development	Notch and its analogs, key proteins in development
Hemostasis (and thrombosis)	Specific glycoproteins on the surface membranes of platelets

11.2.8 Analysis

A variety of methods used in detection, purification, and structural analysis of glycoproteins are

Table 5: Some Important Methods Used to Study Glycoproteins

Method	Use
Periodic acid-Schiff stain	Detects glycoproteins as pink bands after electrophoretic separation.
Incubation of cultured cells with glycoproteins as radioactive decay bands	Leads to detection of a radioactive sugar after electrophoretic separation.
Treatment with appropriate endo- or exoglycosidase or phospholipases	Resultant shifts in electrophoretic migration help distinguish among proteins with N-glycan, O-glycan, or GPI linkages and also between high mannose and complex N-glycans.
Agarose-lectin column chromatography, lectin affinity chromatography	To purify glycoproteins or glycopeptides that bind the particular lectin used.

Method	Use
Lectin affinity electrophoresis	Resultant shifts in electrophoretic migration help distinguish and characterize glycoforms, i.e. variants of a glycoprotein differing in carbohydrate.
Compositional analysis following acid hydrolysis	Identifies sugars that the glycoprotein contains and their stoichiometry.
Mass spectrometry	Provides information on molecular mass, composition, sequence, and sometimes branching of a glycan chain. It can also be used for site-specific glycosylation profiling.
NMR spectroscopy	To identify specific sugars, their sequence, linkages, and the anomeric nature of glycosidic chain.
Multi-angle light scattering	In conjunction with size-exclusion chromatography, UV/Vis absorption and differential refractometry, provides information on molecular mass, protein-carbohydrate ratio, aggregation state, size, and sometimes branching of a glycan chain. In conjunction with composition-gradient analysis, analyzes self- and hetero-association to determine binding affinity and stoichiometry with proteins or carbohydrates in solution without labeling.
Dual Polarisation Interferometry	Measures the mechanisms underlying the biomolecular interactions, including reaction rates, affinities and associated conformational changes.
Methylation (linkage) analysis	To determine linkage between sugars.
Amino acid or cDNA sequencing	Determination of amino acid sequence.

11.2.9 Synthesis

The glycosylation of proteins has an array of different applications from influencing cell to cell communication to changing the thermal stability and the folding of proteins. Due to the unique abilities of glycoproteins, they can be used in many therapies. By understanding glycoproteins and their synthesis, they can be made to treat cancer, Crohn's Disease, high cholesterol, and more.

The process of glycosylation (binding a carbohydrate to a protein) is a post-translational modification, meaning it happens after the production of the protein. Glycosylation is a process that roughly half of all human proteins undergo and heavily influences the properties and functions of the protein. Within the cell, glycosylation occurs in the endoplasmic reticulum.

Recombination

There are several techniques for the assembly of glycoproteins. One technique utilizes recombination. The first consideration for this method is the choice of host, as there are many different factors that can influence the success of glycoprotein recombination such as cost, the host environment, the efficacy of the process, and other considerations. Some examples of host cells include E. coli, yeast, plant cells, insect cells, and mammalian cells. Of these options, mammalian cells are the most common because their use does not face the same challenges that other host cells do such as different glycan structures, shorter half life, and potential unwanted immune responses in humans. Of mammalian cells, the most common cell line used for recombinant glycoprotein production is the Chinese hamster ovary line. However, as technologies develop, the most promising cell lines for recombinant glycoprotein production are human cell lines.

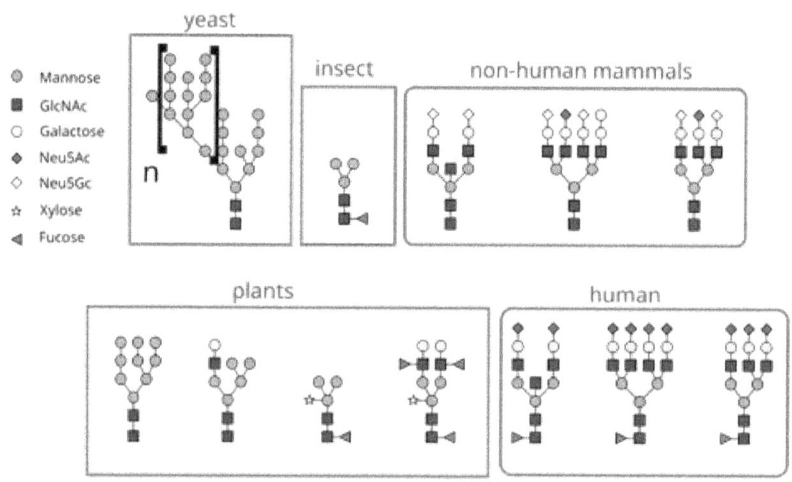

Fig 6: Depiction of Differences in Glycans Amongst Different Animals

Glycosylation

The formation of the link between the glycan and the protein is key element of the synthesis of glycoproteins. The most common method of glycosylation of N-linked glycoproteins is through the reaction between a protected glycan and a protected Asparagine. Similarly, an O-linked glycoprotein can be formed through the addition of a glycosyl donor with a protected Serine or Threonine. These two methods are examples of natural linkage. However, there are also methods of unnatural linkages. Some methods include ligation and a reaction between a serine-derived sulfamidate and thiohexoses in water. Once this linkage is complete, the amino acid sequence can be expanded upon using solid-phase peptide synthesis.

CHAPTER 12

LIPIDS AND BIOLOGICAL EMBRANCES

12.1 LIPIDS

Lipids are a broad group of organic compounds which include fats, waxes, sterols, fat-soluble vitamins (such as vitamins A, D, E and K), monoglycerides, diglycerides, phospholipids, and others. The functions of lipids include storing energy, signaling, and acting as structural components of cell membranes. Lipids have applications in the cosmetic and food industries, and in nanotechnology.

Lipids may be broadly defined as hydrophobic or amphiphilic small molecules; the amphiphilic nature of some lipids allows them to form structures such as vesicles, multilamellar/unilamellar liposomes, or membranes in an aqueous environment. Biological lipids originate entirely or in part from two distinct types of biochemical subunits or "building-blocks": ketoacyl and isoprene groups. Using this approach, lipids may be divided into eight categories: fatty acyls, glycerolipids, glycerophospholipids, sphingolipids, saccharolipids, and polyketides (derived from condensation of ketoacyl subunits); and sterol lipids and prenol lipids (derived from condensation of isoprene subunits).

Although the term "lipid" is sometimes used as a synonym for fats, fats are a subgroup of lipids called triglycerides. Lipids also encompass molecules such as fatty acids and their derivatives (including tri-, di-, monoglycerides, and phospholipids), as well as other sterol-containing metabolites such as cholesterol. Although humans and other mammals use various biosynthetic pathways both to break down and to synthesize lipids, some essential lipids cannot be made this way and must be obtained from the diet.

12.1.1 History

In 1815, Henri Braconnot classified lipids (*graisses*) in two categories, *suifs* (solid greases or tallow) and *huiles* (fluid oils). In 1823, Michel Eugène Chevreul developed a more detailed classification, including oils, greases, tallow, waxes, resins, balsams and volatile oils (or essential oils).

The first synthetic triglyceride was reported by Théophile-Jules Pelouze in 1844, when he produced tributyrin by treating butyric acid with glycerin in the presence of concentrated sulfuric acid. Several years later, Marcellin Berthelot, one of Pelouze's students, synthesized tristearin and tripalmitin by reaction of the analogous fatty acids with glycerin in the presence of gaseous hydrogen chloride at high temperature.

In 1827, William Prout recognized fat ("oily" alimentary matters), along with protein ("albuminous") and carbohydrate ("saccharine"), as an important nutrient for humans and animals.

For a century, chemists regarded "fats" as only simple lipids made of fatty acids and glycerol (glycerides), but new forms were described later. Theodore Gobley (1847) discovered phospholipids in mammalian brain and hen egg, called by him as "lecithins". Thudichum discovered in human brain some phospholipids (cephalin), glycolipids (cerebroside) and sphingolipids (sphingomyelin).

The terms lipoid, lipin, lipide and lipid have been used with varied meanings from author to author. In 1912, Rosenbloom and Gies proposed the substitution of "lipoid" by "lipin". In 1920, Bloor introduced a new classification for "lipoids": simple lipoids (greases and waxes), compound lipoids (phospholipoids and glycolipoids), and the derived lipoids (fatty acids, alcohols, sterols).

In 1947, T. P. Hilditch defined "simple lipids" as greases and waxes (true waxes, sterols, alcohols).

12.1.2 Fatty Acyls

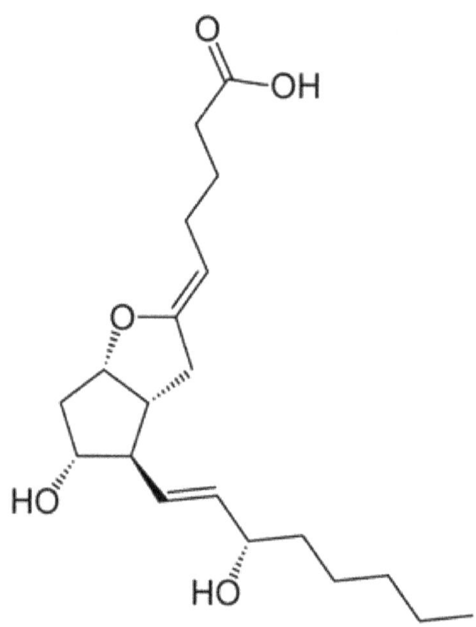

Fig 1: I_2 – Prostacyclin (an Example of a Prostaglandin, an Eicosanoid Fatty Acid)

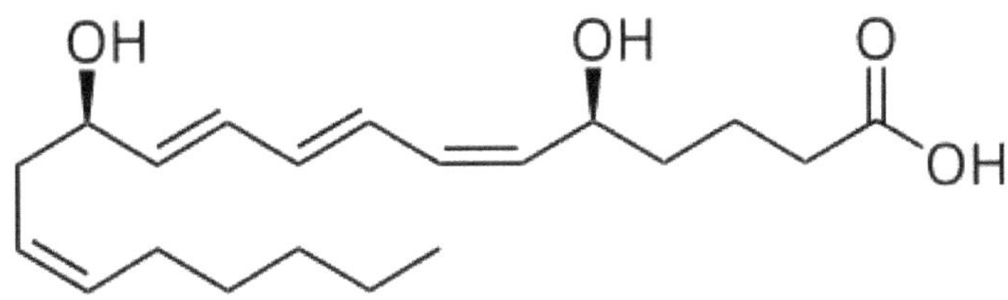

Fig 2: LTB_4 (An Example of a Leukotriene, an Eicosanoid Fatty Acid)

Fatty acyls, a generic term for describing fatty acids, their conjugates and derivatives, are a diverse group of molecules synthesized by chain-elongation of an acetyl-CoA primer with malonyl-CoA or methylmalonyl-CoA groups in a process called fatty acid synthesis.

They are made of a hydrocarbon chain that terminates with a carboxylic acid group; this arrangement confers the molecule with a polar, hydrophilic end, and a nonpolar, hydrophobic end that is insoluble in water. The fatty acid structure is one of the most fundamental categories of biological lipids and is commonly used as a building-block of more structurally complex lipids. The carbon chain, typically between four and 24 carbons long, may be saturated or unsaturated, and may be attached to functional groups containing oxygen, halogens, nitrogen, and sulfur. If a fatty acid contains a double bond, there is the possibility of either a *cis* or *trans* geometric isomerism, which significantly affects the molecule's configuration. *Cis*-double bonds cause the fatty acid chain to bend, an effect that is compounded with more double bonds in the chain.

Three double bonds in 18-carbon *linolenic acid*, the most abundant fatty-acyl chains of plant *thylakoid membranes*, render these membranes highly *fluid* despite environmental low-temperatures, and also makes linolenic acid give dominating sharp peaks in high resolution 13-C NMR spectra of chloroplasts.

This in turn plays an important role in the structure and function of cell membranes. Most naturally occurring fatty acids are of the *cis* configuration, although the *trans* form does exist in some natural and partially hydrogenated fats and oils.

Examples of biologically important fatty acids include the eicosanoids, derived primarily from arachidonic acid and eicosapentaenoic acid, that include prostaglandins, leukotrienes, and thromboxanes.

Docosahexaenoic acid is also important in biological systems, particularly with respect to sight. Other major lipid classes in the fatty acid category are the fatty esters and fatty amides. Fatty esters include important biochemical intermediates such as wax esters, fatty acid thioester coenzyme A derivatives, fatty acid thioester ACP derivatives and fatty acid carnitines. The fatty amides include N-acyl ethanolamines, such as the cannabinoid neurotransmitter anandamide.

12.1.3 Glycerolipids

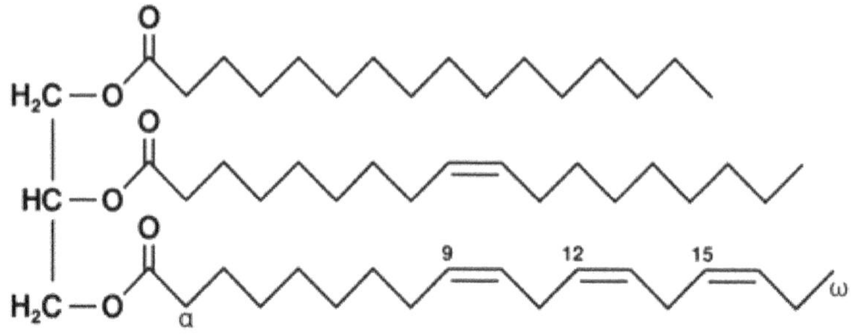

Fig 3: Example of an Unsaturated Fat Triglyceride ($C_{55}H_{98}O_6$). Left Part: Glycerol; Right Part, from Top to Bottom: Palmitic Acid, Oleic Acid, Alpha-Linolenic Acid

Glycerolipids are composed of mono-, di-, and tri-substituted glycerols, the best-known being the fatty acid triesters of glycerol, called triglycerides. The word "triacylglycerol" is sometimes used synonymously with "triglyceride". In these compounds, the three hydroxyl groups of glycerol are each esterified, typically by different fatty acids. Because they function as an energy store, these lipids comprise the bulk of storage fat in animal tissues. The hydrolysis of the ester bonds of triglycerides and the release of glycerol and fatty acids from adipose tissue are the initial steps in metabolizing fat.

Additional subclasses of glycerolipids are represented by glycosylglycerols, which are characterized by the presence of one or more sugar residues attached to glycerol via a glycosidic linkage. Examples of structures in this category are the digalactosyldiacylglycerols found in plant membranes and seminolipid from mammalian sperm cells.

12.1.4 Glycerophospholipids

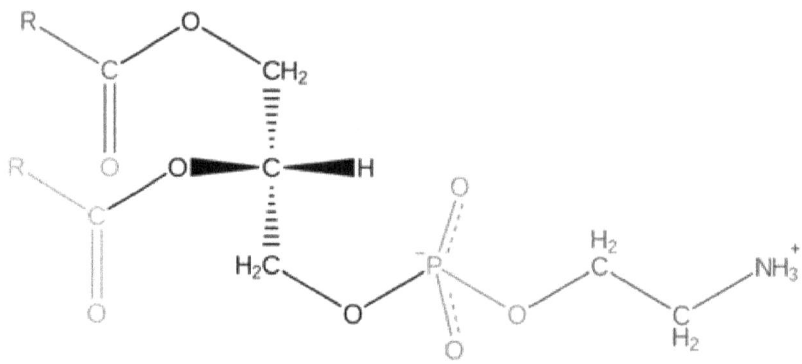

Fig 4: Phosphatidylethanolamine

Glycerophospholipids, usually referred to as phospholipids (though sphingomyelins are also classified as phospholipids), are ubiquitous in nature and are key components of the lipid bilayer of cells, as well as being involved in metabolism and cell signaling. Neural tissue (including the brain) contains relatively high amounts of glycerophospholipids, and alterations in their composition has been implicated in various neurological disorders. Glycerophospholipids may be subdivided into distinct classes, based on the nature of the polar headgroup at the *sn*-3 position of the glycerol backbone in eukaryotes and eubacteria, or the *sn*-1 position in the case of archaebacteria.

Examples of glycerophospholipids found in biological membranes are phosphatidylcholine (also known as PC, GPCho or lecithin), phosphatidylethanolamine (PE or GPEtn) and phosphatidylserine (PS or GPSer). In addition to serving as a primary component of cellular membranes and binding sites for intra- and intercellular proteins, some glycerophospholipids in eukaryotic cells, such as phosphatidylinositols and phosphatidic acids are either precursors of or, themselves, membrane-derived second messengers. Typically, one or both of these hydroxyl groups are acylated with long-chain fatty acids, but there are also alkyl-linked and 1Z-alkenyl-linked (plasmalogen) glycerophospholipids, as well as dialkylether variants in archaebacteria.

12.1.5 Sphingolipids

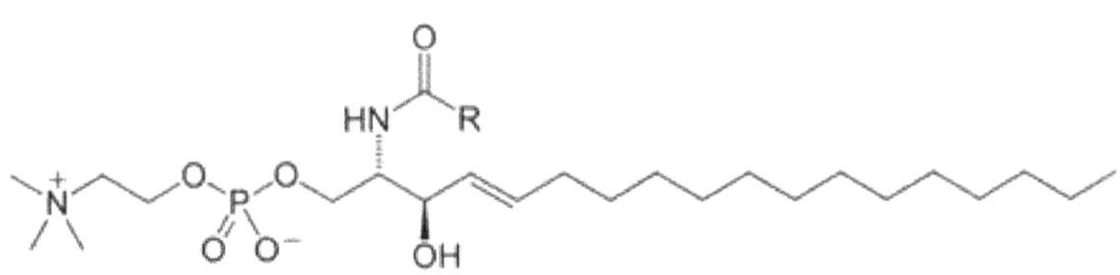

Fig 5: Sphingomyelin

Sphingolipids are a complicated family of compounds that share a common structural feature, a sphingoid base backbone that is synthesized *de novo* from the amino acid serine and a long-chain fatty acyl CoA, then converted into ceramides, phosphosphingolipids, glycosphingolipids and other compounds. The major sphingoid base of mammals is commonly referred to as sphingosine. Ceramides (N-acyl-sphingoid bases) are a major subclass of sphingoid base derivatives with an amide-linked fatty acid. The fatty acids are typically saturated or mono-unsaturated with chain lengths from 16 to 26 carbon atoms.

The major phosphosphingolipids of mammals are sphingomyelins (ceramide phosphocholines), whereas insects contain mainly ceramide phosphoethanolamines and fungi have phytoceramide phosphoinositols and mannose-containing headgroups. The glycosphingolipids are a diverse family of molecules composed of one or more sugar residues linked via a glycosidic bond to the sphingoid base. Examples of these are the simple and complex glycosphingolipids such as cerebrosides and gangliosides.

12.1.6 Sterols

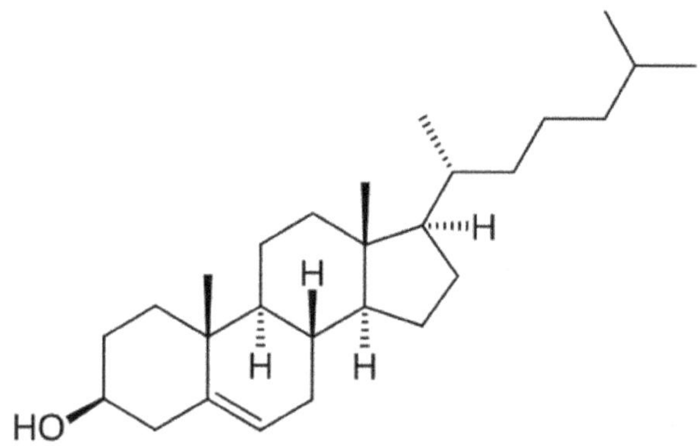

Fig 6: Chemical Structure of Cholesterol

Sterols, such as cholesterol and its derivatives, are an important component of membrane lipids, along with the glycerophospholipids and sphingomyelins. Other examples of sterols are the bile acids and their conjugates, which in mammals are oxidized derivatives of cholesterol and are synthesized in the liver. The plant equivalents are the phytosterols, such as β-sitosterol, stigmasterol, and brassicasterol; the latter compound is also used as a biomarker for algal growth. The predominant sterol in fungal cell membranes is ergosterol.

Sterols are steroids in which one of the hydrogen atoms is substituted with a hydroxyl group, at position 3 in the carbon chain. They have in common with steroids the same fused four-ring core structure. Steroids have different biological roles as hormones and signaling molecules. The eighteen-carbon (C18) steroids include the estrogen family whereas the C19 steroids comprise the androgens such as testosterone and androsterone. The C21 subclass includes the progestogens as well as the glucocorticoids and mineralocorticoids. The secosteroids, comprising various forms of vitamin D, are characterized by cleavage of the B ring of the core structure.

12.1.7 Prenols

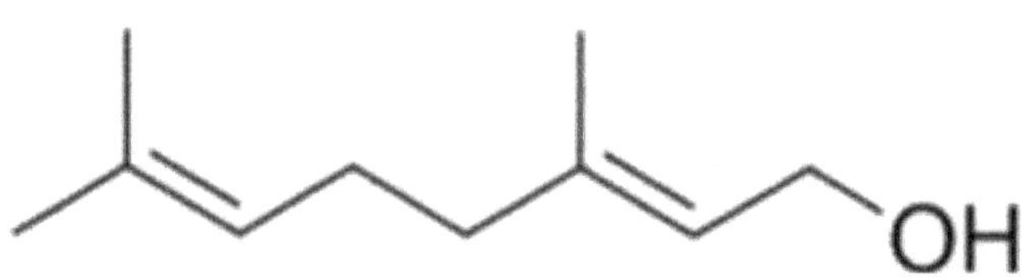

Fig 7: Prenol Lipid (2*E*-Geraniol)

Prenol lipids are synthesized from the five-carbon-unit precursors isopentenyl diphosphate and dimethylallyl diphosphate, which are produced mainly via the mevalonic acid (MVA) pathway. The simple isoprenoids (linear alcohols, diphosphates, etc.) are formed by the successive addition of C5 units, and are classified according to number of these terpene units. Structures containing greater than 40 carbons are known as polyterpenes. Carotenoids are important simple isoprenoids that function as antioxidants and as precursors of vitamin A. Another biologically important class of molecules is exemplified by the quinones and hydroquinones, which contain an isoprenoid tail attached to a quinonoid core of non-isoprenoid origin. Vitamin E and vitamin K, as well as the ubiquinones, are examples of this class. Prokaryotes synthesize polyprenols (called bactoprenols) in which the terminal isoprenoid unit attached to oxygen remains unsaturated, whereas in animal polyprenols (dolichols) the terminal isoprenoid is reduced.

12.1.8 Saccharolipids

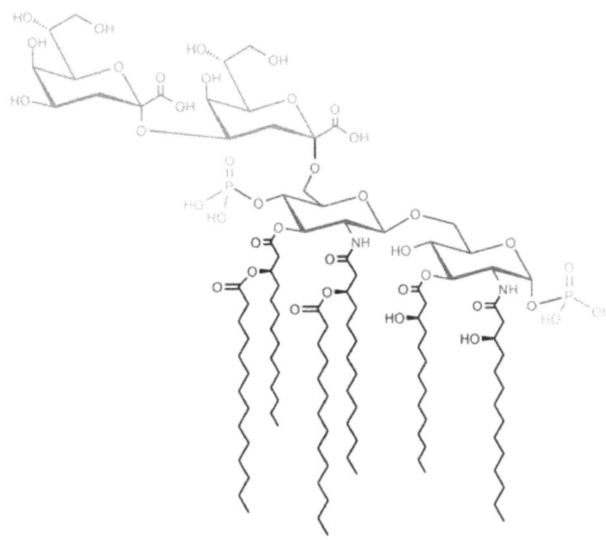

Fig 8: Structure of the Saccharolipid Kdo$_2$-Lipid A. Glucosamine Residues in Blue, Kdo Residues in Red, Acyl Chains in Black and Phosphate Groups in Green

Saccharolipids describe compounds in which fatty acids are linked to a sugar backbone, forming structures that are compatible with membrane bilayers. In the saccharolipids, a monosaccharide substitutes for the glycerol backbone present in glycerolipids and glycerophospholipids.

The most familiar saccharolipids are the acylated glucosamine precursors of the Lipid A component of the lipopolysaccharides in Gram-negative bacteria. Typical lipid A molecules are disaccharides of glucosamine, which are derivatized with as many as seven fatty-acyl chains.

The minimal lipopolysaccharide required for growth in *E. coli* is Kdo$_2$-Lipid A, a hexa-acylated disaccharide of glucosamine that is glycosylated with two 3-deoxy-D-manno-octulosonic acid (Kdo) residues.

12.1.9 Polyketides

Polyketides are synthesized by polymerization of acetyl and propionyl subunits by classic enzymes as well as iterative and multimodular enzymes that share mechanistic features with the fatty acid synthases.

They comprise many secondary metabolites and natural products from animal, plant, bacterial, fungal and marine sources, and have great structural diversity. Many polyketides are cyclic molecules whose backbones are often further modified by glycosylation, methylation, hydroxylation, oxidation, or other processes. Many commonly used antimicrobial, antiparasitic, and anticancer agents are polyketides or polyketide derivatives, such as erythromycins, tetracyclines, avermectins, and antitumor epothilones.

12.1.10 Biological Functions

Component of Biological Membranes

Eukaryotic cells feature the compartmentalized membrane-bound organelles that carry out different biological functions. The glycerophospholipids are the main structural component of biological membranes, as the cellular plasma membrane and the intracellular membranes of organelles; in animal cells, the plasma membrane physically separates the intracellular components from the extracellular environment.

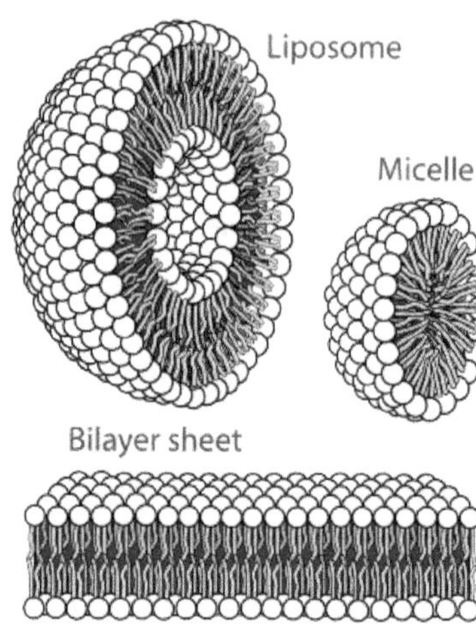

Fig 9: Self-Organization of Phospholipids: A Spherical Liposome, a Micelle, and a Lipid Bilayer

The glycerophospholipids are amphipathic molecules (containing both hydrophobic and hydrophilic regions) that contain a glycerol core linked to two fatty acid-derived "tails" by ester linkages and to one "head" group by a phosphate ester linkage. While glycerophospholipids are the major component of biological membranes, other non-glyceride lipid components such as sphingomyelin and sterols (mainly cholesterol in animal cell membranes) are also found in biological membranes. In plants and algae, the galactosyldiacylglycerols, and sulfoquinovosyldiacylglycerol, which lack a phosphate group, are important components of membranes of chloroplasts and related organelles and are among the most abundant lipids in photosynthetic tissues, including those of higher plants, algae and certain bacteria.

Plant thylakoid membranes have the largest lipid component of a non-bilayer forming monogalactosyl diglyceride (MGDG), and little phospholipids; despite this unique lipid composition, chloroplast thylakoid membranes have been shown to contain a dynamic lipid-bilayer matrix as revealed by magnetic resonance and electron microscope studies.

A biological membrane is a form of lamellar phase lipid bilayer. The formation of lipid bilayers is an energetically preferred process when the glycerophospholipids described above are in an aqueous environment. This is known as the hydrophobic effect. In an aqueous system, the polar heads of lipids align towards the polar, aqueous environment, while the hydrophobic tails minimize their contact with water and tend to cluster together, forming a vesicle; depending on the concentration of the lipid, this biophysical interaction may result in the formation of micelles, liposomes, or lipid bilayers. Other aggregations are also observed and form part of the polymorphism of amphiphile (lipid) behavior. Phase behavior is an area of study within biophysics. Micelles and bilayers form in the polar medium by a process known as the hydrophobic effect. When dissolving a lipophilic or amphiphilic substance in a polar environment, the polar molecules (i.e., water in an aqueous solution) become more ordered around the dissolved lipophilic substance, since the polar molecules cannot form hydrogen bonds to the lipophilic areas of the amphiphile. So in an aqueous environment, the water molecules form an ordered "clathrate" cage around the dissolved lipophilic molecule.

The formation of lipids into protocell membranes represents a key step in models of abiogenesis, the origin of life.

Energy Storage

Triglycerides, stored in adipose tissue, are a major form of energy storage both in animals and plants. They are a major source of energy in aerobic respiration. The complete oxidation of fatty acids releases about 38 kJ/g (9 kcal/g), compared with only 17 kJ/g (4 kcal/g) for the oxidative breakdown of carbohydrates and proteins. The adipocyte, or fat cell, is designed for continuous synthesis and breakdown of triglycerides in animals, with breakdown controlled mainly by the activation of hormone-sensitive enzyme lipase. Migratory birds that must fly long distances without eating use triglycerides to fuel their flights.

Signaling

Evidence has emerged showing that lipid signaling is a vital part of the cell signaling. Lipid signaling may occur via activation of G protein-coupled or nuclear receptors, and members of several different lipid categories have been identified as signaling molecules and cellular messengers.

These include sphingosine-1-phosphate, a sphingolipid derived from ceramide that is a potent messenger molecule involved in regulating calcium mobilization, cell growth, and apoptosis; diacylglycerol and the phosphatidylinositol phosphates (PIPs), involved in calcium-mediated activation of protein kinase C; the prostaglandins, which are one type of fatty-acid derived eicosanoid involved in inflammation and immunity; the steroid hormones such as estrogen, testosterone and cortisol, which modulate a host of functions such as reproduction, metabolism and blood pressure; and the oxysterols such as 25-hydroxy-cholesterol that are liver X receptor agonists.

Phosphatidylserine lipids are known to be involved in signaling for the phagocytosis of apoptotic cells or pieces of cells. They accomplish this by being exposed to the extracellular face of the cell membrane after the inactivation of flippases which place them exclusively on the cytosolic side and the activation of scramblases, which scramble the orientation of the phospholipids. After this occurs, other cells recognize the phosphatidylserines and phagocytosize the cells or cell fragments exposing them.

Other Functions

The "fat-soluble" vitamins (A, D, E and K) – which are isoprene-based lipids – are essential nutrients stored in the liver and fatty tissues, with a diverse range of functions. Acyl-carnitines are involved in the transport and metabolism of fatty acids in and out of mitochondria, where they undergo beta oxidation. Polyprenols and their phosphorylated derivatives also play important transport roles, in this case the transport of oligosaccharides across membranes. Polyprenol phosphate sugars and polyprenol diphosphate sugars function in extra-cytoplasmic glycosylation reactions, in extracellular polysaccharide biosynthesis (for instance, peptidoglycan polymerization in bacteria), and in eukaryotic protein N-glycosylation. Cardiolipins are a subclass of glycerophospholipids containing four acyl chains and three glycerol groups that are particularly abundant in the inner mitochondrial membrane. They are believed to activate enzymes involved with oxidative phosphorylation. Lipids also form the basis of steroid hormones.

12.1.11 Metabolism

The major dietary lipids for humans and other animals are animal and plant triglycerides, sterols, and membrane phospholipids. The process of lipid metabolism synthesizes and degrades the lipid stores and produces the structural and functional lipids characteristic of individual tissues.

Biosynthesis

In animals, when there is an oversupply of dietary carbohydrate, the excess carbohydrate is converted to triglycerides. This involves the synthesis of fatty acids from acetyl-CoA and the esterification of fatty acids in the production of triglycerides, a process called lipogenesis. Fatty acids are made by fatty acid synthases that polymerize and then reduce acetyl-CoA units. The acyl chains in the fatty acids are extended by a cycle of reactions that add the acetyl group, reduce it to an alcohol, dehydrate it to an alkene group and then reduce it again to an alkane group. The enzymes of fatty acid biosynthesis are divided into two groups, in animals and fungi all these fatty acid synthase reactions are carried out by a single multifunctional protein, while in plant plastids and bacteria separate enzymes perform each step in the pathway. The fatty acids may be subsequently converted to triglycerides that are packaged in lipoproteins and secreted from the liver.

The synthesis of unsaturated fatty acids involves a desaturation reaction, whereby a double bond is introduced into the fatty acyl chain. For example, in humans, the desaturation of stearic acid by stearoyl-CoA desaturase-1 produces oleic acid. The doubly unsaturated fatty acid linoleic acid as well as the triply unsaturated α-linolenic acid cannot be synthesized in mammalian tissues, and are therefore essential fatty acids and must be obtained from the diet.

Triglyceride synthesis takes place in the endoplasmic reticulum by metabolic pathways in which acyl groups in fatty acyl-CoAs are transferred to the hydroxyl groups of glycerol-3-phosphate and diacylglycerol.

Terpenes and isoprenoids, including the carotenoids, are made by the assembly and modification of isoprene units donated from the reactive precursors isopentenyl pyrophosphate and dimethylallyl pyrophosphate. These precursors can be made in different ways. In animals and archaea, the mevalonate pathway produces these compounds from acetyl-CoA, while in plants and bacteria the non-mevalonate pathway uses pyruvate and glyceraldehyde 3-phosphate as substrates. One important reaction that uses these activated isoprene donors is steroid biosynthesis. Here, the isoprene units are joined together to make squalene and then folded up and formed into a set of rings to make lanosterol. Lanosterol can then be converted into other steroids such as cholesterol and ergosterol.

Degradation

Beta oxidation is the metabolic process by which fatty acids are broken down in the mitochondria or in peroxisomes to generate acetyl-CoA. For the most part, fatty acids are oxidized by a mechanism that is similar to, but not identical with, a reversal of the process of fatty acid synthesis. That is, two-carbon fragments are removed sequentially from the carboxyl end of the acid after steps of dehydrogenation, hydration, and oxidation to form a beta-keto acid, which is split by thiolysis. The acetyl-CoA is then ultimately converted into adenosine triphosphate (ATP), CO_2, and H_2O using the citric acid cycle and the electron transport chain. Hence the citric acid cycle can start at acetyl-CoA when fat is being broken down for energy if there is little or no glucose available. The energy yield of the complete oxidation of the fatty acid palmitate is 106 ATP. Unsaturated and odd-chain fatty acids require additional enzymatic steps for degradation.

12.1.12 Nutrition and Health

Most of the fat found in food is in the form of triglycerides, cholesterol, and phospholipids. Some dietary fat is necessary to facilitate absorption of fat-soluble vitamins (A, D, E, and K) and carotenoids. Humans and other mammals have a dietary requirement for certain essential fatty acids, such as linoleic acid (an omega-6 fatty acid) and alpha-linolenic acid (an omega-3 fatty acid) because they cannot be synthesized from simple precursors in the diet. Both of these fatty acids are 18-carbon polyunsaturated fatty acids differing in the number and position of the double bonds. Most vegetable oils are rich in linoleic acid (safflower, sunflower, and corn oils). Alpha-linolenic acid is found in the green leaves of plants and in some seeds, nuts, and legumes (in particular flax, rapeseed, walnut, and soy). Fish oils are particularly rich in the longer-chain omega-3 fatty acids eicosapentaenoic acid and docosahexaenoic acid. Many studies have shown positive health benefits associated with consumption of omega-3 fatty acids on infant development, cancer, cardiovascular diseases, and various mental illnesses (such as depression, attention-deficit hyperactivity disorder, and dementia).

In contrast, it is now well-established that consumption of trans fats, such as those present in partially hydrogenated vegetable oils, are a risk factor for cardiovascular disease. Fats that are good for one may be turned into trans fats by improper cooking methods that result in overcooking the lipids.

A few studies have suggested that total dietary fat intake is linked to an increased risk of obesity. and diabetes; Others, including the Women's Health Initiative Dietary Modification Trial, an eight-year study of 49,000 women, the Nurses' Health Study, and the Health Professionals Follow-up Study, revealed no such links. None of these studies suggested any connection between percentage of calories from fat and risk of cancer, heart disease, or weight gain. The Nutrition Source, a website maintained by the department of nutrition at the T. H. Chan School of Public Health at Harvard University, summarizes the current evidence on the effect of dietary fat: "Detailed research—much of it done at Harvard—shows that the total amount of fat in the diet isn't really linked with weight or disease."

12.2 BIOLOGICAL MEMBRANCES

A biological membrane, biomembrane or cell membrane is a selectively permeable membrane that separates the interior of a cell from the external environment or creates intracellular compartments by serving as a boundary between one part of the cell and another. Biological membranes, in the form of eukaryotic cell membranes, consist of a phospholipid bilayer with embedded, integral and peripheral proteins used in communication and transportation of chemicals and ions. The bulk of lipids in a cell membrane provides a fluid matrix for proteins to rotate and laterally diffuse for physiological functioning. Proteins are adapted to high membrane fluidity environment of the lipid bilayer with the presence of an annular lipid shell, consisting of lipid molecules bound tightly to the surface of integral membrane proteins. The cell membranes are different from the isolating tissues formed by layers of cells, such as mucous membranes, basement membranes, and serous membranes.

12.2.1 Composition

Asymmetry

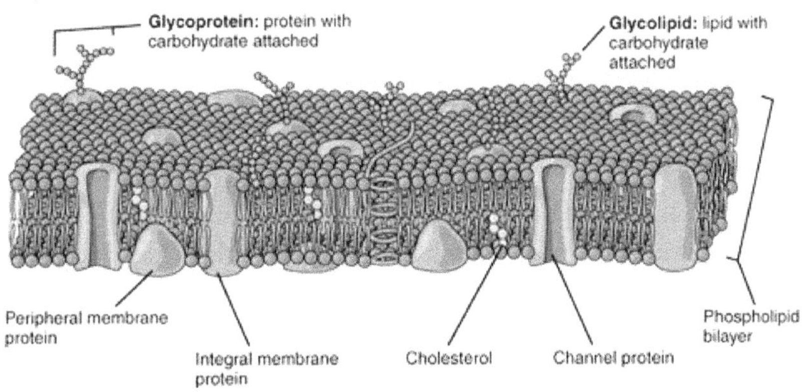

Fig 10: A Fluid Membrane Model of the Phospholipid Bilayer

The lipid bilayer consists of two layers- an outer leaflet and an inner leaflet. The components of bilayers are distributed unequally between the two surfaces to create asymmetry between the outer and inner surfaces. This asymmetric organization is important for cell functions such as cell signaling. The asymmetry of the biological membrane reflects the different functions of the two leaflets of the membrane. As seen in the fluid membrane model of the phospholipid bilayer, the outer leaflet and inner leaflet of the membrane are asymmetrical in their composition. Certain proteins and lipids rest only on one surface of the membrane and not the other.

Both the plasma membrane and internal membranes have cytosolic and exoplasmic faces. This orientation is maintained during membrane trafficking - proteins, lipids, glycoconjugates facing the lumen of the ER and Golgi get expressed on the extracellular side of the plasma membrane. In eukaryotic cells, new phospholipids are manufactured by enzymes bound to the part of the endoplasmic reticulum membrane that faces the cytosol.

These enzymes, which use free fatty acids as substrates, deposit all newly made phospholipids into the cytosolic half of the bilayer. To enable the membrane as a whole to grow evenly, half of the new phospholipid molecules then have to be transferred to the opposite monolayer.

This transfer is catalyzed by enzymes called flippases. In the plasma membrane, flippases transfer specific phospholipids selectively, so that different types become concentrated in each monolayer.

Using selective flippases is not the only way to produce asymmetry in lipid bilayers, however. In particular, a different mechanism operates for glycolipids—the lipids that show the most striking and consistent asymmetric distribution in animal cells.

Lipids

The biological membrane is made up of lipids with hydrophobic tails and hydrophilic heads. The hydrophobic tails are hydrocarbon tails whose length and saturation is important in characterizing the cell. Lipid rafts occur when lipid species and proteins aggregate in domains in the membrane.

These help organize membrane components into localized areas that are involved in specific processes, such as signal transduction.

Red blood cells, or erythrocytes, have a unique lipid composition. The bilayer of red blood cells is composed of cholesterol and phospholipids in equal proportions by weight. Erythrocyte membrane plays a crucial role in blood clotting. In the bilayer of red blood cells is phosphatidylserine.

This is usually in the cytoplasmic side of the membrane. However, it is flipped to the outer membrane to be used during blood clotting.

Proteins

Phospholipid bilayers contain different proteins. These membrane proteins have various functions and characteristics and catalyze different chemical reactions. Integral proteins span the membranes with different domains on either side. Integral proteins hold strong association with the lipid bilayer and cannot easily become detached.

They will dissociate only with chemical treatment that breaks the membrane. Peripheral proteins are unlike integral proteins in that they hold weak interactions with the surface of the bilayer and can easily become dissociated from the membrane. Peripheral proteins are located on only one face of a membrane and create membrane asymmetry.

Table 1: Some Examples of Plasma Membrane Proteins and their Functions

Functional Class	Protein Example	Specific Function
Transporters	Na+ Pump	actively pumps Na+ out of cells and K+ in
Anchors	integrins	link intracellular actin filaments to extracellular matrix proteins
Receptors	platelet-derived growth factor receptor	binds extracellular PDGF and, as a consequence, generates intracellular signals that cause the cell to grow and divide
Enzymes	adenylyl cyclase	catalyzes the production of intracellular signaling molecule cyclic AMP in response to extracellular signals

Oligosaccharides

Oligosaccharides are sugar containing polymers. In the membrane, they can be covalently bound to lipids to form glycolipids or covalently bound to proteins to form glycoproteins. Membranes contain sugar-containing lipid molecules known as glycolipids. In the bilayer, the sugar groups of glycolipids are exposed at the cell surface, where they can form hydrogen bonds. Glycolipids provide the most extreme example of asymmetry in the lipid bilayer. Glycolipids perform a vast number of functions in the biological membrane that are mainly communicative, including cell recognition and cell-cell adhesion. Glycoproteins are integral proteins. They play an important role in the immune response and protection.

12.2.2 Formation

The phospholipid bilayer is formed due to the aggregation of membrane lipids in aqueous solutions. Aggregation is caused by the hydrophobic effect, where hydrophobic ends come into contact with each other and are sequestered away from water. This arrangement maximises hydrogen bonding between hydrophilic heads and water while minimising unfavorable contact between hydrophobic tails and water. The increase in available hydrogen bonding increases the entropy of the system, creating a spontaneous process.

12.2.3 Function

Biological molecules are amphiphilic or amphipathic, i.e. are simultaneously hydrophobic and hydrophilic. The phospholipid bilayer contains charged hydrophilic headgroups, which interact with polar water. The layers also contain hydrophobic tails, which meet with the hydrophobic tails of the complementary layer.

The hydrophobic tails are usually fatty acids that differ in lengths. The interactions of lipids, especially the hydrophobic tails, determine the lipid bilayer physical properties such as fluidity.

Membranes in cells typically define enclosed spaces or compartments in which cells may maintain a chemical or biochemical environment that differs from the outside. For example, the membrane around peroxisomes shields the rest of the cell from peroxides, chemicals that can be toxic to the cell, and the cell membrane separates a cell from its surrounding medium. Peroxisomes are one form of vacuole found in the cell that contain by-products of chemical reactions within the cell. Most organelles are defined by such membranes, and are called membrane-bound organelles.

Selective Permeability

Probably the most important feature of a biomembrane is that it is a selectively permeable structure. This means that the size, charge, and other chemical properties of the atoms and molecules attempting to cross it will determine whether they succeed in doing so. Selective permeability is essential for effective separation of a cell or organelle from its surroundings. Biological membranes also have certain mechanical or elastic properties that allow them to change shape and move as required.

Generally, small hydrophobic molecules can readily cross phospholipid bilayers by simple diffusion.

Particles that are required for cellular function but are unable to diffuse freely across a membrane enter through a membrane transport protein or are taken in by means of endocytosis, where the membrane allows for a vacuole to join onto it and push its contents into the cell. Many types of specialized plasma membranes can separate cell from external environment: apical, basolateral, presynaptic and postsynaptic ones, membranes of flagella, cilia, microvillus, filopodia and lamellipodia, the sarcolemma of muscle cells, as well as specialized myelin and dendritic spine membranes of neurons. Plasma membranes can also form different types of "supramembrane" structures such as caveolae, postsynaptic density, podosome, invadopodium, desmosome, hemidesmosome, focal adhesion, and cell junctions. These types of membranes differ in lipid and protein composition.

Distinct types of membranes also create intracellular organelles: endosome; smooth and rough endoplasmic reticulum; sarcoplasmic reticulum; Golgi apparatus; lysosome; mitochondrion (inner and outer membranes); nucleus (inner and outer membranes); peroxisome; vacuole; cytoplasmic granules; cell vesicles (phagosome, autophagosome, clathrin-coated vesicles, COPI-coated and COPII-coated vesicles) and secretory vesicles (including synaptosome, acrosomes, melanosomes, and chromaffin granules). Different types of biological membranes have diverse lipid and protein compositions. The content of membranes defines their physical and biological properties. Some components of membranes play a key role in medicine, such as the efflux pumps that pump drugs out of a cell.

Fluidity

The hydrophobic core of the phospholipid bilayer is constantly in motion because of rotations around the bonds of lipid tails. Hydrophobic tails of a bilayer bend and lock together. However, because of hydrogen bonding with water, the hydrophilic head groups exhibit less movement as their rotation and mobility are constrained. This results in increasing viscosity of the lipid bilayer closer to the hydrophilic heads.

Below a transition temperature, a lipid bilayer loses fluidity when the highly mobile lipids exhibits less movement becoming a gel-like solid. The transition temperature depends on such components of the lipid bilayer as the hydrocarbon chain length and the saturation of its fatty acids. Temperature-dependence fluidity constitutes an important physiological attribute for bacteria and cold-blooded organisms. These organisms maintain a constant fluidity by modifying membrane lipid fatty acid composition in accordance with differing temperatures.

In animal cells, membrane fluidity is modulated by the inclusion of the sterol cholesterol. This molecule is present in especially large amounts in the plasma membrane, where it constitutes approximately 20% of the lipids in the membrane by weight. Because cholesterol molecules are short and rigid, they fill the spaces between neighboring phospholipid molecules left by the kinks in their unsaturated hydrocarbon tails. In this way, cholesterol tends to stiffen the bilayer, making it more rigid and less permeable.

For all cells, membrane fluidity is important for many reasons. It enables membrane proteins to diffuse rapidly in the plane of the bilayer and to interact with one another, as is crucial, for example, in cell signaling. It permits membrane lipids and proteins to diffuse from sites where they are inserted into the bilayer after their synthesis to other regions of the cell. It allows membranes to fuse with one another and mix their molecules, and it ensures that membrane molecules are distributed evenly between daughter cells when a cell divides. If biological membranes were not fluid, it is hard to imagine how cells could live, grow, and reproduce.

The fluidity property is at the center of the Helfrich model which allows for calculating the energy cost of an elastic deformation to the membrane.

CHAPTER 13
MEMBRANE-CHANNELS AND PUMPS

Cellular membranes are vital structures that regulate the movement of substances into and out of cells.

They act as selective barriers, maintaining the internal environment of the cell and allowing it to communicate with its surroundings. Within these membranes, specialized proteins known as membrane channels and pumps play crucial roles in facilitating the transport of ions, molecules, and other substances across the lipid bilayer.

13.1 THE STRUCTURE OF CELL MEMBRANES

Cell membranes consist of a phospholipid bilayer interspersed with proteins, cholesterol, and carbohydrates. The phospholipids create a semi-permeable barrier that restricts the free passage of polar and charged molecules, such as ions and large organic compounds.

Membrane proteins, including channels and pumps, are embedded within this bilayer, providing pathways for specific substances to cross the membrane. Membrane channels are typically composed of multiple subunits that form a pore through which ions or small molecules can flow.

These channels can be selective, allowing only certain ions to pass based on size or charge. Pumps, on the other hand, are integral membrane proteins that actively transport substances against their concentration gradients using energy, usually derived from ATP hydrolysis.

The specific arrangement and composition of these proteins contribute to the overall function of the cell membrane, regulating various physiological processes.

13.2 MEMBRANE CHANNELS: FACILITATING PASSIVE TRANSPORT

Membrane channels facilitate passive transport, a process where substances move across the membrane along their concentration gradient without the expenditure of energy. This movement can occur through various types of channels, including ion channels and aquaporins.

13.2.1 Ion Channels

Ion channels are specialized proteins that allow the selective passage of ions across the cell membrane. These channels can be classified based on their gating mechanisms:

- **Voltage-gated channels** open or close in response to changes in the membrane potential.
- **Ligand-gated channels** respond to the binding of specific molecules (ligands), such as neurotransmitters.
- **Mechanically gated channels** open in response to mechanical forces, such as stretching or pressure.

Ion channels are crucial for numerous physiological functions, including the generation of action potentials in neurons, muscle contraction, and the regulation of cellular excitability. For instance, the opening of voltage-gated sodium channels during an action potential allows sodium ions to flood into the neuron, leading to depolarization. Similarly, calcium channels play a critical role in muscle contraction by facilitating the influx of calcium ions.

13.2.2 Aquaporins

Aquaporins are a specific class of membrane channels that facilitate the rapid transport of water across the cell membrane. These channels are essential for maintaining cellular hydration and regulating osmotic balance. Aquaporins enable water molecules to pass through the lipid bilayer much faster than would be possible by diffusion alone. They selectively filter water while excluding ions and other solutes, thus preserving the osmotic gradient essential for cellular function. Aquaporins are particularly abundant in tissues with high water permeability, such as the kidneys and plant root cells, where they play vital roles in water reabsorption and uptake.

13.3 MEMBRANE PUMPS: ACTIVE TRANSPORT MECHANISMS

Unlike channels, membrane pumps engage in **active transport**, which requires energy input to move substances against their concentration gradients. Active transport is crucial for maintaining concentration differences across the membrane, essential for various cellular functions.

13.3.1 The Sodium-Potassium Pump

The **sodium-potassium pump** (Na^+/K^+ ATPase) is one of the most studied membrane pumps. It actively transports sodium ions out of the cell and potassium ions into the cell, maintaining the electrochemical gradient essential for cellular function. For every three sodium ions pumped out, two potassium ions are transported into the cell, a process that consumes one molecule of ATP.

This pump is vital for several physiological processes, including the regulation of cell volume, the generation of action potentials in neurons, and the reabsorption of nutrients in the kidneys. The electrochemical gradient established by the sodium-potassium pump is also critical for secondary active transport, where the movement of one substance down its gradient drives the transport of another substance against its gradient.

13.3.2 Other Active Transporters

In addition to the sodium-potassium pump, various other membrane pumps exist that transport specific ions or molecules. For instance, the **calcium pump** (Ca^{2+} ATPase) actively transports calcium ions out of the cell, maintaining low intracellular calcium levels essential for muscle relaxation and neurotransmitter release. Similarly, **proton pumps** (H^+ ATPases) transport protons across the membrane, playing key roles in processes such as pH regulation and ATP synthesis in mitochondria.

13.3.3 Secondary Active Transport

Secondary active transport, also known as cotransport, involves the coupling of the transport of one substance to the movement of another down its concentration gradient. This mechanism does not directly use ATP; instead, it relies on the gradients established by primary active transporters. There are two main types of secondary active transport:

- **Symporters** transport two different substances in the same direction across the membrane.
- **Antiporters** transport two substances in opposite directions.

For example, the sodium-glucose cotransporter (SGLT) utilizes the sodium gradient established by the sodium-potassium pump to transport glucose into the cell against its concentration gradient, demonstrating the interdependence of channels and pumps in cellular transport.

13.4 REGULATION OF MEMBRANE CHANNELS AND PUMPS

The activity of membrane channels and pumps is tightly regulated to ensure proper cellular function. Various factors influence their regulation, including changes in voltage, ligand binding, and post-translational modifications.

13.4.1 Gating Mechanisms

For ion channels, gating mechanisms allow for the precise control of ion flow in response to physiological signals. Voltage-gated channels undergo conformational changes in response to alterations in membrane potential, allowing for the rapid propagation of action potentials. Ligand-gated channels can be modulated by the presence or absence of specific ligands, ensuring that ion flow occurs only when needed.

13.4.2 Phosphorylation and Modulation

The activity of pumps can also be regulated through phosphorylation, a post-translational modification that can activate or inhibit their function. For example, the sodium-potassium pump's activity can be influenced by hormones such as insulin, which enhances glucose uptake by modulating the activity of glucose transporters and the sodium-potassium pump.

13.5 PHYSIOLOGICAL IMPLICATIONS OF MEMBRANE TRANSPORT

The proper functioning of membrane channels and pumps is crucial for maintaining cellular homeostasis and overall organismal health. Dysregulation of these transport mechanisms can lead to various pathophysiological conditions.

13.5.1 Cystic Fibrosis

One example is cystic fibrosis, a genetic disorder caused by mutations in the CFTR gene, which encodes a chloride ion channel. This condition leads to impaired chloride transport and thickened mucus in the lungs, resulting in respiratory complications and increased susceptibility to infections.

13.5.2 Cardiac Function

In the context of cardiac function, abnormalities in ion channels can lead to arrhythmias.

For instance, mutations in potassium channels can disrupt the repolarization phase of the cardiac action potential, leading to conditions such as Long QT syndrome, which increases the risk of sudden cardiac death.

13.5.3 Hypertension

Dysregulation of the sodium-potassium pump can contribute to hypertension (high blood pressure) by affecting the balance of sodium and potassium in the body. Overactivity of this pump may lead to increased sodium reabsorption, resulting in elevated blood volume and blood pressure.

13.6 CONCLUSION

Membrane channels and pumps are essential components of cellular membranes, facilitating the transport of ions and molecules that underpin vital physiological processes.

Through passive and active transport mechanisms, these proteins maintain the electrochemical gradients and concentrations necessary for cellular function, signaling, and homeostasis. Understanding the mechanisms of membrane transport provides insights into various biological functions and highlights the importance of these proteins in health and disease.

As research advances, continued exploration of membrane channels and pumps will enhance our understanding of their roles in cellular physiology and may lead to the development of targeted therapies for conditions arising from their dysregulation.

The intricate balance maintained by these transport proteins is critical for the overall health of cells and organisms, emphasizing their significance in biological systems.

CHAPTER 14
SIGNAL-TRANSDUCTION PATHWAYS

14.1 AN INTRODUCTION

Signal transduction is the process by which a chemical or physical signal is transmitted through a cell as a series of molecular events. Proteins responsible for detecting stimuli are generally termed receptors, although in some cases the term sensor is used. The changes elicited by ligand binding (or signal sensing) in a receptor give rise to a biochemical cascade, which is a chain of biochemical events known as a signaling pathway.

When signaling pathways interact with one another they form networks, which allow cellular responses to be coordinated, often by combinatorial signaling events. At the molecular level, such responses include changes in the transcription or translation of genes, and post-translational and conformational changes in proteins, as well as changes in their location.

These molecular events are the basic mechanisms controlling cell growth, proliferation, metabolism and many other processes. In multicellular organisms, signal transduction pathways regulate cell communication in a wide variety of ways.

Each component (or node) of a signaling pathway is classified according to the role it plays with respect to the initial stimulus. Ligands are termed *first messengers*, while receptors are the *signal transducers*, which then activate *primary effectors*. Such effectors are typically proteins and are often linked to second messengers, which can activate *secondary effectors*, and so on.

Depending on the efficiency of the nodes, a signal can be amplified (a concept known as signal gain), so that one signaling molecule can generate a response involving hundreds to millions of molecules. As with other signals, the transduction of biological signals is characterised by delay, noise, signal feedback and feedforward and interference, which can range from negligible to pathological. With the advent of computational biology, the analysis of signaling pathways and networks has become an essential tool to understand cellular functions and disease, including signaling rewiring mechanisms underlying responses to acquired drug resistance.

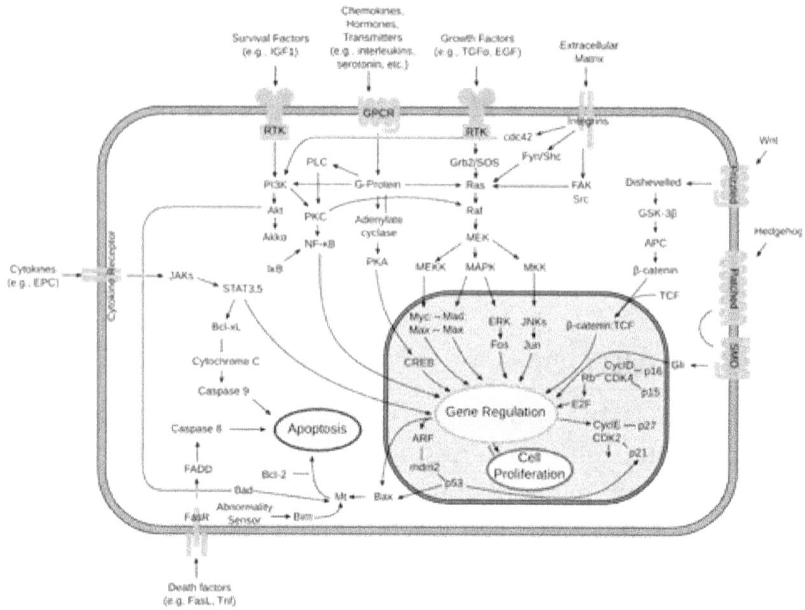

Fig 1: Simplified Representation of Major Signal Transduction Pathways in Mammals

14.2 HISTORY

The earliest notion of signal transduction can be traced back to 1855, when Claude Bernard proposed that ductless glands such as the spleen, the thyroid and adrenal glands, were responsible for the release of "internal secretions" with physiological effects. Bernard's "secretions" were later named "hormones" by Ernest Starling in 1905. Together with William Bayliss, Starling had discovered secretin in 1902. Although many other hormones, most notably insulin, were discovered in the following years, the mechanisms remained largely unknown.

The discovery of nerve growth factor by Rita Levi-Montalcini in 1954, and epidermal growth factor by Stanley Cohen in 1962, led to more detailed insights into the molecular basis of cell signaling, in particular growth factors. Their work, together with Earl Wilbur Sutherland's discovery of cyclic AMP in 1956, prompted the redefinition of endocrine signaling to include only signaling from glands, while the terms autocrine and paracrine began to be used. Sutherland was awarded the 1971 Nobel Prize in Physiology or Medicine, while Levi-Montalcini and Cohen shared it in 1986.

In 1970, Martin Rodbell examined the effects of glucagon on a rat's liver cell membrane receptor. He noted that guanosine triphosphate disassociated glucagon from this receptor and stimulated the G-protein, which strongly influenced the cell's metabolism. Thus, he deduced that the G-protein is a transducer that accepts glucagon molecules and affects the cell. For this, he shared the 1994 Nobel Prize in Physiology or Medicine with Alfred G. Gilman. Thus, the characterization of RTKs and GPCRs led to the formulation of the concept of "signal transduction", a word first used in 1972.

14.2.1 Signal Transduction in Immunology

The relevant events begin with the sequencing of myeloma protein light chains, which are found in abundance in the urine of individuals with multiple myeloma. Biochemical experiments revealed that these so-called Bence Jones proteins consisted of 2 discrete domains -one that varied from one molecule to the next (the V domain) and one that did not (the Fc domain or the Fragment crystallizable region). An analysis of multiple V region sequences by Wu and Kabat identified locations within the V region that were hypervariable and which, they hypothesized, combined in the folded protein to form the antigen recognition site. Thus, within a relatively short time a plausible model was developed for the molecular basis of immunological specificity, and for mediation of biological function through the Fc domain. Crystallization of an IgG molecule soon followed) confirming the inferences based on sequencing, and providing an understanding of immunological specificity at the highest level of resolution.

The biological significance of these developments was encapsulated in the theory of clonal selection which holds that a B cell has on its surface immunoglobulin receptors whose antigen-binding site is identical to that of antibodies that are secreted by the cell when it encounters an antigen, and more specifically a particular B cell clone secretes antibodies with identical sequences. The final piece of the story, the Fluid mosaic model of the plasma membrane provided all the ingredients for a new model for the initiation of signal transduction; viz, receptor dimerization.

The first hints of this were obtained by Becker et al who demonstrated that the extent to which human basophils—for which bivalent Immunoglobulin E (IgE) functions as a surface receptor - degranulate, depends on the concentration of anti IgE antibodies to which they are exposed, and results in a redistribution of surface molecules, which is absent when monovalent ligand is used. The latter observation was consistent with earlier findings by Fanger et al. These observations tied a biological response to events and structural details of molecules on the cell surface. A preponderance of evidence soon developed that receptor dimerization initiates responses (reviewed in) in a variety of cell types, including B cells.

Such observations led to a number of theoretical (mathematical) developments. The first of these was a simple model proposed by Bell which resolved an apparent paradox: clustering forms stable networks; i.e. binding is essentially irreversible, whereas the affinities of antibodies secreted by B cells increase as the immune response progresses. A theory of the dynamics of cell surface clustering on lymphocyte membranes was developed by DeLisi and Perelson who found the size distribution of clusters as a function of time, and its dependence on the affinity and valence of the ligand. Subsequent theories for basophils and mast cells were developed by Goldstein and Sobotka and their collaborators, all aimed at the analysis of dose-response patterns of immune cells and their biological correlates. For a recent review of clustering in immunological systems see.

Ligand binding to cell surface receptors is also critical to motility, a phenomenon that is best understood in single-celled organisms. An example is a detection and response to concentration gradients by bacteria --the classic mathematical theory appearing in. A recent account can be found in

14.3 STIMULI

The basis for signal transduction is the transformation of a certain stimulus into a biochemical signal. The nature of such stimuli can vary widely, ranging from extracellular cues, such as the presence of EGF, to intracellular events, such as the DNA damage resulting from replicative telomere attrition. Traditionally, signals that reach the central nervous system are classified as senses. These are transmitted from neuron to neuron in a process called synaptic transmission. Many other intercellular signal relay mechanisms exist in multicellular organisms, such as those that govern embryonic development.

14.3.1 Ligands

The majority of signal transduction pathways involve the binding of signaling molecules, known as ligands, to receptors that trigger events inside the cell. The binding of a signaling molecule with a receptor causes a change in the conformation of the receptor, known as *receptor activation.* Most ligands are soluble molecules from the extracellular medium which bind to cell surface receptors. These include growth factors, cytokines and neurotransmitters. Components of the extracellular matrix such as fibronectin and hyaluronan can also bind to such receptors (integrins and CD44, respectively).

In addition, some molecules such as steroid hormones are lipid-soluble and thus cross the plasma membrane to reach cytoplasmic or nuclear receptors. In the case of steroid hormone receptors, their stimulation leads to binding to the promoter region of steroid-responsive genes.

Not all classifications of signaling molecules take into account the molecular nature of each class member. For example, odorants belong to a wide range of molecular classes, as do neurotransmitters, which range in size from small molecules such as dopamine to neuropeptides such as endorphins. Moreover, some molecules may fit into more than one class, e.g. epinephrine is a neurotransmitter when secreted by the central nervous system and a hormone when secreted by the adrenal medulla.

Some receptors such as HER2 are capable of ligand-independent activation when overexpressed or mutated. This leads to constitutive activation of the pathway, which may or may not be overturned by compensation mechanisms. In the case of HER2, which acts as a dimerization partner of other EGFRs, constitutive activation leads to hyperproliferation and cancer.

14.3.2 Mechanical Forces

The prevalence of basement membranes in the tissues of Eumetazoans means that most cell types require attachment to survive. This requirement has led to the development of complex mechanotransduction pathways, allowing cells to sense the stiffness of the substratum. Such signaling is mainly orchestrated in focal adhesions, regions where the integrin-bound actin cytoskeleton detects changes and transmits them downstream through YAP1. Calcium-dependent cell adhesion molecules such as cadherins and selectins can also mediate mechanotransduction. Specialised forms of mechanotransduction within the nervous system are responsible for mechanosensation: hearing, touch, proprioception and balance.

14.3.3 Osmolarity

Cellular and systemic control of osmotic pressure (the difference in osmolarity between the cytosol and the extracellular medium) is critical for homeostasis. There are three ways in which cells can detect osmotic stimuli: as changes in macromolecular crowding, ionic strength, and changes in the properties of the plasma membrane or cytoskeleton (the latter being a form of mechanotransduction). These changes are detected by proteins known as osmosensors or osmoreceptors. In humans, the best characterised osmosensors are transient receptor potential channels present in the primary cilium of human cells. In yeast, the HOG pathway has been extensively characterised.

14.3.4 Temperature

The sensing of temperature in cells is known as thermoception and is primarily mediated by transient receptor potential channels. Additionally, animal cells contain a conserved mechanism to prevent high temperatures from causing cellular damage, the heat-shock response. Such response is triggered when high temperatures cause the dissociation of inactive HSF1 from complexes with heat shock proteins Hsp40/Hsp70 and Hsp90. With help from the ncRNA *hsr1*, HSF1 then trimerizes, becoming active and upregulating the expression of its target genes. Many other thermosensory mechanisms exist in both prokaryotes and eukaryotes.

14.3.5 Light

In mammals, light controls the sense of sight and the circadian clock by activating light-sensitive proteins in photoreceptor cells in the eye's retina. In the case of vision, light is detected by rhodopsin in rod and cone cells. In the case of the circadian clock, a different photopigment, melanopsin, is responsible for detecting light in intrinsically photosensitive retinal ganglion cells.

14.4 RECEPTORS

Receptors can be roughly divided into two major classes: intracellular and extracellular receptors.

14.4.1 Extracellular Receptors

Extracellular receptors are integral transmembrane proteins and make up most receptors. They span the plasma membrane of the cell, with one part of the receptor on the outside of the cell and the other on the inside. Signal transduction occurs as a result of a ligand binding to the outside region of the receptor (the ligand does not pass through the membrane). Ligand-receptor binding induces a change in the conformation of the inside part of the receptor, a process sometimes called "receptor activation". This results in either the activation of an enzyme domain of the receptor or the exposure of a binding site for other intracellular signaling proteins within the cell, eventually propagating the signal through the cytoplasm.

In eukaryotic cells, most intracellular proteins activated by a ligand/receptor interaction possess an enzymatic activity; examples include tyrosine kinase and phosphatases. Often such enzymes are covalently linked to the receptor. Some of them create second messengers such as cyclic AMP and IP_3, the latter controlling the release of intracellular calcium stores into the cytoplasm. Other activated proteins interact with adaptor proteins that facilitate signaling protein interactions and coordination of signaling complexes necessary to respond to a particular stimulus. Enzymes and adaptor proteins are both responsive to various second messenger molecules.

Many adaptor proteins and enzymes activated as part of signal transduction possess specialized protein domains that bind to specific secondary messenger molecules. For example, calcium ions bind to the EF hand domains of calmodulin, allowing it to bind and activate calmodulin-dependent kinase. PIP_3 and other phosphoinositides do the same thing to the Pleckstrin homology domains of proteins such as the kinase protein AKT.

G Protein–Coupled Receptors

G protein-coupled receptors (GPCRs), also known as seven-(pass)-transmembrane domain receptors, 7TM receptors, heptahelical receptors, serpentine receptors, and G protein-linked receptors (GPLR), form a large group of evolutionarily related proteins that are cell surface receptors that detect molecules outside the cell and activate cellular responses. They are coupled with G proteins. They pass through the cell membrane seven times in the form of six loops (three extracellular loops interacting with ligand molecules, three intracellular loops interacting with G proteins, an N-terminal extracellular region and a C-terminal intracellular region) of amino acid residues, which is why they are sometimes referred to as seven-transmembrane receptors. Ligands can bind either to the extracellular N-terminus and loops (e.g. glutamate receptors) or to the binding site within transmembrane helices (rhodopsin-like family). They are all activated by agonists, although a spontaneous auto-activation of an empty receptor has also been observed.

G protein-coupled receptors are found only in eukaryotes, including yeast, and choanoflagellates. The ligands that bind and activate these receptors include light-sensitive compounds, odors, pheromones, hormones, and neurotransmitters, and vary in size from small molecules to peptides to large proteins. G protein-coupled receptors are involved in many diseases.

There are two principal signal transduction pathways involving the G protein-coupled receptors:

- The cAMP signal pathway.
- The phosphatidylinositol signal pathway.

When a ligand binds to the GPCR it causes a conformational change in the GPCR, which allows it to act as a guanine nucleotide exchange factor (GEF). The GPCR can then activate an associated G protein by exchanging the GDP bound to the G protein for a GTP. The G protein's α subunit, together with the bound GTP, can then dissociate from the β and γ subunits to further affect intracellular signaling proteins or target functional proteins directly depending on the α subunit type ($G_{\alpha s}$, $G_{\alpha i/o}$, $G_{\alpha q/11}$, $G_{\alpha 13/12}$).

GPCRs are an important drug target and approximately 34% of all Food and Drug Administration (FDA) approved drugs target 108 members of this family. The global sales volume for these drugs is estimated to be 180 billion US dollars as of 2018. It is estimated that GPCRs are targets for about 50% of drugs currently on the market, mainly due to their involvement in signaling pathways related to many diseases i.e. mental, metabolic including endocrinological disorders, immunological including viral infections, cardiovascular, inflammatory, senses disorders, and cancer. The long ago discovered association between GPCRs and many endogenous and exogenous substances, resulting in e.g. analgesia, is another dynamically developing field of the pharmaceutical research.

History and Significance

With the determination of the first structure of the complex between a G-protein coupled receptor (GPCR) and a G-protein trimer ($G\alpha\beta\gamma$) in 2011 a new chapter of GPCR research was opened for structural investigations of global switches with more than one protein being investigated. The previous breakthroughs involved determination of the crystal structure of the first GPCR, rhodopsin, in 2000 and the crystal structure of the first GPCR with a diffusible ligand (β_2AR) in 2007. The way in which the seven transmembrane helices of a GPCR are arranged into a bundle was suspected based on the low-resolution model of frog rhodopsin from cryogenic electron microscopy studies of the two-dimensional crystals. The crystal structure of rhodopsin, that came up three years later, was not a surprise apart from the presence of an additional cytoplasmic helix H8 and a precise location of a loop covering retinal binding site. However, it provided a scaffold which was hoped to be a universal template for homology modeling and drug design for other GPCRs – a notion that proved to be too optimistic.

Seven years later, the crystallization of β_2-adrenergic receptor (β_2AR) with a diffusible ligand brought surprising results because it revealed quite a different shape of the receptor extracellular side than that of rhodopsin. This area is important because it is responsible for the ligand binding and is targeted by many drugs. Moreover, the ligand binding site was much more spacious than in the rhodopsin structure and was open to the exterior. In the other receptors crystallized shortly afterwards the binding side was even more easily accessible to the ligand. New structures complemented with biochemical investigations uncovered mechanisms of action of molecular switches which modulate the structure of the receptor leading to activation states for agonists or to complete or partial inactivation states for inverse agonists.

The 2012 Nobel Prize in Chemistry was awarded to Brian Kobilka and Robert Lefkowitz for their work that was "crucial for understanding how G protein-coupled receptors function". There have been at least seven other Nobel Prizes awarded for some aspect of G protein–mediated signaling. As of 2012, two of the top ten global best-selling drugs (Advair Diskus and Abilify) act by targeting G protein-coupled receptors.

Classification

The exact size of the GPCR superfamily is unknown, but at least 831 different human genes (or about 4% of the entire protein-coding genome) have been predicted to code for them from genome sequence analysis. Although numerous classification schemes have been proposed, the superfamily was classically divided into three main classes (A, B, and C) with no detectable shared sequence homology between classes.

The largest class by far is class A, which accounts for nearly 85% of the GPCR genes. Of class A GPCRs, over half of these are predicted to encode olfactory receptors, while the remaining receptors are liganded by known endogenous compounds or are classified as orphan receptors. Despite the lack of sequence homology between classes, all GPCRs have a common structure and mechanism of signal transduction. The very large rhodopsin A group has been further subdivided into 19 subgroups (A1-A19).

According to the classical A-F system, GPCRs can be grouped into six classes based on sequence homology and functional similarity:

- Class A (or 1) (Rhodopsin-like).
- Class B (or 2) (Secretin receptor family).
- Class C (or 3) (Metabotropic glutamate/pheromone).
- Class D (or 4) (Fungal mating pheromone receptors).
- Class E (or 5) (Cyclic AMP receptors).
- Class F (or 6) (Frizzled/Smoothened).

More recently, an alternative classification system called GRAFS (Glutamate, Rhodopsin, *Adhesion*, Frizzled/Taste2, Secretin) has been proposed for vertebrate GPCRs. They correspond to classical classes C, A, B2, F, and B.

An early study based on available DNA sequence suggested that the human genome encodes roughly 750 G protein-coupled receptors, about 350 of which detect hormones, growth factors, and other endogenous ligands. Approximately 150 of the GPCRs found in the human genome have unknown functions.

Some web-servers and bioinformatics prediction methods have been used for predicting the classification of GPCRs according to their amino acid sequence alone, by means of the pseudo amino acid composition approach.

Physiological Roles

GPCRs are involved in a wide variety of physiological processes. Some examples of their physiological roles include:

- **The Visual Sense:** The opsins use a photoisomerization reaction to translate electromagnetic radiation into cellular signals. Rhodopsin, for example, uses the conversion of *11-cis*-retinal to *all-trans*-retinal for this purpose.
- **The Gustatory Sense (Taste):** GPCRs in taste cells mediate release of gustducin in response to bitter-, umami- and sweet-tasting substances.
- **The Sense of Smell:** Receptors of the olfactory epithelium bind odorants (olfactory receptors) and pheromones (vomeronasal receptors).
- **Behavioral and Mood Regulation:** Receptors in the mammalian brain bind several different neurotransmitters, including serotonin, dopamine, histamine, GABA, and glutamate.

- **Regulation of Immune System Activity and Inflammation:** Chemokine receptors bind ligands that mediate intercellular communication between cells of the immune system; receptors such as histamine receptors bind inflammatory mediators and engage target cell types in the inflammatory response. GPCRs are also involved in immune-modulation, e. g., regulating interleukin induction or suppressing TLR-induced immune responses from T cells.
- **Autonomic Nervous System Transmission:** Both the sympathetic and parasympathetic nervous systems are regulated by GPCR pathways, responsible for control of many automatic functions of the body such as blood pressure, heart rate, and digestive processes.
- **Cell Density Sensing:** A novel GPCR role in regulating cell density sensing.
- Homeostasis modulation (e.g., water balance).
- Involved in growth and metastasis of some types of tumors.
- Used in the endocrine system for peptide and amino-acid derivative hormones that bind to GCPRs on the cell membrane of a target cell. This activates cAMP, which in turn activates several kinases, allowing for a cellular response, such as transcription.

Receptor Structure

GPCRs are integral membrane proteins that possess seven membrane-spanning domains or transmembrane helices. The extracellular parts of the receptor can be glycosylated.

These extracellular loops also contain two highly conserved cysteine residues that form disulfide bonds to stabilize the receptor structure. Some seven-transmembrane helix proteins (channelrhodopsin) that resemble GPCRs may contain ion channels, within their protein.

In 2000, the first crystal structure of a mammalian GPCR, that of bovine rhodopsin (1F88), was solved. In 2007, the first structure of a human GPCR was solved This human β_2-adrenergic receptor GPCR structure proved highly similar to the bovine rhodop sin. The structures of activated or agonist-bound GPCRs have also been determ ined. These structures indicate how ligand binding at the extracellular side of a receptor leads to conformational changes in the cytoplasmic side of the receptor. The biggest change is an outward movement of the cytoplasmic part of the 5th and 6th transmembrane helix (TM5 and TM6). The structure of activated beta-2 adrenergic receptor in complex with G_s confirmed that the Gα binds to a cavity created by this movement.

GPCRs exhibit a similar structure to some other proteins with seven transmembrane domains, such as microbial rhodopsins and adiponectin receptors 1 and 2 (ADIPOR1 and ADIPOR2). However, these 7TMH (7-transmembrane helices) receptors and channels do not associate with G proteins. In addition, ADIPOR1 and ADIPOR2 are oriented oppositely to GPCRs in the membrane (i.e. GPCRs usually have an extracellular N-terminus, cytoplasmic C-terminus, whereas ADIPORs are inverted).

Structure–Function Relationships

In terms of structure, GPCRs are characterized by an extracellular N-terminus, followed by seven transmembrane (7-TM) α-helices (TM-1 to TM-7) connected by three intracellular (IL-1 to IL-3) and three extracellular loops (EL-1 to EL-3), and finally an intracellular C-terminus. The GPCR arranges itself into a tertiary structure resembling a barrel, with the seven transmembrane helices forming a cavity within the plasma membrane that serves a ligand-binding domain that is often covered by EL-2. Ligands may also bind elsewhere, however, as is the case for bulkier ligands (e.g., proteins or large peptides), which instead interact with the extracellular loops, or, as illustrated by the class C metabotropic glutamate receptors (mGluRs), the N-terminal tail.

The class C GPCRs are distinguished by their large N-terminal tail, which also contains a ligand-binding domain. Upon glutamate-binding to an mGluR, the N-terminal tail undergoes a conformational change that leads to its interaction with the residues of the extracellular loops and TM domains. The eventual effect of all three types of agonist-induced activation is a change in the relative orientations of the TM helices (likened to a twisting motion) leading to a wider intracellular surface and "revelation" of residues of the intracellular helices and TM domains crucial to signal transduction function (i.e., G-protein coupling). Inverse agonists and antagonists may also bind to a number of different sites, but the eventual effect must be prevention of this TM helix reorientation.

The structure of the N- and C-terminal tails of GPCRs may also serve important functions beyond ligand-binding. For example, The C-terminus of M_3 muscarinic receptors is sufficient, and the six-amino-acid polybasic (KKKRRK) domain in the C-terminus is necessary for its preassembly with G_q proteins. In particular, the C-terminus often contains serine (Ser) or threonine (Thr) residues that, when phosphorylated, increase the affinity of the intracellular surface for the binding of scaffolding proteins called β-arrestins (β-arr). Once bound, β-arrestins both sterically prevent G-protein coupling and may recruit other proteins, leading to the creation of signaling complexes involved in extracellular-signal regulated kinase (ERK) pathway activation or receptor endocytosis (internalization). As the phosphorylation of these Ser and Thr residues often occurs as a result of GPCR activation, the β-arr-mediated G-protein-decoupling and internalization of GPCRs are important mechanisms of desensitization. In addition, internalized "mega-complexes" consisting of a single GPCR, β-arr(in the tail conformation), and heterotrimeric G protein exist and may account for protein signaling from endosomes.

A final common structural theme among GPCRs is palmitoylation of one or more sites of the C-terminal tail or the intracellular loops. Palmitoylation is the covalent modification of cysteine (Cys) residues via addition of hydrophobic acyl groups, and has the effect of targeting the receptor to cholesterol- and sphingolipid-rich microdomains of the plasma membrane called lipid rafts. As many of the downstream transducer and effector molecules of GPCRs (including those involved in negative feedback pathways) are also targeted to lipid rafts, this has the effect of facilitating rapid receptor signaling.

GPCRs respond to extracellular signals mediated by a huge diversity of agonists, ranging from proteins to biogenic amines to protons, but all transduce this signal via a mechanism of G-protein coupling. This is made possible by a guanine-nucleotide exchange factor (GEF) domain primarily formed by a combination of IL-2 and IL-3 along with adjacent residues of the associated TM helices.

Mechanism

The G protein-coupled receptor is activated by an external signal in the form of a ligand or other signal mediator. This creates a conformational change in the receptor, causing activation of a G protein. Further effect depends on the type of G protein. G proteins are subsequently inactivated by GTPase activating proteins, known as RGS proteins.

- **Ligand Binding:**
  - GPCRs include one or more receptors for the following ligands: sensory signal mediators (e.g., light and olfactory stimulatory molecules); adenosine, bombesin, bradykinin, endothelin, γ-aminobutyric acid (GABA), hepatocyte growth factor (HGF), melanocortins, neuropeptide Y, opioid peptides, opsins, somatostatin, GH, tachykinins, members of the vasoactive intestinal peptide family, and vasopressin; biogenic amines (e.g., dopamine, epinephrine, norepinephrine, histamine, serotonin, and melatonin); glutamate (metabotropic effect); glucagon; acetylcholine (muscarinic effect); chemokines; lipid mediators of inflammation (e.g., prostaglandins, prostanoids, platelet-activating factor, and leukotrienes); peptide hormones (e.g., calcitonin, C5a anaphylatoxin, follicle-stimulating hormone [FSH], gonadotropin-releasing hormone [GnRH], neurokinin, thyrotropin-releasing hormone [TRH], and oxytocin); and endocannabinoids.
  - GPCRs that act as receptors for stimuli that have not yet been identified are known as orphan receptors.

- However, in contrast to other types of receptors that have been studied, wherein ligands bind externally to the membrane, the ligands of GPCRs typically bind within the transmembrane domain. However, protease-activated receptors are activated by cleavage of part of their extracellular domain.

- **Conformational Change:**
  - The transduction of the signal through the membrane by the receptor is not completely understood. It is known that in the inactive state, the GPCR is bound to a heterotrimeric G protein complex. Binding of an agonist to the GPCR results in a conformational change in the receptor that is transmitted to the bound G_α subunit of the heterotrimeric G protein via protein domain dynamics. The activated G_α subunit exchanges GTP in place of GDP which in turn triggers the dissociation of G_α subunit from the $G_{\beta\gamma}$ dimer and from the receptor. The dissociated G_α and $G_{\beta\gamma}$ subunits interact with other intracellular proteins to continue the signal transduction cascade while the freed GPCR is able to rebind to another heterotrimeric G protein to form a new complex that is ready to initiate another round of signal transduction.
  - It is believed that a receptor molecule exists in a conformational equilibrium between active and inactive biophysical states. The binding of ligands to the receptor may shift the equilibrium toward the active receptor states. Three types of ligands exist: Agonists are ligands that shift the equilibrium in favour of active states; inverse agonists are ligands that shift the equilibrium in favour of inactive states; and neutral antagonists are ligands that do not affect the equilibrium. It is not yet known how exactly the active and inactive states differ from each other.

- **G-Protein Activation/Deactivation Cycle:**
  - When the receptor is inactive, the GEF domain may be bound to an also inactive α-subunit of a heterotrimeric G-protein. These “G-proteins” are a trimer of α, β, and γ subunits (known as Gα, Gβ, and Gγ, respectively) that is rendered inactive when reversibly bound to Guanosine diphosphate (GDP) (or, alternatively, no guanine nucleotide) but active when bound to guanosine triphosphate (GTP). Upon receptor activation, the GEF domain, in turn, allosterically activates the G-protein by facilitating the exchange of a molecule of GDP for GTP at the G-protein’s α-subunit. The cell maintains a 10:1 ratio of cytosolic GTP:GDP so exchange for GTP is ensured. At this point, the subunits of the G-protein dissociate from the receptor, as well as each other, to yield a Gα-GTP monomer and a tightly interacting Gβγ dimer, which are now free to modulate the activity of other intracellular proteins. The extent to which they may diffuse, however, is limited due to the palmitoylation of Gα and the presence of an isoprenoid moiety that has been covalently added to the C-termini of Gγ.

- Because Gα also has slow GTP → GDP hydrolysis capability, the inactive form of the α-subunit (Gα-GDP) is eventually regenerated, thus allowing reassociation with a Gβγ dimer to form the "resting" G-protein, which can again bind to a GPCR and await activation. The rate of GTP hydrolysis is often accelerated due to the actions of another family of allosteric modulating proteins called regulators of G-protein signaling, or RGS proteins, which are a type of GTPase-activating protein, or GAP. In fact, many of the primary effector proteins (e.g., adenylate cyclases) that become activated/inactivated upon interaction with Gα-GTP also have GAP activity. Thus, even at this early stage in the process, GPCR-initiated signaling has the capacity for self-termination.

- **Crosstalk:**
  - GPCRs downstream signals have been shown to possibly interact with integrin signals, such as FAK. Integrin signaling will phosphorylate FAK, which can then decrease GPCR $G_{\alpha s}$ activity.

Signaling

If a receptor in an active state encounters a G protein, it may activate it. Some evidence suggests that receptors and G proteins are actually pre-coupled. For example, binding of G proteins to receptors affects the receptor's affinity for ligands. Activated G proteins are bound to GTP.

Further signal transduction depends on the type of G protein. The enzyme adenylate cyclase is an example of a cellular protein that can be regulated by a G protein, in this case the G protein G_s. Adenylate cyclase activity is activated when it binds to a subunit of the activated G protein. Activation of adenylate cyclase ends when the G protein returns to the GDP-bound state.

Adenylate cyclases (of which 9 membrane-bound and one cytosolic forms are known in humans) may also be activated or inhibited in other ways (e.g., Ca2+/calmodulin binding), which can modify the activity of these enzymes in an additive or synergistic fashion along with the G proteins.

The signaling pathways activated through a GPCR are limited by the primary sequence and tertiary structure of the GPCR itself but ultimately determined by the particular conformation stabilized by a particular ligand, as well as the availability of transducer molecules. Currently, GPCRs are considered to utilize two primary types of transducers: G-proteins and β-arrestins. Because β-arr's have high affinity only to the phosphorylated form of most GPCRs, the majority of signaling is ultimately dependent upon G-protein activation. However, the possibility for interaction does allow for G-protein-independent signaling to occur.

- **G-Protein-Dependent Signaling:**
  - There are three main G-protein-mediated signaling pathways, mediated by four sub-classes of G-proteins distinguished from each other by sequence

homology ($G_{\alpha s}$, $G_{\alpha i/o}$, $G_{\alpha q/11}$, and $G_{\alpha 13/12}$). Each sub-class of G-protein consists of multiple proteins, each the product of multiple genes or splice variations that may imbue them with differences ranging from subtle to distinct with regard to signaling properties, but in general they appear reasonably grouped into four classes. Because the signal transducing properties of the various possible βγ combinations do not appear to radically differ from one another, these classes are defined according to the isoform of their α-subunit.

- While most GPCRs are capable of activating more than one Gα-subtype, they also show a preference for one subtype over another. When the subtype activated depends on the ligand that is bound to the GPCR, this is called functional selectivity (also known as agonist-directed trafficking, or conformation-specific agonism). However, the binding of any single particular agonist may also initiate activation of multiple different G-proteins, as it may be capable of stabilizing more than one conformation of the GPCR's GEF domain, even over the course of a single interaction. In addition, a conformation that preferably activates one isoform of Gα may activate another if the preferred is less available. Furthermore, feedback pathways may result in receptor modifications (e.g., phosphorylation) that alter the G-protein preference. Regardless of these various nuances, the GPCR's preferred coupling partner is usually defined according to the G-protein most obviously activated by the endogenous ligand under most physiological or experimental conditions.

- **Gα Signaling:**
  - The effector of both the $G_{\alpha s}$ and $G_{\alpha i/o}$ pathways is the cyclic-adenosine monophosphate (cAMP)-generating enzyme adenylate cyclase, or AC. While there are ten different AC gene products in mammals, each with subtle differences in tissue distribution or function, all catalyze the conversion of cytosolic adenosine triphosphate (ATP) to cAMP, and all are directly stimulated by G-proteins of the $G_{\alpha s}$ class. In contrast, however, interaction with Gα subunits of the $G_{\alpha i/o}$ type inhibits AC from generating cAMP. Thus, a GPCR coupled to $G_{\alpha s}$ counteracts the actions of a GPCR coupled to $G_{\alpha i/o}$, and vice versa. The level of cytosolic cAMP may then determine the activity of various ion channels as well as members of the ser/thr-specific protein kinase A (PKA) family. Thus cAMP is considered a second messenger and PKA a secondary effector.
  - The effector of the $G_{\alpha q/11}$ pathway is phospholipase C-β (PLCβ), which catalyzes the cleavage of membrane-bound phosphatidylinositol 4,5-bisphosphate (PIP2) into the second messengers inositol (1,4,5) trisphosphate (IP3) and diacylglycerol (DAG). IP3 acts on IP3 receptors found in the membrane of the endoplasmic reticulum (ER) to elicit Ca^{2+} release from the ER, while DAG diffuses along the plasma membrane where it may activate any membrane localized forms of a second ser/thr kinase called protein ki-

nase C (PKC). Since many isoforms of PKC are also activated by increases in intracellular Ca^{2+}, both these pathways can also converge on each other to signal through the same secondary effector. Elevated intracellular Ca^{2+} also binds and allosterically activates proteins called calmodulins, which in turn tosolic small GTPase, Rho. Once bound to GTP, Rho can then go on to activate various proteins responsible for cytoskeleton regulation such as Rho-kinase (ROCK). Most GPCRs that couple to $G_{\alpha 13/12}$ also couple to other sub-classes, often $G_{\alpha q/11}$.

- **Gβγ Signaling:**
  - The above descriptions ignore the effects of Gβγ–signalling, which can also be important, in particular in the case of activated $G_{\alpha i/o}$-coupled GPCRs. The primary effectors of Gβγ are various ion channels, such as G-protein-regulated inwardly rectifying K^+ channels (GIRKs), P/Q- and N-type voltage-gated Ca^{2+} channels, as well as some isoforms of AC and PLC, along with some phosphoinositide-3-kinase (PI3K) isoforms.
- **G-Protein-Independent Signaling:**
  - Although they are classically thought of working only together, GPCRs may signal through G-protein-independent mechanisms, and heterotrimeric G-proteins may play functional roles independent of GPCRs. GPCRs may signal independently through many proteins already mentioned for their roles in G-protein-dependent signaling such as β-arrs, GRKs, and Srcs. Such signaling has been shown to be physiologically relevant, for example, β-arrestin signaling mediated by the chemokine receptor CXCR3 was necessary for full efficacy chemotaxis of activated T cells. In addition, further scaffolding proteins involved in subcellular localization of GPCRs (e.g., PDZ-domain-containing proteins) may also act as signal transducers. Most often the effector is a member of the MAPK family.
  - GPCR-Independent Signaling by Heterotrimeric G-Proteins:
  - Although it is a relatively immature area of research, it appears that heterotrimeric G-proteins may also take part in non-GPCR signaling. There is evidence for roles as signal transducers in nearly all other types of receptor-mediated signaling, including integrins, receptor tyrosine kinases (RTKs), cytokine receptors (JAK/STATs), as well as modulation of various other "accessory" proteins such as GEFs, guanine-nucleotide dissociation inhibitors (GDIs) and protein phosphatases. There may even be specific proteins of these classes whose primary function is as part of GPCR-independent pathways, termed activators of G-protein signalling (AGS). Both the ubiquity of these interactions and the importance of Gα vs. Gβγ subunits to these processes are still unclear.

cAMP Signal Pathway

The cAMP signal transduction contains five main characters: stimulative hormone receptor (Rs) or inhibitory hormone receptor (Ri); stimulative regulative G-protein

(Gs) or inhibitory regulative G-protein (Gi); adenylyl cyclase; protein kinase A (PKA); and cAMP phosphodiesterase.

Stimulative hormone receptor (Rs) is a receptor that can bind with stimulative signal molecules, while inhibitory hormone receptor (Ri) is a receptor that can bind with inhibitory signal molecules.

Stimulative regulative G-protein is a G-protein linked to stimulative hormone receptor (Rs), and its α subunit upon activation could stimulate the activity of an enzyme or other intracellular metabolism. On the contrary, inhibitory regulative G-protein is linked to an inhibitory hormone receptor, and its α subunit upon activation could inhibit the activity of an enzyme or other intracellular metabolism.

Adenylyl cyclase is a 12-transmembrane glycoprotein that catalyzes the conversion of ATP to cAMP with the help of cofactor Mg^{2+} or Mn^{2+}. The cAMP produced is a second messenger in cellular metabolism and is an allosteric activator of protein kinase A.

Protein kinase A is an important enzyme in cell metabolism due to its ability to regulate cell metabolism by phosphorylating specific committed enzymes in the metabolic pathway. It can also regulate specific gene expression, cellular secretion, and membrane permeability. The protein enzyme contains two catalytic subunits and two regulatory subunits. When there is no cAMP, the complex is inactive. When cAMP binds to the regulatory subunits, their conformation is altered, causing the dissociation of the regulatory subunits, which activates protein kinase A and allows further biological effects.

These signals then can be terminated by cAMP phosphodiesterase, which is an enzyme that degrades cAMP to 5'-AMP and inactivates protein kinase A.

Phosphatidylinositol Signal Pathway

In the phosphatidylinositol signal pathway, the extracellular signal molecule binds with the G-protein receptor (G_q) on the cell surface and activates phospholipase C, which is located on the plasma membrane. The lipase hydrolyzes phosphatidylinositol 4,5-bisphosphate (PIP2) into two second messengers: inositol 1,4,5-trisphosphate (IP3) and diacylglycerol (DAG). IP3 binds with the IP3 receptor in the membrane of the smooth endoplasmic reticulum and mitochondria to open Ca^{2+} channels. DAG helps activate protein kinase C (PKC), which phosphorylates many other proteins, changing their catalytic activities, leading to cellular responses.

The effects of Ca^{2+} are also remarkable: it cooperates with DAG in activating PKC and can activate the CaM kinase pathway, in which calcium-modulated protein calmodulin (CaM) binds Ca^{2+}, undergoes a change in conformation, and activates CaM kinase II, which has unique ability to increase its binding affinity to CaM by autophosphorylation, making CaM unavailable for the activation of other enzymes.

The kinase then phosphorylates target enzymes, regulating their activities. The two signal pathways are connected together by Ca^{2+}-CaM, which is also a regulatory subunit of adenylyl cyclase and phosphodiesterase in the cAMP signal pathway.

Phosphorylation by cAMP-Dependent Protein Kinases

Cyclic AMP-dependent protein kinases (protein kinase A) are activated by the signal chain coming from the G protein (that was activated by the receptor) via adenylate cyclase and cyclic AMP (cAMP). In a *feedback mechanism*, these activated kinases phosphorylate the receptor. The longer the receptor remains active the more kinases are activated and the more receptors are phosphorylated. In β_2-adrenoceptors, this phosphorylation results in the switching of the coupling from the G_s class of G-protein to the G_i class. cAMP-dependent PKA mediated phosphorylation can cause heterologous desensitisation in receptors other than those activated.

Phosphorylation by GRKs

The G protein-coupled receptor kinases (GRKs) are protein kinases that phosphorylate only active GPCRs. G-protein-coupled receptor kinases (GRKs) are key modulators of G-protein-coupled receptor (GPCR) signaling. They constitute a family of seven mammalian serine-threonine protein kinases that phosphorylate agonist-bound receptor. GRKs-mediated receptor phosphorylation rapidly initiates profound impairment of receptor signaling and desensitization. Activity of GRKs and subcellular targeting is tightly regulated by interaction with receptor domains, G protein subunits, lipids, anchoring proteins and calcium-sensitive proteins.

Phosphorylation of the receptor can have two consequences:

- **Translocation:** The receptor is, along with the part of the membrane it is embedded in, brought to the inside of the cell, where it is dephosphorylated within the acidic vesicular environment and then brought back. This mechanism is used to regulate long-term exposure, for example, to a hormone, by allowing resensitisation to follow desensitisation. Alternatively, the receptor may undergo lysozomal degradation, or remain internalised, where it is thought to participate in the initiation of signalling events, the nature of which depending on the internalised vesicle's subcellular localisation.
- **Arrestin Linking:** The phosphorylated receptor can be linked to *arrestin* molecules that prevent it from binding (and activating) G proteins, in effect switching it off for a short period of time. This mechanism is used, for example, with rhodopsin in retina cells to compensate for exposure to bright light. In many cases, arrestin's binding to the receptor is a prerequisite for translocation. For example, beta-arrestin bound to β_2-

adrenoreceptors acts as an adaptor for binding with clathrin, and with the beta-subunit of AP2 (clathrin adaptor molecules); thus, the arrestin here acts as a scaffold assembling the components needed for clathrin-mediated endocytosis of β_2-adrenoreceptors.

Mechanisms of GPCR Signal Termination

As mentioned above, G-proteins may terminate their own activation due to their intrinsic GTP $\rightarrow$ GDP hydrolysis capability. However, this reaction proceeds at a slow rate (≈0.02 times/sec) and, thus, it would take around 50 seconds for any single G-protein to deactivate if other factors did not come into play. Indeed, there are around 30 isoforms of RGS proteins that, when bound to Gα through their GAP domain, accelerate the hydrolysis rate to ≈30 times/sec. This 1500-fold increase in rate allows for the cell to respond to external signals with high speed, as well as spatial resolution due to limited amount of second messenger that can be generated and limited distance a G-protein can diffuse in 0.03 seconds. For the most part, the RGS proteins are promiscuous in their ability to deactivate G-proteins, while which RGS is involved in a given signaling pathway seems more determined by the tissue and GPCR involved than anything else. In addition, RGS proteins have the additional function of increasing the rate of GTP-GDP exchange at GPCRs, (i.e., as a sort of co-GEF) further contributing to the time resolution of GPCR signaling.

In addition, the GPCR may be desensitized itself. This can occur as:

- A direct result of ligand occupation, wherein the change in conformation allows recruitment of GPCR-Regulating Kinases (GRKs), which go on to phosphorylate various serine/threonine residues of IL-3 and the C-terminal tail. Upon GRK phosphorylation, the GPCR's affinity for β-arrestin (β-arrestin-1/2 in most tissues) is increased, at which point β-arrestin may bind and act to both sterically hinder G-protein coupling as well as initiate the process of receptor internalization through clathrin-mediated endocytosis. Because only the liganded receptor is desensitized by this mechanism, it is called homologous desensitization.
- The affinity for β-arrestin may be increased in a ligand occupation and GRK-independent manner through phosphorylation of different ser/thr sites (but also of IL-3 and the C-terminal tail) by PKC and PKA. These phosphorylations are often sufficient to impair G-protein coupling on their own as well.
- PKC/PKA may, instead, phosphorylate GRKs, which can also lead to GPCR phosphorylation and β-arrestin binding in an occupation-independent manner. These latter two mechanisms allow for desensitization of one GPCR due to the activities of others, or heterologous desensitization.

GRKs may also have GAP domains and so may contribute to inactivation through non-kinase mechanisms as well. A combination of these mechanisms may also occur.

Once β-arrestin is bound to a GPCR, it undergoes a conformational change allowing it to serve as a scaffolding protein for an adaptor complex termed AP-2, which in turn recruits another protein called clathrin. If enough receptors in the local area recruit clathrin in this manner, they aggregate and the membrane buds inwardly as a result of interactions between the molecules of clathrin, in a process called opsonization.

Once the pit has been pinched off the plasma membrane due to the actions of two other proteins called amphiphysin and dynamin, it is now an endocytic vesicle. At this point, the adapter molecules and clathrin have dissociated, and the receptor is either trafficked back to the plasma membrane or targeted to lysosomes for degradation.

At any point in this process, the β-arrestins may also recruit other proteins—such as the non-receptor tyrosine kinase (nRTK), c-SRC—which may activate ERK1/2, or other mitogen-activated protein kinase (MAPK) signaling through, for example, phosphorylation of the small GTPase, Ras, or recruit the proteins of the ERK cascade directly at which point signaling is initiated due to their close proximity to one another.

Another target of c-SRC are the dynamin molecules involved in endocytosis. Dynamins polymerize around the neck of an incoming vesicle, and their phosphorylation by c-SRC provides the energy necessary for the conformational change allowing the final "pinching off" from the membrane.

GPCR Cellular Regulation

Receptor desensitization is mediated through a combination phosphorylation, β-arr binding, and endocytosis as described above. Downregulation occurs when endocytosed receptor is embedded in an endosome that is trafficked to merge with an organelle called a lysosome. Because lysosomal membranes are rich in proton pumps, their interiors have low pH (≈4.8 vs. the pH≈7.2 cytosol), which acts to denature the GPCRs.

In addition, lysosomes contain many degradative enzymes, including proteases, which can function only at such low pH, and so the peptide bonds joining the residues of the GPCR together may be cleaved. Whether or not a given receptor is trafficked to a lysosome, detained in endosomes, or trafficked back to the plasma membrane depends on a variety of factors, including receptor type and magnitude of the signal. GPCR regulation is additionally mediated by gene transcription factors. These factors can increase or decrease gene transcription and thus increase or decrease the generation of new receptors (up- or down-regulation) that travel to the cell membrane.

Receptor Oligomerization

G-protein-coupled receptor oligomerisation is a widespread phenomenon. One of the best-studied examples is the metabotropic $GABA_B$ receptor. This so-called constitutive receptor is formed by heterodimerization of $GABA_BR1$ and $GABA_BR2$ subunits. Expression of the $GABA_BR1$ without the $GABA_BR2$ in heterologous systems leads to retention of the subunit in the endoplasmic reticulum. Expression of the $GABA_BR2$ subunit alone, meanwhile, leads to surface expression of the subunit, although with no functional activity (i.e., the receptor does not bind agonist and cannot initiate a response following exposure to agonist). Expression of the two subunits together leads to plasma membrane expression of functional receptor. It has been shown that $GABA_BR2$ binding to $GABA_BR1$ causes masking of a retention signal of functional receptors.

Origin and Diversification of the Superfamily

Signal transduction mediated by the superfamily of GPCRs dates back to the origin of multicellularity. Mammalian-like GPCRs are found in fungi, and have been classified according to the GRAFS classification system based on GPCR fingerprints. Identification of the superfamily members across the eukaryotic domain, and comparison of the family-specific motifs, have shown that the superfamily of GPCRs have a common origin. Characteristic motifs indicate that three of the five GRAFS families, *Rhodopsin*, *Adhesion*, and *Frizzled*, evolved from the *Dictyostelium discoideum* cAMP receptors before the split of opisthokonts. Later, the *Secretin* family evolved from the *Adhesion* GPCR receptor family before the split of nematodes. Insect GPCRs appear to be in their own group and Taste2 is identified as descending from *Rhodopsin*. Note that the *Secretin/Adhesion* split is based on presumed function rather than signature, as the classical Class B (7tm_2, Pfam PF00002) is used to identify both in the studies.

Tyrosine, Ser/Thr and Histidine-Specific Protein Kinases

Receptor tyrosine kinases (RTKs) are transmembrane proteins with an intracellular kinase domain and an extracellular domain that binds ligands; examples include growth factor receptors such as the insulin receptor. To perform signal transduction, RTKs need to form dimers in the plasma membrane; the dimer is stabilized by ligands binding to the receptor. The interaction between the cytoplasmic domains stimulates the autophosphorylation of tyrosine residues within the intracellular kinase domains of the RTKs, causing conformational changes. Subsequent to this, the receptors' kinase domains are activated, initiating phosphorylation signaling cascades of downstream cytoplasmic molecules that facilitate various cellular processes such as cell differentiation and metabolism. Many Ser/Thr and dual-specificity protein kinases are important for signal transduction, either acting downstream of [receptor tyrosine kinases], or as membrane-embedded or cell-soluble versions in their own right. The process of signal transduction involves around 560 known protein kinases and pseudokinases, encoded by the human kinome

As is the case with GPCRs, proteins that bind GTP play a major role in signal transduction from the activated RTK into the cell. In this case, the G proteins are members of the Ras, Rho, and Raf families, referred to collectively as small G proteins. They act as molecular switches usually tethered to membranes by isoprenyl groups linked to their carboxyl ends. Upon activation, they assign proteins to specific membrane subdomains where they participate in signaling. Activated RTKs in turn activate small G proteins that activate guanine nucleotide exchange factors such as SOS1. Once activated, these exchange factors can activate more small G proteins, thus amplifying the receptor's initial signal. The mutation of certain RTK genes, as with that of GPCRs, can result in the expression of receptors that exist in a constitutively activated state; such mutated genes may act as oncogenes.

Histidine-specific protein kinases are structurally distinct from other protein kinases and are found in prokaryotes, fungi, and plants as part of a two-component signal transduction mechanism: a phosphate group from ATP is first added to a histidine residue within the kinase, then transferred to an aspartate residue on a receiver domain on a different protein or the kinase itself, thus activating the aspartate residue.

Integrins

Integrins are produced by a wide variety of cells; they play a role in cell attachment to other cells and the extracellular matrix and in the transduction of signals from extracellular matrix components such as fibronectin and collagen. Ligand binding to the extracellular domain of integrins changes the protein's conformation, clustering it at the cell membrane to initiate signal transduction. Integrins lack kinase activity; hence, integrin-mediated signal transduction is achieved through a variety of intracellular protein kinases and adaptor molecules, the main coordinator being integrin-linked kinase. As shown in the adjacent picture, cooperative integrin-RTK signaling determines the timing of cellular survival, apoptosis, proliferation, and differentiation.

Important differences exist between integrin-signaling in circulating blood cells and non-circulating cells such as epithelial cells; integrins of circulating cells are normally inactive. For example, cell membrane integrins on circulating leukocytes are maintained in an inactive state to avoid epithelial cell attachment; they are activated only in response to stimuli such as those received at the site of an inflammatory response. In a similar manner, integrins at the cell membrane of circulating platelets are normally kept inactive to avoid thrombosis. Epithelial cells (which are non-circulating) normally have active integrins at their cell membrane, helping maintain their stable adhesion to underlying stromal cells that provide signals to maintain normal functioning.

In plants, there are no bona fide integrin receptors identified to date; nevertheless, several integrin-like proteins were proposed based on structural homology with the metazoan receptors. Plants contain integrin-linked kinases that are very similar in their primary structure with the animal ILKs. In the experimental model plant *Arabidopsis thaliana*, one of the integrin-linked kinase genes, *ILK1*, has been shown to be a critical element in the plant immune response to signal molecules from bacterial pathogens and plant sensitivity to salt and osmotic stress. ILK1 protein interacts with the high-affinity potassium transporter HAK5 and with the calcium sensor CML9.

Toll-Like Receptors

When activated, toll-like receptors (TLRs) take adapter molecules within the cytoplasm of cells in order to propagate a signal. Four adaptor molecules are known to be involved in signaling, which are Myd88, TIRAP, TRIF, and TRAM. These adapters activate other intracellular molecules such as IRAK1, IRAK4, TBK1, and IKKi that amplify the signal, eventually leading to the induction or suppression of genes that cause certain responses. Thousands of genes are activated by TLR signaling, implying that this method constitutes an important gateway for gene modulation.

Ligand-Gated Ion Channels

A ligand-gated ion channel, upon binding with a ligand, changes conformation to open a channel in the cell membrane through which ions relaying signals can pass. An example of this mechanism is found in the receiving cell of a neural synapse. The influx of ions that occurs in response to the opening of these channels induces action potentials, such as those that travel along nerves, by depolarizing the membrane of post-synaptic cells, resulting in the opening of voltage-gated ion channels. An example of an ion allowed into the cell during a ligand-gated ion channel opening is Ca^{2+}; it acts as a second messenger initiating signal transduction cascades and altering the physiology of the responding cell. This results in amplification of the synapse response between synaptic cells by remodelling the dendritic spines involved in the synapse.

14.4.2 Intracellular Receptors

Intracellular receptors, such as nuclear receptors and cytoplasmic receptors, are soluble proteins localized within their respective areas. The typical ligands for nuclear receptors are non-polar hormones like the steroid hormones testosterone and progesterone and derivatives of vitamins A and D. To initiate signal transduction, the ligand must pass through the plasma membrane by passive diffusion. On binding with the receptor, the ligands pass through the nuclear membrane into the nucleus, altering gene expression.

Activated nuclear receptors attach to the DNA at receptor-specific hormone-responsive element (HRE) sequences, located in the promoter region of the genes activated by the hormone-receptor complex. Due to their enabling gene transcription, they are alternatively called inductors of gene expression. All hormones that act by regulation of gene expression have two consequences in their mechanism of action; their effects are produced after a characteristically long period of time and their effects persist for another long period of time, even after their concentration has been reduced to zero, due to a relatively slow turnover of most enzymes and proteins that would either deactivate or terminate ligand binding onto the receptor.

Nucleic receptors have DNA-binding domains containing zinc fingers and a ligand-binding domain; the zinc fingers stabilize DNA binding by holding its phosphate backbone. DNA sequences that match the receptor are usually hexameric repeats of any kind; the sequences are similar but their orientation and distance differentiate them.

The ligand-binding domain is additionally responsible for dimerization of nucleic receptors prior to binding and providing structures for transactivation used for communication with the translational apparatus.

Steroid receptors are a subclass of nuclear receptors located primarily within the cytosol. In the absence of steroids, they associate in an aporeceptor complex containing chaperone or heatshock proteins (HSPs).

The HSPs are necessary to activate the receptor by assisting the protein to fold in a way such that the signal sequence enabling its passage into the nucleus is accessible. Steroid receptors, on the other hand, may be repressive on gene expression when their transactivation domain is hidden. Receptor activity can be enhanced by phosphorylation of serine residues at their N-terminal as a result of another signal transduction pathway, a process called crosstalk.

Retinoic acid receptors are another subset of nuclear receptors. They can be activated by an endocrine-synthesized ligand that entered the cell by diffusion, a ligand synthesised from a precursor like retinol brought to the cell through the bloodstream or a completely intracellularly synthesised ligand like prostaglandin. These receptors are located in the nucleus and are not accompanied by HSPs. They repress their gene by binding to their specific DNA sequence when no ligand binds to them, and vice versa.

Certain intracellular receptors of the immune system are cytoplasmic receptors; recently identified NOD-like receptors (NLRs) reside in the cytoplasm of some eukaryotic cells and interact with ligands using a leucine-rich repeat (LRR) motif similar to TLRs. Some of these molecules like NOD2 interact with RIP2 kinase that activates NF-κB signaling, whereas others like NALP3 interact with inflammatory caspases and initiate processing of particular cytokines like interleukin-1β.

14.5 SECOND MESSENGERS

First messengers are the signaling molecules (hormones, neurotransmitters, and paracrine/autocrine agents) that reach the cell from the extracellular fluid and bind to their specific receptors. Second messengers are the substances that enter the cytoplasm and act within the cell to trigger a response. In essence, second messengers serve as chemical relays from the plasma membrane to the cytoplasm, thus carrying out intracellular signal transduction.

14.5.1 Calcium

The release of calcium ions from the endoplasmic reticulum into the cytosol results in its binding to signaling proteins that are then activated; it is then sequestered in the smooth endoplasmic reticulum and the mitochondria. Two combined receptor/ion channel proteins control the transport of calcium: the $InsP_3$-receptor that transports calcium upon interaction with inositol triphosphate on its cytosolic side; and the ryanodine receptor named after the alkaloid ryanodine, similar to the $InsP_3$ receptor but having a feedback mechanism that releases more calcium upon binding with it. The nature of calcium in the cytosol means that it is active for only a very short time, meaning its free state concentration is very low and is mostly bound to organelle molecules like calreticulin when inactive.

Calcium is used in many processes including muscle contraction, neurotransmitter release from nerve endings, and cell migration. The three main pathways that lead to its activation are GPCR pathways, RTK pathways, and gated ion channels; it regulates proteins either directly or by binding to an enzyme.

14.5.2 Lipid Messengers

Lipophilic second messenger molecules are derived from lipids residing in cellular membranes; enzymes stimulated by activated receptors activate the lipids by modifying them. Examples include diacylglycerol and ceramide, the former required for the activation of protein kinase C.

14.5.3 Nitric Oxide

Nitric oxide (NO) acts as a second messenger because it is a free radical that can diffuse through the plasma membrane and affect nearby cells. It is synthesised from arginine and oxygen by the NO synthase and works through activation of soluble guanylyl cyclase, which when activated produces another second messenger, cGMP. NO can also act through covalent modification of proteins or their metal co-factors; some have a redox mechanism and are reversible. It is toxic in high concentrations and causes damage during stroke, but is the cause of many other functions like the relaxation of blood vessels, apoptosis, and penile erections.

14.5.4 Redox Signaling

In addition to nitric oxide, other electronically activated species are also signal-transducing agents in a process called redox signaling. Examples include superoxide, hydrogen peroxide, carbon monoxide, and hydrogen sulfide. Redox signaling also includes active modulation of electronic flows in semiconductive biological macromolecules.

14.6 CELLULAR RESPONSES

Gene activations and metabolism alterations are examples of cellular responses to extracellular stimulation that require signal transduction. Gene activation leads to further cellular effects, since the products of responding genes include instigators of activation; transcription factors produced as a result of a signal transduction cascade can activate even more genes. Hence, an initial stimulus can trigger the expression of a large number of genes, leading to physiological events like the increased uptake of glucose from the blood stream and the migration of neutrophils to sites of infection. The set of genes and their activation order to certain stimuli is referred to as a genetic program. Mammalian cells require stimulation for cell division and survival; in the absence of growth factor, apoptosis ensues. Such requirements for extracellular stimulation are necessary for controlling cell behavior in unicellular and multicellular organisms; signal transduction pathways are perceived to be so central to biological processes that a large number of diseases are attributed to their dysregulation. Three basic signals determine cellular growth:

- **Stimulatory (Growth Factors):**
  - *Transcription Dependent Response:* For example, steroids act directly as transcription factor (gives slow response, as transcription factor must bind DNA, which needs to be transcribed. Produced mRNA needs to be translated, and the produced protein/peptide can undergo posttranslational modification (PTM))
  - *Transcription Independent Response:* For example, epidermal growth factor (EGF) binds the epidermal growth factor receptor (EGFR), which causes dimerization and autophosphorylation of the EGFR, which in turn activates the intracellular signaling pathway.
- **Inhibitory** (cell-cell contact).
- **Permissive** (cell-matrix interactions).

The combination of these signals is integrated into altered cytoplasmic machinery which leads to altered cell behaviour.

14.7 MAJOR PATHWAYS

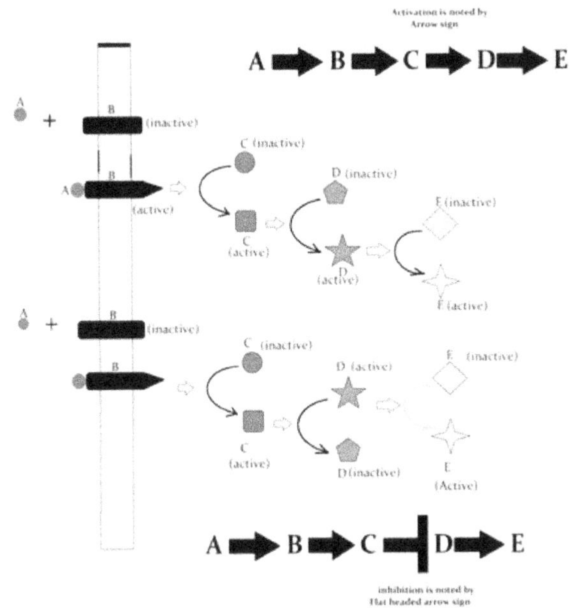

Fig 2: How to Read Signal Transduction Diagrams, what does Normal Arrow and Flathead Arrow Means

Following are some major signaling pathways, demonstrating how ligands binding to their receptors can affect second messengers and eventually result in altered cellular responses.

- **MAPK/ERK Pathway:** A pathway that couples intracellular responses to the binding of growth factors to cell surface receptors. This pathway is very complex and includes many protein components. In many cell types, activation of this pathway promotes cell division, and many forms of cancer are associated with aberrations in it.
- **cAMP-Dependent Pathway:** In humans, cAMP works by activating protein kinase A (PKA, cAMP-dependent protein kinase) and, thus, further effects depend mainly on cAMP-dependent protein kinase, which vary based on the type of cell.
- **IP_3/DAG Pathway:** PLC cleaves the phospholipid phosphatidylinositol 4,5-bisphosphate (PIP2), yielding diacyl glycerol (DAG) and inositol 1,4,5-triphosphate (IP_3). DAG remains bound to the membrane, and IP_3 is released as a soluble structure into the cytosol. IP_3 then diffuses through the cytosol to bind to IP_3 receptors, particular calcium channels in the

endoplasmic reticulum (ER). These channels are specific to calcium and allow the passage of only calcium to move through. This causes the cytosolic concentration of Calcium to increase, causing a cascade of intracellular changes and activity. In addition, calcium and DAG together works to activate PKC, which goes on to phosphorylate other molecules, leading to altered cellular activity. End-effects include taste, manic depression, tumor promotion, etc.

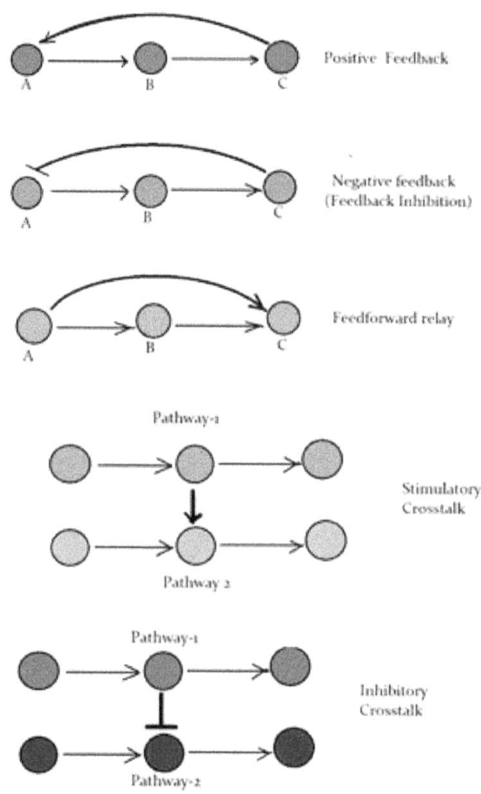

Fig 3: Elements of Signal Transduction Cascade Networking

CHAPTER 15

METABOLISM: BASIC CONCEPTS AND THEMES

15.1 AN INTRODUCTION

Metabolism is the set of life-sustaining chemical reactions in organisms. The three main functions of metabolism are: the conversion of the energy in food to energy available to run cellular processes; the conversion of food to building blocks of proteins, lipids, nucleic acids, and some carbohydrates; and the elimination of metabolic wastes.

These enzyme-catalyzed reactions allow organisms to grow and reproduce, maintain their structures, and respond to their environments. The word *metabolism* can also refer to the sum of all chemical reactions that occur in living organisms, including digestion and the transportation of substances into and between different cells, in which case the above described set of reactions within the cells is called intermediary (or intermediate) metabolism.

Metabolic reactions may be categorized as *catabolic*—the *breaking down* of compounds (for example, of glucose to pyruvate by cellular respiration); or *anabolic*—the *building up* (synthesis) of compounds (such as proteins, carbohydrates, lipids, and nucleic acids). Usually, catabolism releases energy, and anabolism consumes energy.

The chemical reactions of metabolism are organized into metabolic pathways, in which one chemical is transformed through a series of steps into another chemical, each step being facilitated by a specific enzyme. Enzymes are crucial to metabolism because they allow organisms to drive desirable reactions that require energy and will not occur by themselves, by coupling them to spontaneous reactions that release energy. Enzymes act as catalysts—they allow a reaction to proceed more rapidly—and they also allow the regulation of the rate of a metabolic reaction, for example in response to changes in the cell's environment or to signals from other cells.

The metabolic system of a particular organism determines which substances it will find nutritious and which poisonous. For example, some prokaryotes use hydrogen sulfide as a nutrient, yet this gas is poisonous to animals. The basal metabolic rate of an organism is the measure of the amount of energy consumed by all of these chemical reactions.

A striking feature of metabolism is the similarity of the basic metabolic pathways among vastly different species. For example, the set of carboxylic acids that are best known as the intermediates in the citric acid cycle are present in all known organisms, being found in species as diverse as the unicellular bacterium *Escherichia coli* and huge multicellular organisms like elephants. These similarities in metabolic pathways are likely due to their early appearance in evolutionary history, and their retention is likely due to their efficacy. In various diseases, such as type II diabetes, metabolic syndrome, and cancer, normal metabolism is disrupted. The metabolism of cancer cells is also different from the metabolism of normal cells, and these differences can be used to find targets for therapeutic intervention in cancer.

15.2 HISTORY

15.2.1 Greek Philosophy

Aristotle's *The Parts of Animals* sets out enough details of his views on metabolism for an open flow model to be made. He believed that at each stage of the process, materials from food were transformed, with heat being released as the classical element of fire, and residual materials being excreted as urine, bile, or faeces.

Ibn al-Nafis described metabolism in his 1260 AD work titled Al-Risalah al-Kamiliyyah fil Siera al-Nabawiyyah (The Treatise of Kamil on the Prophet's Biography) which included the following phrase "Both the body and its parts are in a continuous state of dissolution and nourishment, so they are inevitably undergoing permanent change."

15.2.2 Application of the Scientific Method and Modern Metabolic Theories

The history of the scientific study of metabolism spans several centuries and has moved from examining whole animals in early studies, to examining individual metabolic reactions in modern biochemistry. The first controlled experiments in human metabolism were published by Santorio Santorio in 1614 in his book *Ars de statica medicina*. He described how he weighed himself before and after eating, sleep, working, sex, fasting, drinking, and excreting. He found that most of the food he took in was lost through what he called "insensible perspiration".

In these early studies, the mechanisms of these metabolic processes had not been identified and a vital force was thought to animate living tissue. In the 19th century, when studying the fermentation of sugar to alcohol by yeast, Louis Pasteur concluded that fermentation was catalyzed by substances within the yeast cells he called "ferments". He wrote that "alcoholic fermentation is an act correlated with the life and organization of the yeast cells, not with the death or putrefaction of the cells." This discovery, along with the publication by Friedrich Wöhler in 1828 of a paper on the chemical synthesis of urea, and is notable for being the first organic compound prepared from wholly inorganic precursors. This proved that the organic compounds and chemical reactions found in cells were no different in principle than any other part of chemistry.

It was the discovery of enzymes at the beginning of the 20th century by Eduard Buchner that separated the study of the chemical reactions of metabolism from the biological study of cells, and marked the beginnings of biochemistry. The mass of biochemical knowledge grew rapidly throughout the early 20th century. One of the most prolific of these modern biochemists was Hans Krebs who made huge contributions to the study of metabolism.

He discovered the urea cycle and later, working with Hans Kornberg, the citric acid cycle and the glyoxylate cycle. Modern biochemical research has been greatly aided by the development of new techniques such as chromatography, X-ray diffraction, NMR spectroscopy, radioisotopic labelling, electron microscopy and molecular dynamics simulations. These techniques have allowed the discovery and detailed analysis of the many molecules and metabolic pathways in cells.

15.3 KEY BIOCHEMICALS

Most of the structures that make up animals, plants and microbes are made from four basic classes of molecules: amino acids, carbohydrates, nucleic acid and lipids (often called fats). As these molecules are vital for life, metabolic reactions either focus on making these molecules during the construction of cells and tissues, or on breaking them down and using them to obtain energy, by their digestion. These biochemicals can be joined to make polymers such as DNA and proteins, essential macromolecules of life.

Type of Molecule	Name of Monomer Forms	Name of Polymer Forms	Examples of Polymer Forms
Amino acids	Amino acids	Proteins (made of polypeptides)	Fibrous proteins and globular proteins
Carbohydrates	Monosaccharides	Polysaccharides	Starch, glycogen and cellulose
Nucleic acids	Nucleotides	Polynucleotides	DNA and RNA

15.3.1 Amino Acids and Proteins

Proteins are made of amino acids arranged in a linear chain joined by peptide bonds. Many proteins are enzymes that catalyze the chemical reactions in metabolism. Other proteins have structural or mechanical functions, such as those that form the cytoskeleton, a system of scaffolding that maintains the cell shape.

Proteins are also important in cell signaling, immune responses, cell adhesion, active transport across membranes, and the cell cycle. Amino acids also contribute to cellular energy metabolism by providing a carbon source for entry into the citric acid cycle (tricarboxylic acid cycle), especially when a primary source of energy, such as glucose, is scarce, or when cells undergo metabolic stress.

15.3.2 Lipids

Lipids are the most diverse group of biochemicals. Their main structural uses are as part of internal and external biological membranes, such as the cell membrane. Their chemical energy can also be used. Lipids contain a long, non-polar hydrocarbon chain with a small polar region containing oxygen. Lipids are usually defined as hydrophobic or amphipathic biological molecules but will dissolve in organic solvents such as ethanol, benzene or chloroform. The fats are a large group of compounds that contain fatty acids and glycerol; a glycerol molecule attached to three fatty acids by ester linkages is called a triacylglyceride. Several variations of the basic structure exist, including backbones such as sphingosine in sphingomyelin, and hydrophilic groups such as phosphate in phospholipids. Steroids such as sterol are another major class of lipids.

15.3.3 Carbohydrates

Carbohydrates are aldehydes or ketones, with many hydroxyl groups attached, that can exist as straight chains or rings. Carbohydrates are the most abundant biological molecules, and fill numerous roles, such as the storage and transport of energy (starch, glycogen) and structural components (cellulose in plants, chitin in animals). The basic carbohydrate units are called monosaccharides and include galactose, fructose, and most importantly glucose. Monosaccharides can be linked together to form polysaccharides in almost limitless ways.

15.3.4 Nucleotides

The two nucleic acids, DNA and RNA, are polymers of nucleotides. Each nucleotide is composed of a phosphate attached to a ribose or deoxyribose sugar group which is attached to a nitrogenous base. Nucleic acids are critical for the storage and use of genetic information, and its interpretation through the processes of transcription and protein biosynthesis. This information is protected by DNA repair mechanisms and propagated through DNA replication. Many viruses have an RNA genome, such as HIV, which uses reverse transcription to create a DNA template from its viral RNA genome. RNA in ribozymes such as spliceosomes and ribosomes is similar to enzymes as it can catalyze chemical reactions. Individual nucleosides are made by attaching a nucleobase to a ribose sugar. These bases are heterocyclic rings containing nitrogen, classified as purines or pyrimidines. Nucleotides also act as coenzymes in metabolic-group-transfer reactions.

15.3.5 Coenzymes

Metabolism involves a vast array of chemical reactions, but most fall under a few basic types of reactions that involve the transfer of functional groups of atoms and their bonds within molecules. This common chemistry allows cells to use a small set of metabolic intermediates to carry chemical groups between different reactions. These group-transfer intermediates are called coenzymes. Each class of group-transfer reactions is carried out by a particular coenzyme, which is the substrate for a set of enzymes that produce it, and a set of enzymes that consume it. These coenzymes are therefore continuously made, consumed and then recycled.

One central coenzyme is adenosine triphosphate (ATP), the energy currency of cells. This nucleotide is used to transfer chemical energy between different chemical reactions. There is only a small amount of ATP in cells, but as it is continuously regenerated, the human body can use about its own weight in ATP per day. ATP acts as a bridge between catabolism and anabolism. Catabolism breaks down molecules, and anabolism puts them together. Catabolic reactions generate ATP, and anabolic reactions consume it. It also serves as a carrier of phosphate groups in phosphorylation reactions.

A vitamin is an organic compound needed in small quantities that cannot be made in cells. In human nutrition, most vitamins function as coenzymes after modification; for example, all water-soluble vitamins are phosphorylated or are coupled to nucleotides when they are used in cells. Nicotinamide adenine dinucleotide (NAD^+), a derivative of vitamin B_3 (niacin), is an important coenzyme that acts as a hydrogen acceptor. Hundreds of separate types of dehydrogenases remove electrons from their substrates and reduce NAD^+ into NADH. This reduced form of the coenzyme is then a substrate for any of the reductases in the cell that need to transfer hydrogen atoms to their substrates. Nicotinamide adenine dinucleotide exists in two related forms in the cell, NADH and NADPH. The NAD^+/NADH form is more important in catabolic reactions, while $NADP^+$/NADPH is used in anabolic reactions.

15.3.6 Mineral and Cofactors

Inorganic elements play critical roles in metabolism; some are abundant (e.g. sodium and potassium) while others function at minute concentrations. About 99% of a human's body weight is made up of the elements carbon, nitrogen, calcium, sodium, chlorine, potassium, hydrogen, phosphorus, oxygen and sulfur. Organic compounds (proteins, lipids and carbohydrates) contain the majority of the carbon and nitrogen; most of the oxygen and hydrogen is present as water.

The abundant inorganic elements act as electrolytes. The most important ions are sodium, potassium, calcium, magnesium, chloride, phosphate and the organic ion bicarbonate. The maintenance of precise ion gradients across cell membranes maintains osmotic pressure and pH. Ions are also critical for nerve and muscle function, as action potentials in these tissues are produced by the exchange of electrolytes between the extracellular fluid and the cell's fluid, the cytosol.

Electrolytes enter and leave cells through proteins in the cell membrane called ion channels. For example, muscle contraction depends upon the movement of calcium, sodium and potassium through ion channels in the cell membrane and T-tubules.

Transition metals are usually present as trace elements in organisms, with zinc and iron being most abundant of those. Metal cofactors are bound tightly to specific sites in proteins; although enzyme cofactors can be modified during catalysis, they always return to their original state by the end of the reaction catalyzed. Metal micronutrients are taken up into organisms by specific transporters and bind to storage proteins such as ferritin or metallothionein when not in use.

15.4 CATABOLISM

Catabolism is the set of metabolic processes that break down large molecules. These include breaking down and oxidizing food molecules.

Table: Classification of Organisms Based on their Metabolism

<table>
<tr><td rowspan="2">Energy source</td><td>sunlight</td><td>photo-</td><td></td><td></td><td rowspan="6">-troph</td></tr>
<tr><td>molecules</td><td>chemo-</td><td></td><td></td></tr>
<tr><td rowspan="2">Hydrogen or electron donor</td><td>organic compound</td><td></td><td>organo-</td><td></td></tr>
<tr><td>inorganic compound</td><td></td><td>litho-</td><td></td></tr>
<tr><td rowspan="2">Carbon source</td><td>organic compound</td><td></td><td></td><td>hetero-</td></tr>
<tr><td>inorganic compound</td><td></td><td></td><td>auto-</td></tr>
</table>

The purpose of the catabolic reactions is to provide the energy and components needed by anabolic reactions which build molecules. The exact nature of these catabolic reactions differ from organism to organism, and organisms can be classified based on their sources of energy, hydrogen, and carbon (their primary nutritional groups).

Organic molecules are used as a source of hydrogen atoms or electrons by organotrophs, while lithotrophs use inorganic substrates. Whereas phototrophs convert sunlight to chemical energy, chemotrophs depend on redox reactions that involve the transfer of electrons from reduced donor molecules such as organic molecules, hydrogen, hydrogen sulfide or ferrous ions to oxygen, nitrate or sulfate. In animals, these reactions involve complex organic molecules that are broken down to simpler molecules, such as carbon dioxide and water. Photosynthetic organisms, such as plants and cyanobacteria, use similar electron-transfer reactions to store energy absorbed from sunlight.

The most common set of catabolic reactions in animals can be separated into three main stages. In the first stage, large organic molecules, such as proteins, polysaccharides or lipids, are digested into their smaller components outside cells. Next, these smaller molecules are taken up by cells and converted to smaller molecules, usually acetyl coenzyme A (acetyl-CoA), which releases some energy. Finally, the acetyl group on acetyl-CoA is oxidized to water and carbon dioxide in the citric acid cycle and electron transport chain, releasing more energy while reducing the coenzyme nicotinamide adenine dinucleotide (NAD^+) into NADH.

15.4.1 Digestion

Macromolecules cannot be directly processed by cells. Macromolecules must be broken into smaller units before they can be used in cell metabolism.

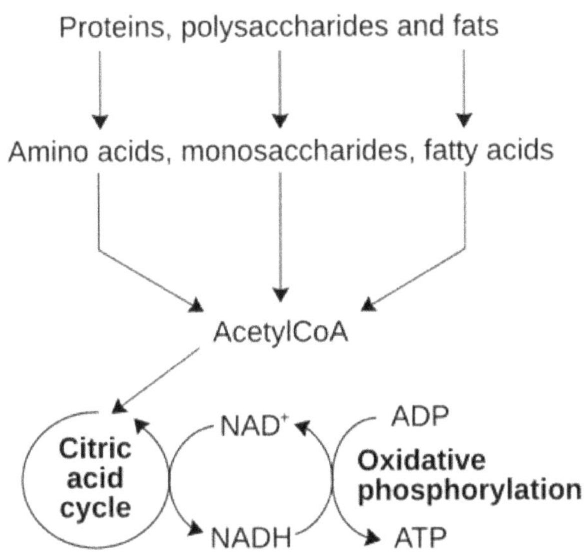

Fig 1: A Simplified Outline of the Catabolism of Proteins, Carbohydrates and Fats

Different classes of enzymes are used to digest these polymers. These digestive enzymes include proteases that digest proteins into amino acids, as well as glycoside hydrolases that digest polysaccharides into simple sugars known as monosaccharides.

Microbes simply secrete digestive enzymes into their surroundings, while animals only secrete these enzymes from specialized cells in their guts, including the stomach and pancreas, and in salivary glands. The amino acids or sugars released by these extracellular enzymes are then pumped into cells by active transport proteins.

15.4.2 Energy from Organic Compounds

Carbohydrate catabolism is the breakdown of carbohydrates into smaller units. Carbohydrates are usually taken into cells after they have been digested into monosaccharides such as glucose and fructose.

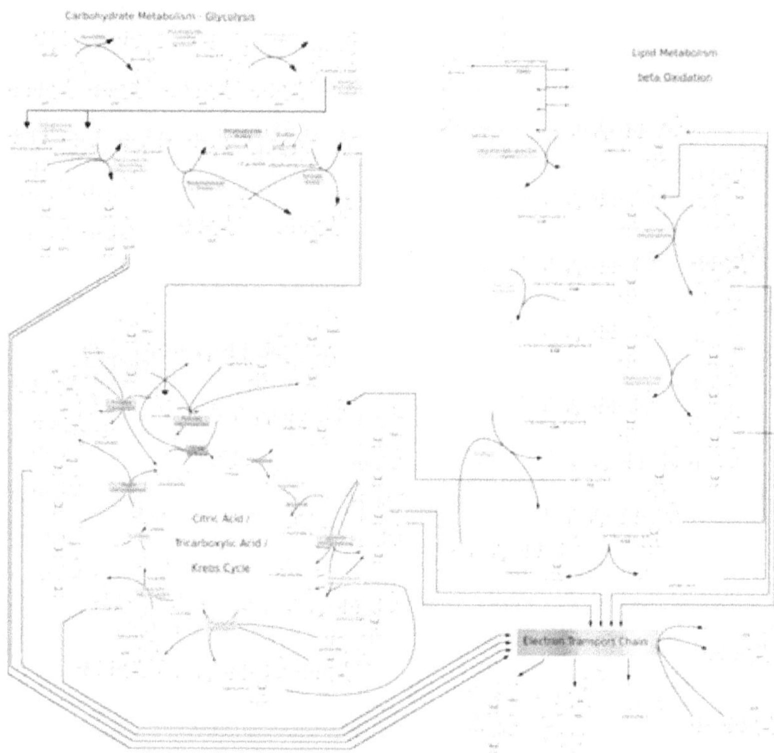

Fig 2: Carbon Catabolism Pathway Map for Free Energy Including Carbohydrate and Lipid Sources of Energy

Once inside, the major route of breakdown is glycolysis, in which glucose is converted into pyruvate. This process generates the energy-conveying molecule NADH from NAD^+, and generates ATP from ADP for use in powering many processes within the cell.

Pyruvate is an intermediate in several metabolic pathways, but the majority is converted to acetyl-CoA and fed into the citric acid cycle, which enables more ATP production by means of oxidative phosphorylation.

This oxidation consumes molecular oxygen and releases water and the waste product carbon dioxide. When oxygen is lacking, or when pyruvate is temporarily produced faster than it can be consumed by the citric acid cycle (as in intense muscular exertion), pyruvate is converted to lactate by the enzyme lactate dehydrogenase, a process that also oxidizes NADH back to NAD^+ for re-use in further glycolysis, allowing energy production to continue. The lactate is later converted back to pyruvate for ATP production where energy is needed, or back to glucose in the Cori cycle. An alternative route for glucose breakdown is the pentose phosphate pathway, which produces less energy but supports anabolism (biomolecule synthesis). This pathway reduces the coenzyme $NADP^+$ to NADPH and produces pentose compounds such as ribose 5-phosphate for synthesis of many biomolecules such as nucleotides and aromatic amino acids.

Fats are catabolized by hydrolysis to free fatty acids and glycerol. The glycerol enters glycolysis and the fatty acids are broken down by beta oxidation to release acetyl-CoA, which then is fed into the citric acid cycle. Fatty acids release more energy upon oxidation than carbohydrates. Steroids are also broken down by some bacteria in a process similar to beta oxidation, and this breakdown process involves the release of significant amounts of acetyl-CoA, propionyl-CoA, and pyruvate, which can all be used by the cell for energy. *M. tuberculosis* can also grow on the lipid cholesterol as a sole source of carbon, and genes involved in the cholesterol-use pathway(s) have been validated as important during various stages of the infection lifecycle of *M. tuberculosis.*

Amino acids are either used to synthesize proteins and other biomolecules, or oxidized to urea and carbon dioxide to produce energy. The oxidation pathway starts with the removal of the amino group by a transaminase. The amino group is fed into the urea cycle, leaving a deaminated carbon skeleton in the form of a keto acid. Several of these keto acids are intermediates in the citric acid cycle, for example α-ketoglutarate formed by deamination of glutamate. The glucogenic amino acids can also be converted into glucose, through gluconeogenesis.

15.5 ENERGY TRANSFORMATIONS

15.5.1 Oxidative Phosphorylation

In oxidative phosphorylation, the electrons removed from organic molecules in areas such as the citric acid cycle are transferred to oxygen and the energy released is used to make ATP.

This is done in eukaryotes by a series of proteins in the membranes of mitochondria called the electron transport chain. In prokaryotes, these proteins are found in the cell's inner membrane. These proteins use the energy from reduced molecules like NADH to pump protons across a membrane.

Pumping protons out of the mitochondria creates a proton concentration difference across the membrane and generates an electrochemical gradient. This force drives protons back into the mitochondrion through the base of an enzyme called ATP synthase. The flow of protons makes the stalk subunit rotate, causing the active site of the synthase domain to change shape and phosphorylate adenosine diphosphate—turning it into ATP.

15.5.2 Energy from Inorganic Compounds

Chemolithotrophy is a type of metabolism found in prokaryotes where energy is obtained from the oxidation of inorganic compounds. These organisms can use hydrogen, reduced sulfur compounds (such as sulfide, hydrogen sulfide and thiosulfate), ferrous iron (Fe(II)) or ammonia as sources of reducing power and they gain energy from the oxidation of these compounds. These microbial processes are important in global biogeochemical cycles such as acetogenesis, nitrification and denitrification and are critical for soil fertility.

15.5.3 Energy from Light

The energy in sunlight is captured by plants, cyanobacteria, purple bacteria, green sulfur bacteria and some protists. This process is often coupled to the conversion of carbon dioxide into organic compounds, as part of photosynthesis, which is discussed below. The energy capture and carbon fixation systems can, however, operate separately in prokaryotes, as purple bacteria and green sulfur bacteria can use sunlight as a source of energy, while switching between carbon fixation and the fermentation of organic compounds.

In many organisms, the capture of solar energy is similar in principle to oxidative phosphorylation, as it involves the storage of energy as a proton concentration gradient. This proton motive force then drives ATP synthesis. The electrons needed to drive this electron transport chain come from light-gathering proteins called photosynthetic reaction centres. Reaction centers are classified into two types depending on the nature of photosynthetic pigment present, with most photosynthetic bacteria only having one type, while plants and cyanobacteria have two. In plants, algae, and cyanobacteria, photosystem II uses light energy to remove electrons from water, releasing oxygen as a waste product. The electrons then flow to the cytochrome b6f complex, which uses their energy to pump protons across the thylakoid membrane in the chloroplast. These protons move back through the membrane as they drive the ATP synthase, as before. The electrons then flow through photosystem I and can then be used to reduce the coenzyme $NADP^+$.

15.6 ANABOLISM

Anabolism is the set of constructive metabolic processes where the energy released by catabolism is used to synthesize complex molecules. In general, the complex molecules that make up cellular structures are constructed step-by-step from smaller and simpler precursors. Anabolism involves three basic stages. First, the production of precursors such as amino acids, monosaccharides, isoprenoids and nucleotides, secondly, their activation into reactive forms using energy from ATP, and thirdly, the assembly of these precursors into complex molecules such as proteins, polysaccharides, lipids and nucleic acids.

Anabolism in organisms can be different according to the source of constructed molecules in their cells. Autotrophs such as plants can construct the complex organic molecules in their cells such as polysaccharides and proteins from simple molecules like carbon dioxide and water. Heterotrophs, on the other hand, require a source of more complex substances, such as monosaccharides and amino acids, to produce these complex molecules. Organisms can be further classified by ultimate source of their energy: photoautotrophs and photoheterotrophs obtain energy from light, whereas chemoautotrophs and chemoheterotrophs obtain energy from oxidation reactions.

15.6.1 Carbon Fixation

Photosynthesis is the synthesis of carbohydrates from sunlight and carbon dioxide (CO_2). In plants, cyanobacteria and algae, oxygenic photosynthesis splits water, with oxygen produced as a waste product. This process uses the ATP and NADPH produced by the photosynthetic reaction centres, as described above, to convert CO_2 into glycerate 3-phosphate, which can then be converted into glucose. This carbon-fixation reaction is carried out by the enzyme RuBisCO as part of the Calvin–Benson cycle. Three types of photosynthesis occur in plants, C3 carbon fixation, C4 carbon fixation and CAM photosynthesis. These differ by the route that carbon dioxide takes to the Calvin cycle, with C3 plants fixing CO_2 directly, while C4 and CAM photosynthesis incorporate the CO_2 into other compounds first, as adaptations to deal with intense sunlight and dry conditions.

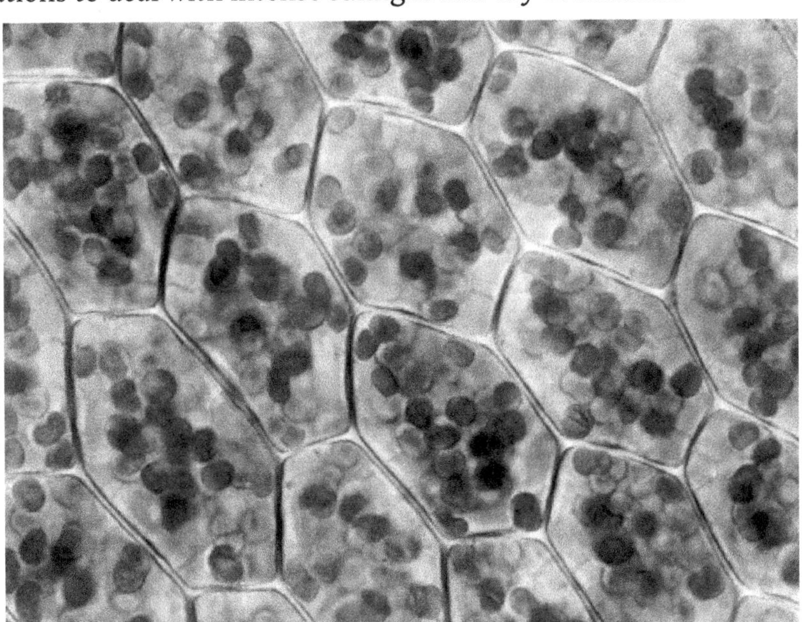

Fig 3: Plant Cells (Bounded by Purple Walls) Filled with Chloroplasts (Green), which are the Site of Photosynthesis

In photosynthetic prokaryotes the mechanisms of carbon fixation are more diverse. Here, carbon dioxide can be fixed by the Calvin–Benson cycle, a reversed citric acid cycle, or the carboxylation of acetyl-CoA. Prokaryotic chemoautotrophs also fix CO_2 through the Calvin–Benson cycle, but use energy from inorganic compounds to drive the reaction.

15.6.2 Carbohydrates and Glycans

In carbohydrate anabolism, simple organic acids can be converted into monosaccharides such as glucose and then used to assemble polysaccharides such as starch.

The generation of glucose from compounds like pyruvate, lactate, glycerol, glycerate 3-phosphate and amino acids is called gluconeogenesis. Gluconeogenesis converts pyruvate to glucose-6-phosphate through a series of intermediates, many of which are shared with glycolysis. However, this pathway is not simply glycolysis run in reverse, as several steps are catalyzed by non-glycolytic enzymes. This is important as it allows the formation and breakdown of glucose to be regulated separately, and prevents both pathways from running simultaneously in a futile cycle.

Although fat is a common way of storing energy, in vertebrates such as humans the fatty acids in these stores cannot be converted to glucose through gluconeogenesis as these organisms cannot convert acetyl-CoA into pyruvate; plants do, but animals do not, have the necessary enzymatic machinery.

As a result, after long-term starvation, vertebrates need to produce ketone bodies from fatty acids to replace glucose in tissues such as the brain that cannot metabolize fatty acids. In other organisms such as plants and bacteria, this metabolic problem is solved using the glyoxylate cycle, which bypasses the decarboxylation step in the citric acid cycle and allows the transformation of acetyl-CoA to oxaloacetate, where it can be used for the production of glucose.

Other than fat, glucose is stored in most tissues, as an energy resource available within the tissue through glycogenesis which was usually being used to maintained glucose level in blood.

Polysaccharides and glycans are made by the sequential addition of monosaccharides by glycosyltransferase from a reactive sugar-phosphate donor such as uridine diphosphate glucose (UDP-Glc) to an acceptor hydroxyl group on the growing polysaccharide. As any of the hydroxyl groups on the ring of the substrate can be acceptors, the polysaccharides produced can have straight or branched structures. The polysaccharides produced can have structural or metabolic functions themselves, or be transferred to lipids and proteins by the enzymes oligosaccharyltransferases.

15.6.3 Fatty Acids, Isoprenoids and Sterol

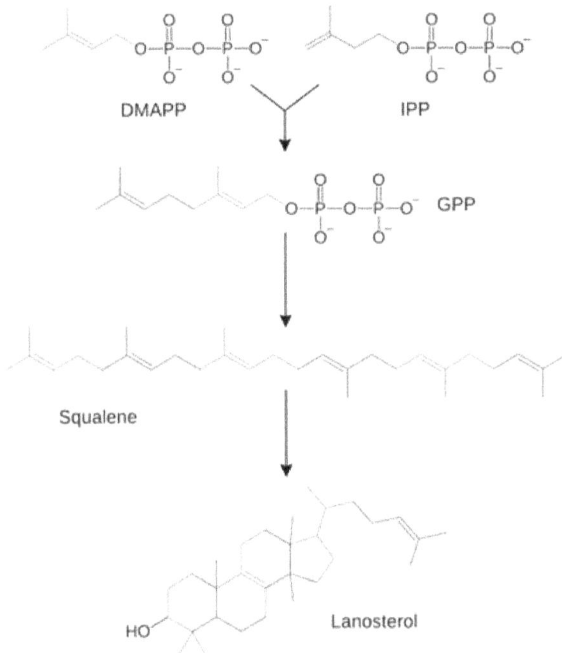

Fig 4: Simplified Version of the Steroid Synthesis Pathway with the Intermediates Isopentenyl Pyrophosphate (IPP), Dimethylallyl Pyrophosphate (DMAPP), Geranyl Pyrophosphate (GPP) and Squalene Shown

Fatty acids are made by fatty acid synthases that polymerize and then reduce acetyl-CoA units. The acyl chains in the fatty acids are extended by a cycle of reactions that add the acyl group, reduce it to an alcohol, dehydrate it to an alkene group and then reduce it again to an alkane group. The enzymes of fatty acid biosynthesis are divided into two groups: in animals and fungi, all these fatty acid synthase reactions are carried out by a single multifunctional type I protein, while in plant plastids and bacteria separate type II enzymes perform each step in the pathway.

Terpenes and isoprenoids are a large class of lipids that include the carotenoids and form the largest class of plant natural products. These compounds are made by the assembly and modification of isoprene units donated from the reactive precursors isopentenyl pyrophosphate and dimethylallyl pyrophosphate.

These precursors can be made in different ways. In animals and archaea, the mevalonate pathway produces these compounds from acetyl-CoA, while in plants and bacteria the non-mevalonate pathway uses pyruvate and glyceraldehyde 3-phosphate as substrates. One important reaction that uses these activated isoprene donors is sterol biosynthesis. Here, the isoprene units are joined to make squalene and then folded up and formed into a set of rings to make lanosterol.

Lanosterol can then be converted into other sterols such as cholesterol and ergosterol.

15.6.4 Proteins

Organisms vary in their ability to synthesize the 20 common amino acids. Most bacteria and plants can synthesize all twenty, but mammals can only synthesize eleven nonessential amino acids, so nine essential amino acids must be obtained from food. Some simple parasites, such as the bacteria *Mycoplasma pneumoniae*, lack all amino acid synthesis and take their amino acids directly from their hosts. All amino acids are synthesized from intermediates in glycolysis, the citric acid cycle, or the pentose phosphate pathway. Nitrogen is provided by glutamate and glutamine. Nonessensial amino acid synthesis depends on the formation of the appropriate alpha-keto acid, which is then transaminated to form an amino acid.

Amino acids are made into proteins by being joined in a chain of peptide bonds. Each different protein has a unique sequence of amino acid residues: this is its primary structure. Just as the letters of the alphabet can be combined to form an almost endless variety of words, amino acids can be linked in varying sequences to form a huge variety of proteins. Proteins are made from amino acids that have been activated by attachment to a transfer RNA molecule through an ester bond. This aminoacyl-tRNA precursor is produced in an ATP-dependent reaction carried out by an aminoacyl tRNA synthetase. This aminoacyl-tRNA is then a substrate for the ribosome, which joins the amino acid onto the elongating protein chain, using the sequence information in a messenger RNA.

15.6.5 Nucleotide Synthesis and Salvage

Nucleotides are made from amino acids, carbon dioxide and formic acid in pathways that require large amounts of metabolic energy. Consequently, most organisms have efficient systems to salvage preformed nucleotides. Purines are synthesized as nucleosides (bases attached to ribose). Both adenine and guanine are made from the precursor nucleoside inosine monophosphate, which is synthesized using atoms from the amino acids glycine, glutamine, and aspartic acid, as well as formate transferred from the coenzyme tetrahydrofolate. Pyrimidines, on the other hand, are synthesized from the base orotate, which is formed from glutamine and aspartate.

15.7 XENOBIOTICS AND REDOX METABOLISM

All organisms are constantly exposed to compounds that they cannot use as foods and that would be harmful if they accumulated in cells, as they have no metabolic function. These potentially damaging compounds are called xenobiotics. Xenobiotics such as synthetic drugs, natural poisons and antibiotics are detoxified by a set of xenobiotic-metabolizing enzymes. In humans, these include cytochrome

P450 oxidases, UDP-glucuronosyltransferases, and glutathione *S*-transferases. This system of enzymes acts in three stages to firstly oxidize the xenobiotic (phase I) and then conjugate water-soluble groups onto the molecule (phase II). The modified water-soluble xenobiotic can then be pumped out of cells and in multicellular organisms may be further metabolized before being excreted (phase III). In ecology, these reactions are particularly important in microbial biodegradation of pollutants and the bioremediation of contaminated land and oil spills. Many of these microbial reactions are shared with multicellular organisms, but due to the incredible diversity of types of microbes these organisms are able to deal with a far wider range of xenobiotics than multicellular organisms, and can degrade even persistent organic pollutants such as organochloride compounds.

A related problem for aerobic organisms is oxidative stress. Here, processes including oxidative phosphorylation and the formation of disulfide bonds during protein folding produce reactive oxygen species such as hydrogen peroxide. These damaging oxidants are removed by antioxidant metabolites such as glutathione and enzymes such as catalases and peroxidases.

15.8 THERMODYNAMICS OF LIVING ORGANISMS

Living organisms must obey the laws of thermodynamics, which describe the transfer of heat and work. The second law of thermodynamics states that in any isolated system, the amount of entropy (disorder) cannot decrease. Although living organisms' amazing complexity appears to contradict this law, life is possible as all organisms are open systems that exchange matter and energy with their surroundings. Living systems are not in equilibrium, but instead are dissipative systems that maintain their state of high complexity by causing a larger increase in the entropy of their environments. The metabolism of a cell achieves this by coupling the spontaneous processes of catabolism to the non-spontaneous processes of anabolism. In thermodynamic terms, metabolism maintains order by creating disorder.

15.9 REGULATION AND CONTROL

As the environments of most organisms are constantly changing, the reactions of metabolism must be finely regulated to maintain a constant set of conditions within cells, a condition called homeostasis. Metabolic regulation also allows organisms to respond to signals and interact actively with their environments. Two closely linked concepts are important for understanding how metabolic pathways are controlled. Firstly, the *regulation* of an enzyme in a pathway is how its activity is increased and decreased in response to signals. Secondly, the *control* exerted by this enzyme is the effect that these changes in its activity have on the overall rate of the pathway (the flux through the pathway). For example, an enzyme may show large changes in activity (i.e. it is highly regulated) but if these changes have little effect on the flux of a metabolic pathway, then this enzyme is not involved in the control of the pathway.

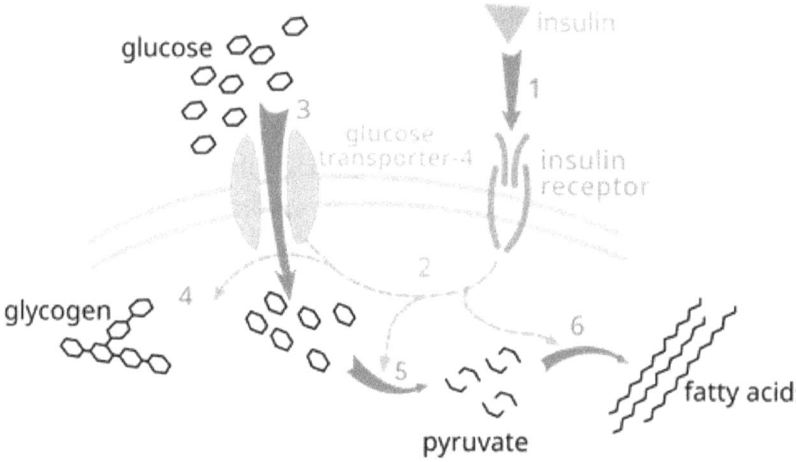

Fig 5: Effect of Insulin on Glucose Uptake and Metabolism: (1) Insulin Binds to its Receptor, (2) which in Turn Starts Many Protein Activation Cascades (3) these include: Translocation of Glut-4 Transporter to the Plasma Membrane and Influx of Glucose (4) Glycogen Synthesis (5) Glycolysis (6) Fatty Acid Synthesis

There are multiple levels of metabolic regulation. In intrinsic regulation, the metabolic pathway self-regulates to respond to changes in the levels of substrates or products; for example, a decrease in the amount of product can increase the flux through the pathway to compensate. This type of regulation often involves allosteric regulation of the activities of multiple enzymes in the pathway. Extrinsic control involves a cell in a multicellular organism changing its metabolism in response to signals from other cells. These signals are usually in the form of water-soluble messengers such as hormones and growth factors and are detected by specific receptors on the cell surface. These signals are then transmitted inside the cell by second messenger systems that often involved the phosphorylation of proteins.

A very well understood example of extrinsic control is the regulation of glucose metabolism by the hormone insulin. Insulin is produced in response to rises in blood glucose levels. Binding of the hormone to insulin receptors on cells then activates a cascade of protein kinases that cause the cells to take up glucose and convert it into storage molecules such as fatty acids and glycogen. The metabolism of glycogen is controlled by activity of phosphorylase, the enzyme that breaks down glycogen, and glycogen synthase, the enzyme that makes it. These enzymes are regulated in a reciprocal fashion, with phosphorylation inhibiting glycogen synthase, but activating phosphorylase. Insulin causes glycogen synthesis by activating protein phosphatases and producing a decrease in the phosphorylation of these enzymes.

15.10 EVOLUTION

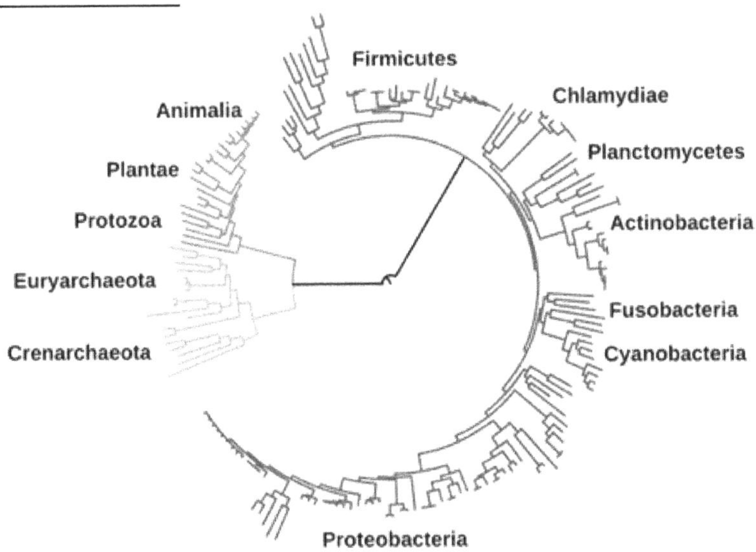

Fig 6: Evolutionary Tree Showing the Common Ancestry of Organisms from all Three Domains of Life. Bacteria are Colored Blue, Eukaryotes Red, and Archaea Green

The central pathways of metabolism described above, such as glycolysis and the citric acid cycle, are present in all three domains of living things and were present in the last universal common ancestor. This universal ancestral cell was prokaryotic and probably a methanogen that had extensive amino acid, nucleotide, carbohydrate and lipid metabolism.

The retention of these ancient pathways during later evolution may be the result of these reactions having been an optimal solution to their particular metabolic problems, with pathways such as glycolysis and the citric acid cycle producing their end products highly efficiently and in a minimal number of steps. The first pathways of enzyme-based metabolism may have been parts of purine nucleotide metabolism, while previous metabolic pathways were a part of the ancient RNA world.

Many models have been proposed to describe the mechanisms by which novel metabolic pathways evolve. These include the sequential addition of novel enzymes to a short ancestral pathway, the duplication and then divergence of entire pathways as well as the recruitment of pre-existing enzymes and their assembly into a novel reaction pathway. The relative importance of these mechanisms is unclear, but genomic studies have shown that enzymes in a pathway are likely to have a shared ancestry, suggesting that many pathways have evolved in a step-by-step fashion with novel functions created from pre-existing steps in the pathway. An alternative model comes from studies that trace the evolution of proteins' structures in metabolic networks, this has suggested that enzymes are pervasively recruited, borrowing enzymes to perform similar functions in different metabolic pathways (evident in the MANET database) These recruitment processes result in an evolutionary enzymatic mosaic. A third possibility is that some parts of metabolism might exist as "modules" that can be reused in different pathways and perform similar functions on different molecules.

As well as the evolution of new metabolic pathways, evolution can also cause the loss of metabolic functions. For example, in some parasites metabolic processes that are not essential for survival are lost and preformed amino acids, nucleotides and carbohydrates may instead be scavenged from the host. Similar reduced metabolic capabilities are seen in endosymbiotic organisms.

15.11 INVESTIGATION AND MANIPULATION

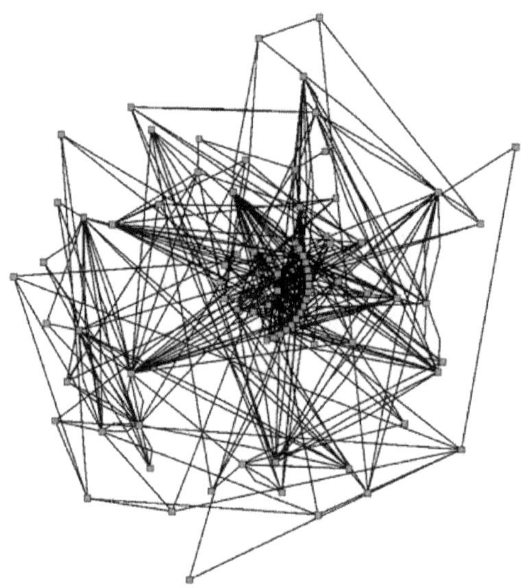

Fig 7: Metabolic Network of the *Arabidopsis Thaliana* Citric Acid Cycle. Enzymes and Metabolites are shown as Red Squares and the Interactions between them as Black Lines

Classically, metabolism is studied by a reductionist approach that focuses on a single metabolic pathway. Particularly valuable is the use of radioactive tracers at the whole-organism, tissue and cellular levels, which define the paths from precursors to final products by identifying radioactively labelled intermediates and products. The enzymes that catalyze these chemical reactions can then be purified and their kinetics and responses to inhibitors investigated. A parallel approach is to identify the small molecules in a cell or tissue; the complete set of these molecules is called the metabolome. Overall, these studies give a good view of the structure and function of simple metabolic pathways, but are inadequate when applied to more complex systems such as the metabolism of a complete cell.

However, it is now possible to use this genomic data to reconstruct complete networks of biochemical reactions and produce more holistic mathematical models that may explain and predict their behavior. These models are especially powerful when used to integrate the pathway and metabolite data obtained through classical methods with data on gene expression from proteomic and DNA microarray studies. Using these techniques, a model of human metabolism has now been produced, which will guide future drug discovery and biochemical research. These models are now used in network analysis, to classify human diseases into groups that share common proteins or metabolites.

Bacterial metabolic networks are a striking example of bow-tie organization, an architecture able to input a wide range of nutrients and produce a large variety of products and complex macromolecules using a relatively few intermediate common currencies.

A major technological application of this information is metabolic engineering. Here, organisms such as yeast, plants or bacteria are genetically modified to make them more useful in biotechnology and aid the production of drugs such as antibiotics or industrial chemicals such as 1,3-propanediol and shikimic acid. These genetic modifications usually aim to reduce the amount of energy used to produce the product, increase yields and reduce the production of wastes.

CHAPTER 16

GLYCOLYSIS AND GLUCONEOGENESIS

16.1 GLYCOLYSIS

Glycolysis is the metabolic pathway that converts glucose ($C_6H_{12}O_6$) into pyruvate and, in most organisms, occurs in the liquid part of cells (the cytosol). The free energy released in this process is used to form the high-energy molecules adenosine triphosphate (ATP) and reduced nicotinamide adenine dinucleotide (NADH). Glycolysis is a sequence of ten reactions catalyzed by enzymes.

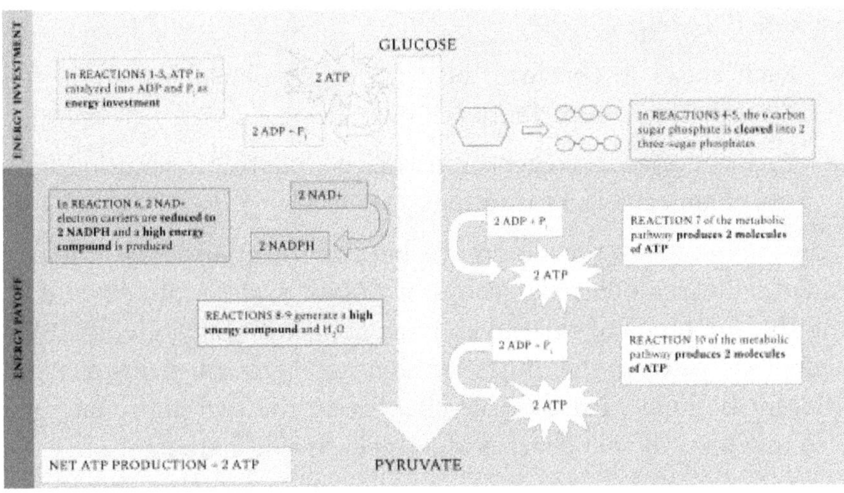

Fig 1: Summary of the 10 Reactions of the Glycolysis Pathway

The wide occurrence of glycolysis in other species indicates that it is an ancient metabolic pathway. Indeed, the reactions that make up glycolysis and its parallel pathway, the pentose phosphate pathway, can occur in the oxygen-free conditions of the Archean oceans, also in the absence of enzymes, catalyzed by metal ions, meaning this is a plausible prebiotic pathway for abiogenesis.

The most common type of glycolysis is the *Embden-Meyerhof-Parnas (EMP) pathway*, which was discovered by Gustav Embden, Otto Meyerhof, and Jakub Karol Parnas. Glycolysis also refers to other pathways, such as the *Entner-Doudoroff pathway* and various heterofermentative and homofermentative pathways. However, the discussion here will be limited to the Embden–Meyerhof–Parnas pathway.

The glycolysis pathway can be separated into two phases:

- **Investment Phase**—wherein ATP is consumed.
- **Yield Phase**—wherein more ATP is produced than originally consumed.

16.1.1 Overview

The overall reaction of glycolysis is:

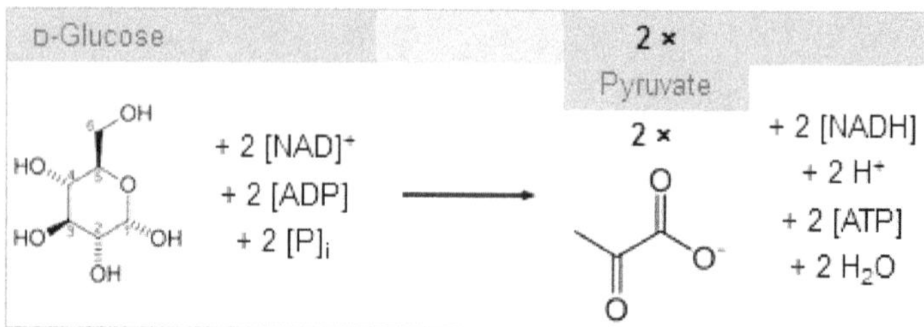

The use of symbols in this equation makes it appear unbalanced with respect to oxygen atoms, hydrogen atoms, and charges. Atom balance is maintained by the two phosphate (P_i) groups:

- Each exists in the form of a hydrogen phosphate anion ($[HPO_4]^{2-}$), dissociating to contribute $2H^+$ overall.
- Each liberates an oxygen atom when it binds to an adenosine diphosphate (ADP) molecule, contributing 2 O overall.

Charges are balanced by the difference between ADP and ATP. In the cellular environment, all three hydroxyl groups of ADP dissociate into $-O^-$ and H^+, giving ADP^{3-}, and this ion tends to exist in an ionic bond with Mg^{2+}, giving $ADPMg^-$. ATP behaves identically except that it has four hydroxyl groups, giving $ATPMg^{2-}$. When these differences along with the true charges on the two phosphate groups are considered together, the net charges of -4 on each side are balanced.

For simple fermentations, the metabolism of one molecule of glucose to two molecules of pyruvate has a net yield of two molecules of ATP. Most cells will then carry out further reactions to "repay" the used NAD^+ and produce a final product of ethanol or lactic acid. Many bacteria use inorganic compounds as hydrogen acceptors to regenerate the NAD^+.

Cells performing aerobic respiration synthesize much more ATP, but not as part of glycolysis. These further aerobic reactions use pyruvate, and NADH + H^+ from glycolysis. Eukaryotic aerobic respiration produces approximately 34 additional molecules of ATP for each glucose molecule, however most of these are produced by a mechanism vastly different from the substrate-level phosphorylation in glycolysis. The lower-energy production, per glucose, of anaerobic respiration relative to aerobic respiration, results in greater flux through the pathway under hypoxic (low-oxygen) conditions, unless alternative sources of anaerobically oxidizable substrates, such as fatty acids, are found.

16.1.2 History

The pathway of glycolysis as it is known today took almost 100 years to fully elucidate. The combined results of many smaller experiments were required in order to understand the intricacies of the entire pathway. The first steps in understanding glycolysis began in the nineteenth century with the wine industry. For economic reasons, the French wine industry sought to investigate why wine sometimes turned distasteful, instead of fermenting into alcohol. French scientist Louis Pasteur researched this issue during the 1850s, and the results of his experiments began the long road to elucidating the pathway of glycolysis. His experiments showed that fermentation occurs by the action of living microorganisms, yeasts, and that yeast's glucose consumption decreased under aerobic conditions of fermentation, in comparison to anaerobic conditions (the Pasteur effect).

Insight into the component steps of glycolysis were provided by the non-cellular fermentation experiments of Eduard Buchner during the 1890s. Buchner demonstrated that the conversion of glucose to ethanol was possible using a non-living extract of yeast, due to the action of enzymes in the extract. This experiment not only revolutionized biochemistry, but also allowed later scientists to analyze this pathway in a more controlled laboratory setting. In a series of experiments (1905–1911), scientists Arthur Harden and William Young discovered more pieces of glycolysis. They discovered the regulatory effects of ATP on glucose consumption during alcohol fermentation. They also shed light on the role of one compound as a glycolysis intermediate: fructose 1,6-bisphosphate.

The elucidation of fructose 1,6-bisphosphate was accomplished by measuring CO_2 levels when yeast juice was incubated with glucose. CO_2 production increased rapidly then slowed down. Harden and Young noted that this process would restart if an inorganic phosphate (Pi) was added to the mixture. Harden and Young deduced that this process produced organic phosphate esters, and further experiments allowed them to extract fructose diphosphate (F-1,6-DP).

Arthur Harden and William Young along with Nick Sheppard determined, in a second experiment, that a heat-sensitive high-molecular-weight subcellular fraction (the enzymes) and a heat-insensitive low-molecular-weight cytoplasm fraction (ADP, ATP and NAD^+ and other cofactors) are required together for fermentation to proceed. This experiment begun by observing that dialyzed (purified) yeast juice could not ferment or even create a sugar phosphate. This mixture was rescued with the addition of undialyzed yeast extract that had been boiled. Boiling the yeast extract renders all proteins inactive (as it denatures them).

The ability of boiled extract plus dialyzed juice to complete fermentation suggests that the cofactors were non-protein in character. In the 1920s Otto Meyerhof was able to link together some of the many individual pieces of glycolysis discovered by Buchner, Harden, and Young. Meyerhof and his team were able to extract different glycolytic enzymes from muscle tissue, and combine them to artificially create the pathway from glycogen to lactic acid.

In one paper, Meyerhof and scientist Renate Junowicz-Kockolaty investigated the reaction that splits fructose 1,6-diphosphate into the two triose phosphates. Previous work proposed that the split occurred via 1,3-diphosphoglyceraldehyde plus an oxidizing enzyme and cozymase. Meyerhoff and Junowicz found that the equilibrium constant for the isomerase and aldoses reaction were not affected by inorganic phosphates or any other cozymase or oxidizing enzymes. They further removed diphosphoglyceraldehyde as a possible intermediate in glycolysis. With all of these pieces available by the 1930s, Gustav Embden proposed a detailed, step-by-step outline of that pathway we now know as glycolysis. The biggest difficulties in determining the intricacies of the pathway were due to the very short lifetime and low steady-state concentrations of the intermediates of the fast glycolytic reactions. By the 1940s, Meyerhof, Embden and many other biochemists had finally completed the puzzle of glycolysis.

16.1.3 Sequence of Reactions

Summary of Reactions

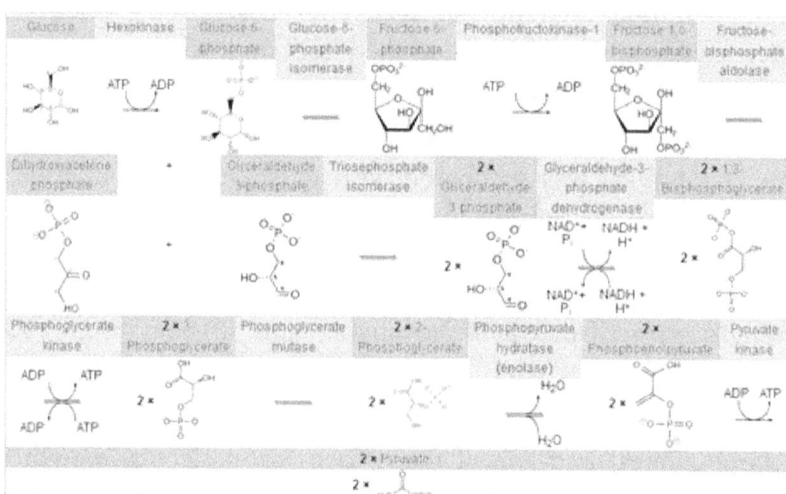

Preparatory Phase

The first five steps of Glycolysis are regarded as the preparatory (or investment) phase, since they consume energy to convert the glucose into two three-carbon sugar phosphates (G3P).

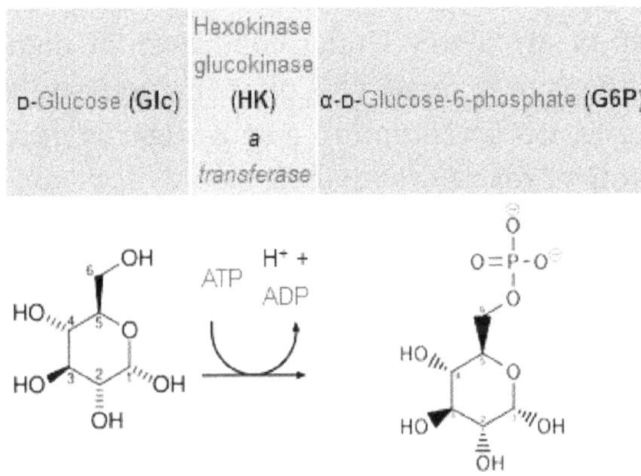

Once glucose enters the cell, the first step is phosphorylation of glucose by a family of enzymes called hexokinases to form glucose 6-phosphate (G6P). This reaction consumes ATP, but it acts to keep the glucose concentration inside the cell low, promoting continuous transport of blood glucose into the cell through the plasma membrane transporters. In addition, phosphorylation blocks the glucose from leaking out – the cell lacks transporters for G6P, and free diffusion out of the cell is prevented due to the charged nature of G6P. Glucose may alternatively be formed from the phosphorolysis or hydrolysis of intracellular starch or glycogen.

In animals, an isozyme of hexokinase called glucokinase is also used in the liver, which has a much lower affinity for glucose (K_m in the vicinity of normal glycemia), and differs in regulatory properties. The different substrate affinity and alternate regulation of this enzyme are a reflection of the role of the liver in maintaining blood sugar levels. G6P is then rearranged into fructose 6-phosphate (F6P) by glucose phosphate isomerase. Fructose can also enter the glycolytic pathway by phosphorylation at this point.

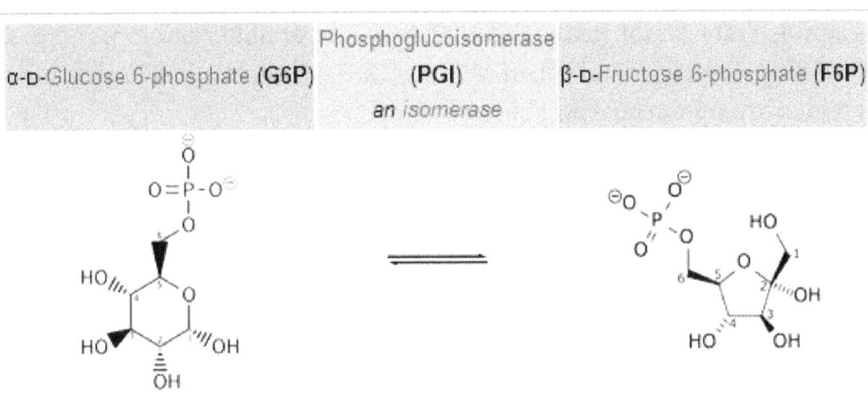

The change in structure is an isomerization, in which the G6P has been converted to F6P. The reaction requires an enzyme, phosphoglucose isomerase, to proceed. This reaction is freely reversible under normal cell conditions. However, it is often driven forward because of a low concentration of F6P, which is constantly consumed during the next step of glycolysis. Under conditions of high F6P concentration, this reaction readily runs in reverse. This phenomenon can be explained through Le Chatelier's Principle. Isomerization to a keto sugar is necessary for carbanion stabilization in the fourth reaction step (below).

The energy expenditure of another ATP in this step is justified in 2 ways: The glycolytic process (up to this step) becomes irreversible, and the energy supplied destabilizes the molecule. Because the reaction catalyzed by phosphofructokinase 1 (PFK-1) is coupled to the hydrolysis of ATP (an energetically favorable step) it is, in essence, irreversible, and a different pathway must be used to do the reverse conversion during gluconeogenesis. This makes the reaction a key regulatory point.

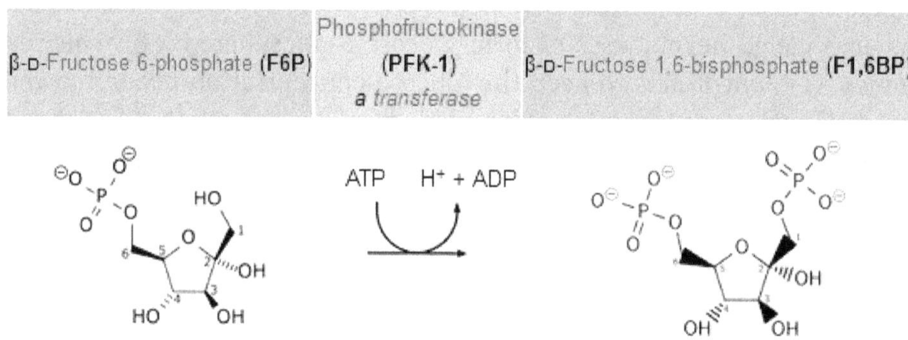

Furthermore, the second phosphorylation event is necessary to allow the formation of two charged groups (rather than only one) in the subsequent step of glycolysis, ensuring the prevention of free diffusion of substrates out of the cell.

The same reaction can also be catalyzed by pyrophosphate-dependent phosphofructokinase (PFP or PPi-PFK), which is found in most plants, some bacteria, archea, and protists, but not in animals. This enzyme uses pyrophosphate (PPi) as a phosphate donor instead of ATP. It is a reversible reaction, increasing the flexibility of glycolytic metabolism. A rarer ADP-dependent PFK enzyme variant has been identified in archaean species.

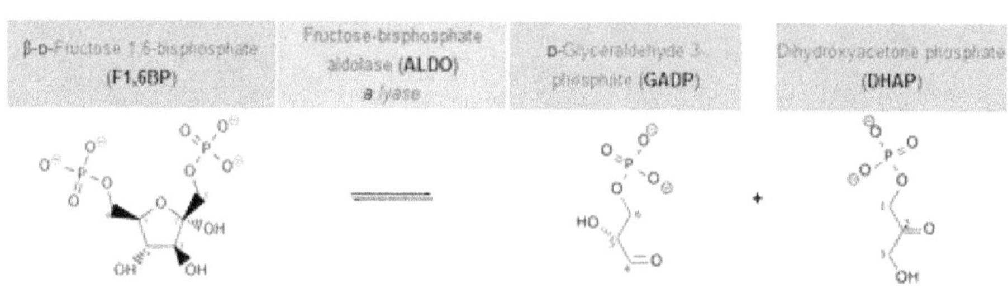

Destabilizing the molecule in the previous reaction allows the hexose ring to be split by aldolase into two triose sugars: dihydroxyacetone phosphate (a ketose), and glyceraldehyde 3-phosphate (an aldose). There are two classes of aldolases: class I aldolases, present in animals and plants, and class II aldolases, present in fungi and bacteria; the two classes use different mechanisms in cleaving the ketose ring. Electrons delocalized in the carbon-carbon bond cleavage associate with the alcohol group. The resulting carbanion is stabilized by the structure of the carbanion itself via resonance charge distribution and by the presence of a charged ion prosthetic group. Triosephosphate isomerase rapidly interconverts dihydroxyacetone phosphate with glyceraldehyde 3-phosphate (GADP) that proceeds further into glycolysis. This is advantageous, as it directs dihydroxyacetone phosphate down the same pathway as glyceraldehyde 3-phosphate, simplifying regulation.

Dihydroxyacetone phosphate **(DHAP)**	Triosephosphate isomerase **(TPI)** *an isomerase*	D-Glyceraldehyde 3-phosphate **(GADP)**

Pay-Off Phase

The second half of glycolysis is known as the pay-off phase, characterised by a net gain of the energy-rich molecules ATP and NADH. Since glucose leads to two triose sugars in the preparatory phase, each reaction in the pay-off phase occurs twice per glucose molecule. This yields 2 NADH molecules and 4 ATP molecules, leading to a net gain of 2 NADH molecules and 2 ATP molecules from the glycolytic pathway per glucose.

Glyceraldehyde 3-phosphate **(GADP)**	Glyceraldehyde phosphate dehydrogenase **(GAPDH)** *an oxidoreductase*	D-1,3-Bisphosphoglycerate **(1,3BPG)**

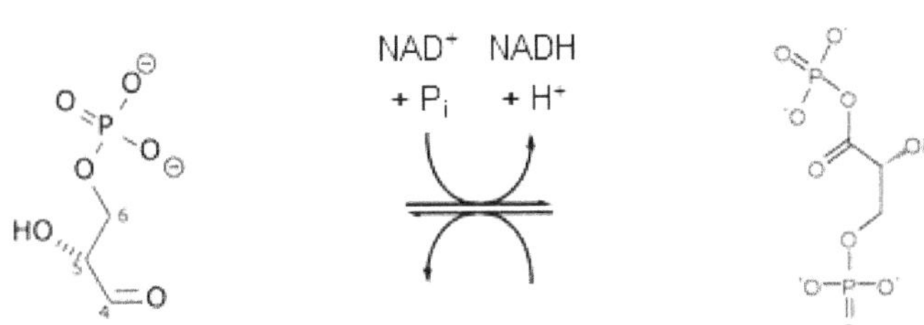

The aldehyde groups of the triose sugars are oxidised, and inorganic phosphate is added to them, forming 1,3-bisphosphoglycerate. The hydrogen is used to reduce two molecules of NAD^+, a hydrogen carrier, to give NADH + H^+ for each triose. Hydrogen atom balance and charge balance are both maintained because the phosphate (P_i) group actually exists in the form of a hydrogen phosphate anion (HPO2−4), which dissociates to contribute the extra H^+ ion and gives a net charge of -3 on both sides.

Here, arsenate ($[AsO_4]^{3-}$), an anion akin to inorganic phosphate may replace phosphate as a substrate to form 1-arseno-3-phosphoglycerate. This, however, is unstable and readily hydrolyzes to form 3-phosphoglycerate, the intermediate in the next step of the pathway.

As a consequence of bypassing this step, the molecule of ATP generated from 1-3 bisphosphoglycerate in the next reaction will not be made, even though the reaction proceeds. As a result, arsenate is an uncoupler of glycolysis.

Thisstepistheenzymatictransferofaphosphategroupfrom1,3-bisphosphoglycerate to ADP by phosphoglycerate kinase, forming ATP and 3-phosphoglycerate. At this step, glycolysis has reached the break-even point: 2 molecules of ATP were consumed, and 2 new molecules have now been synthesized.

This step, one of the two substrate-level phosphorylation steps, requires ADP; thus, when the cell has plenty of ATP (and little ADP), this reaction does not occur.

Because ATP decays relatively quickly when it is not metabolized, this is an important regulatory point in the glycolytic pathway.

1,3-Bisphosphoglycerate (**1,3BPG**)	Phosphoglycerate kinase (**PGK**) *a transferase*	3-Phosphoglycerate (**3PG**)

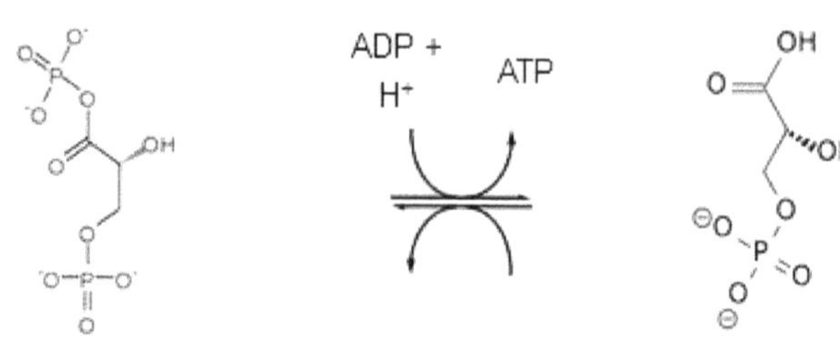

Phosphoglycerate kinase (**PGK**)

ADP actually exists as $ADPMg^-$, and ATP as $ATPMg^{2-}$, balancing the charges at -5 both sides. Phosphoglycerate mutase isomerises 3-phosphoglycerate into 2-phosphoglycerate.

3-Phosphoglycerate (3PG)	Phosphoglycerate mutase (PGM) *a mutase*	2-Phosphoglycerate (2PG)

Enolase next converts 2-phosphoglycerate to phosphoenolpyruvate. This reaction is an elimination reaction involving an E1cB mechanism.

Cofactors: 2 Mg^{2+}, one "conformational" ion to coordinate with the carboxylate group of the substrate, and one "catalytic" ion that participates in the dehydration.

2-Phosphoglycerate (2PG)	Enolase (ENO) *a lyase*	Phosphoenolpyruvate (PEP)

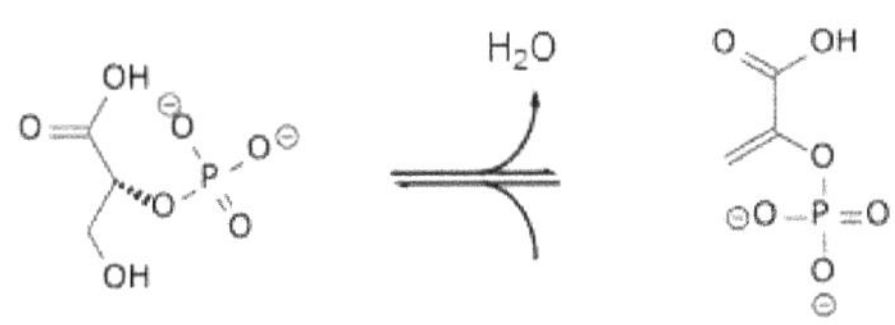

Enolase (ENO)

A final substrate-level phosphorylation now forms a molecule of pyruvate and a molecule of ATP by means of the enzyme pyruvate kinase. This serves as an additional regulatory step, similar to the phosphoglycerate kinase step.

Phosphoenolpyruvate (PEP)	Pyruvate kinase (PK) a *transferase*	Pyruvate (Pyr)

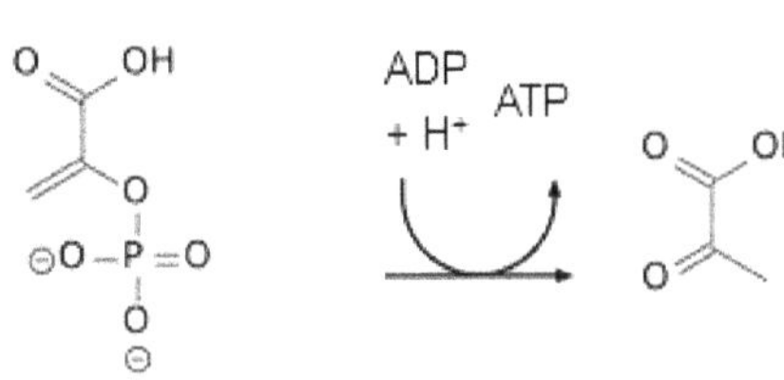

Biochemical Logic

The existence of more than one point of regulation indicates that intermediates between those points enter and leave the glycolysis pathway by other processes. For example, in the first regulated step, hexokinase converts glucose into glucose-6-phosphate. Instead of continuing through the glycolysis pathway, this intermediate can be converted into glucose storage molecules, such as glycogen or starch. The reverse reaction, breaking down, e.g., glycogen, produces mainly glucose-6-phosphate; very little free glucose is formed in the reaction. The glucose-6-phosphate so produced can enter glycolysis *after* the first control point.

In the second regulated step (the third step of glycolysis), phosphofructokinase converts fructose-6-phosphate into fructose-1,6-bisphosphate, which then is converted into glyceraldehyde-3-phosphate and dihydroxyacetone phosphate. The dihydroxyacetone phosphate can be removed from glycolysis by conversion into glycerol-3-phosphate, which can be used to form triglycerides. Conversely, triglycerides can be broken down into fatty acids and glycerol; the latter, in turn, can be converted into dihydroxyacetone phosphate, which can enter glycolysis *after* the second control point.

Free Energy Changes

The change in free energy, ΔG, for each step in the glycolysis pathway can be calculated using $\Delta G = \Delta G^{\circ\prime} + RT\ln Q$, where Q is the reaction quotient. This requires knowing the concentrations of the metabolites. All of these values are available for erythrocytes, with the exception of the concentrations of NAD^+ and NADH. The ratio of NAD^+ to NADH in the cytoplasm is approximately 1000, which makes the oxidation of glyceraldehyde-3-phosphate (step 6) more favourable.

Table 1: Concentrations of Metabolites in Erythrocytes

Compound	Concentration / mM
Glucose	5.0
Glucose-6-phosphate	0.083
Fructose-6-phosphate	0.014
Fructose-1,6-bisphosphate	0.031
Dihydroxyacetone phosphate	0.14
Glyceraldehyde-3-phosphate	0.019
1,3-Bisphosphoglycerate	0.001
2,3-Bisphosphoglycerate	4.0
3-Phosphoglycerate	0.12
2-Phosphoglycerate	0.03
Phosphoenolpyruvate	0.023
Pyruvate	0.051

Compound	Concentration / mM
ATP	1.85
ADP	0.14
P_i	1.0

Using the measured concentrations of each step, and the standard free energy changes, the actual free energy change can be calculated. (Neglecting this is very common—the delta G of ATP hydrolysis in cells is not the standard free energy change of ATP hydrolysis quoted in textbooks).

Table 2: Change in Free Energy for Each Step of Glycolysis

Step	Reaction	$\Delta G°'$ (kJ/mol)		ΔG (kJ/mol)	
1	Glucose + ATP^{4-} ⟶ Glucose-6-phosphate^{2-} + ADP^{3-} + H^+	-16	.7	-34	
2	Glucose-6-phosphate^{2-} ⟶ Fructose-6-phosphate^{2-}	1	.67	-2	.9
3	Fructose-6-phosphate^{2-} + ATP^{4-} ⟶ Fructose-1,6-bisphosphate^{4-} + ADP^{3-} + H^+	-14	.2	-19	
4	Fructose-1,6-bisphosphate^{4-} ⟶ Dihydroxyacetone phosphate^{2-} + Glyceraldehyde-3-phosphate^{2-}	23	.9	-0	.23
5	Dihydroxyacetone phosphate^{2-} ⟶ Glyceraldehyde-3-phosphate^{2-}	7	.56	2	.4
6	Glyceraldehyde-3-phosphate^{2-} + P_i^{2-} + NAD^+ ⟶ 1,3-Bisphosphoglycerate^{4-} + NADH + H^+	6	.30	-1	.29
7	1,3-Bisphosphoglycerate^{4-} + ADP^{3-} ⟶ 3-Phosphoglycerate^{3-} + ATP^{4-}	-18	.9	0	.09
8	3-Phosphoglycerate^{3-} ⟶ 2-Phosphoglycerate^{3-}	4	.4	0	.83
9	2-Phosphoglycerate^{3-} ⟶ Phosphoenolpyruvate^{3-} + H_2O	1	.8	1	.1
10	Phosphoenolpyruvate^{3-} + ADP^{3-} + H^+ ⟶ Pyruvate$^-$ + ATP^{4-}	-31	.7	-23	.0

From measuring the physiological concentrations of metabolites in an erythrocyte it seems that about seven of the steps in glycolysis are in equilibrium for that cell type. Three of the steps—the ones with large negative free energy changes—are not in equilibrium and are referred to as *irreversible*; such steps are often subject to regulation. Step 5, because that step is a side-reaction that can decrease or increase the concentration of the intermediate glyceraldehyde-3-phosphate. That compound is converted to dihydroxyacetone phosphate by the enzyme triose phosphate isomerase, which is a catalytically perfect enzyme; its rate is so fast that the reaction can be assumed to be in equilibrium. The fact that ΔG is not zero indicates that the actual concentrations in the erythrocyte are not accurately known.

16.1.4 Regulation

The enzymes that catalyse glycolysis are regulated via a range of biological mechanisms in order to control overall flux though the pathway. This is vital for both homeostatsis in a static environment, and metabolic adaptation to a changing environment or need. The details of regulation for some enzymes are highly conserved between species, whereas others vary widely.

- **Gene Expression:** Firstly, the cellular concentrations of glycolytic enzymes are modulated via regulation of gene expression via transcription factors, with several glycolysis enzymes themselves acting as regulatory protein kinases in the nucleus.
- **Allosteric Inhibition and Activation by Metabolites:** In particular end-product inhibition of regulated enzymes by metabolites such as ATP serves as negative feedback regulation of the pathway.
- **Allosteric Inhibition and Activation by Protein-Protein Interactions (PPI):** Indeed, some proteins interact with and regulate multiple glycolytic enzymes.
- **Post-Translational Modification (PTM):** In particular, phosphorylation and dephosphorylation is a key mechanism of regulation of pyruvate kinase in the liver.

Regulation by Insulin in Animals

In animals, regulation of blood glucose levels by the pancreas in conjunction with the liver is a vital part of homeostasis. The beta cells in the pancreatic islets are sensitive to the blood glucose concentration. A rise in the blood glucose concentration causes them to release insulin into the blood, which has an effect particularly on the liver, but also on fat and muscle cells, causing these tissues to remove glucose from the blood. When the blood sugar falls the pancreatic beta cells cease insulin production, but, instead, stimulate the neighboring pancreatic alpha cells to release glucagon into the blood. This, in turn, causes the liver to release glucose into the blood by breaking down stored glycogen, and by means of gluconeogenesis. If the fall in the blood glucose level is particularly rapid or severe, other glucose sensors cause the release of epinephrine from the adrenal glands into the blood. This has the same action as glucagon on glucose metabolism, but its effect is more pronounced. In the liver glucagon and epinephrine cause the phosphorylation of the key, regulated enzymes of glycolysis, fatty acid synthesis, cholesterol synthesis, gluconeogenesis, and glycogenolysis. Insulin has the opposite effect on these enzymes. The phosphorylation and dephosphorylation of these enzymes (ultimately in response to the glucose level in the blood) is the dominant manner by which these pathways are controlled in the liver, fat, and muscle cells.

Thus the phosphorylation of phosphofructokinase inhibits glycolysis, whereas its dephosphorylation through the action of insulin stimulates glycolysis.

Regulated Enzymes in Glycolysis

The three regulatory enzymes are hexokinase (or glucokinase in the liver), phosphofructokinase, and pyruvate kinase. The flux through the glycolytic pathway is adjusted in response to conditions both inside and outside the cell. The internal factors that regulate glycolysis do so primarily to provide ATP in adequate quantities for the cell's needs. The external factors act primarily on the liver, fat tissue, and muscles, which can remove large quantities of glucose from the blood after meals (thus preventing hyperglycemia by storing the excess glucose as fat or glycogen, depending on the tissue type). The liver is also capable of releasing glucose into the blood between meals, during fasting, and exercise thus preventing hypoglycemia by means of glycogenolysis and gluconeogenesis. These latter reactions coincide with the halting of glycolysis in the liver.

In addition hexokinase and glucokinase act independently of the hormonal effects as controls at the entry points of glucose into the cells of different tissues. Hexokinase responds to the glucose-6-phosphate (G6P) level in the cell, or, in the case of glucokinase, to the blood sugar level in the blood to impart entirely intracellular controls of the glycolytic pathway in different tissues.

When glucose has been converted into G6P by hexokinase or glucokinase, it can either be converted to glucose-1-phosphate (G1P) for conversion to glycogen, or it is alternatively converted by glycolysis to pyruvate, which enters the mitochondrion where it is converted into acetyl-CoA and then into citrate. Excess citrate is exported from the mitochondrion back into the cytosol, where ATP citrate lyase regenerates acetyl-CoA and oxaloacetate (OAA). The acetyl-CoA is then used for fatty acid synthesis and cholesterol synthesis, two important ways of utilizing excess glucose when its concentration is high in blood. The regulated enzymes catalyzing these reactions perform these functions when they have been dephosphorylated through the action of insulin on the liver cells. Between meals, during fasting, exercise or hypoglycemia, glucagon and epinephrine are released into the blood. This causes liver glycogen to be converted back to G6P, and then converted to glucose by the liver-specific enzyme glucose 6-phosphatase and released into the blood. Glucagon and epinephrine also stimulate gluconeogenesis, which converts non-carbohydrate substrates into G6P, which joins the G6P derived from glycogen, or substitutes for it when the liver glycogen store have been depleted. This is critical for brain function, since the brain utilizes glucose as an energy source under most conditions. The simultaneously phosphorylation of, particularly, phosphofructokinase, but also, to a certain extent pyruvate kinase, prevents glycolysis occurring at the same time as gluconeogenesis and glycogenolysis.

Hexokinase and Glucokinase

All cells contain the enzyme hexokinase, which catalyzes the conversion of glucose that has entered the cell into glucose-6-phosphate (G6P). Since the cell membrane is impervious to G6P, hexokinase essentially acts to transport glucose into the cells from which it can then no longer escape. Hexokinase is inhibited by high levels of G6P in the cell. Thus the rate of entry of glucose into cells partially depends on how fast G6P can be disposed of by glycolysis, and by glycogen synthesis (in the cells which store glycogen, namely liver and muscles).

Glucokinase, unlike hexokinase, is not inhibited by G6P. It occurs in liver cells, and will only phosphorylate the glucose entering the cell to form G6P, when the glucose in the blood is abundant. This being the first step in the glycolytic pathway in the liver, it therefore imparts an additional layer of control of the glycolytic pathway in this organ.

Phosphofructokinase

Phosphofructokinase is an important control point in the glycolytic pathway, since it is one of the irreversible steps and has key allosteric effectors, AMP and fructose 2,6-bisphosphate (F2,6BP).

F2,6BP is a very potent activator of phosphofructokinase (PFK-1) that is synthesized when F6P is phosphorylated by a second phosphofructokinase (PFK2). In the liver, when blood sugar is low and glucagon elevates cAMP, PFK2 is phosphorylated by protein kinase A. The phosphorylation inactivates PFK2, and another domain on this protein becomes active as fructose bisphosphatase-2, which converts F2,6BP back to F6P. Both glucagon and epinephrine cause high levels of cAMP in the liver. The result of lower levels of liver F2,6BP is a decrease in activity of phosphofructokinase and an increase in activity of fructose 1,6-bisphosphatase, so that gluconeogenesis (in essence, "glycolysis in reverse") is favored. This is consistent with the role of the liver in such situations, since the response of the liver to these hormones is to release glucose to the blood.

ATP competes with AMP for the allosteric effector site on the PFK enzyme. ATP concentrations in cells are much higher than those of AMP, typically 100-fold higher, but the concentration of ATP does not change more than about 10% under physiological conditions, whereas a 10% drop in ATP results in a 6-fold increase in AMP. Thus, the relevance of ATP as an allosteric effector is questionable. An increase in AMP is a consequence of a decrease in energy charge in the cell.

Citrate inhibits phosphofructokinase when tested *in vitro* by enhancing the inhibitory effect of ATP. However, it is doubtful that this is a meaningful effect *in vivo*, because citrate in the cytosol is utilized mainly for conversion to acetyl-CoA for fatty acid and cholesterol synthesis.

TIGAR, a p53 induced enzyme, is responsible for the regulation of phosphofructokinase and acts to protect against oxidative stress. TIGAR is a single enzyme with dual function that regulates F2,6BP. It can behave as a phosphatase (fructuose-2,6-bisphosphatase) which cleaves the phosphate at carbon-2 producing F6P. It can also behave as a kinase (PFK2) adding a phosphate onto carbon-2 of F6P which produces F2,6BP. In humans, the TIGAR protein is encoded by *C12orf5* gene. The TIGAR enzyme will hinder the forward progression of glycolysis, by creating a build up of fructose-6-phosphate (F6P) which is isomerized into glucose-6-phosphate (G6P). The accumulation of G6P will shunt carbons into the pentose phosphate pathway.

Pyruvate Kinase

The final step of glycolysis is catalysed by pyruvate kinase to form pyruvate and another ATP. It is regulated by a range of different transcriptional, covalent and non-covalent regulation mechanisms, which can vary widely in different tissues. For example, in the liver, pyruvate kinase is regulated based on glucose availability. During fasting (no glucose available), glucagon activates protein kinase A which phosphorylates pyruvate kinase to inhibit it. An increase in blood sugar leads to secretion of insulin, which activates protein phosphatase 1, leading to dephosphorylation and re-activation of pyruvate kinase. These controls prevent pyruvate kinase from being active at the same time as the enzymes that catalyze the reverse reaction (pyruvate carboxylase and phosphoenolpyruvate carboxykinase), preventing a futile cycle. Conversely, the isoform of pyruvate kinasein found in muscle is not affected by protein kinase A (which is activated by adrenaline in that tissue), so that glycolysis remains active in muscles even during fasting.

16.1.5 Post-Glycolysis Processes

The overall process of glycolysis is:

$$Glucose + 2\ NAD^{+} + 2\ ADP + 2\ P_i \rightarrow 2\ Pyruvate + 2\ NADH + 2\ H^{+} + 2\ ATP + 2\ H_2O$$

If glycolysis were to continue indefinitely, all of the NAD^+ would be used up, and glycolysis would stop. To allow glycolysis to continue, organisms must be able to oxidize NADH back to NAD^+. How this is performed depends on which external electron acceptor is available.

Anoxic Regeneration of NAD+

One method of doing this is to simply have the pyruvate do the oxidation; in this process, pyruvate is converted to lactate (the conjugate base of lactic acid) in a process called lactic acid fermentation:

$$Pyruvate + NADH + H^{+} \rightarrow Lactate + NAD^{+}$$

This process occurs in the bacteria involved in making yogurt (the lactic acid causes the milk to curdle). This process also occurs in animals under hypoxic (or partially anaerobic) conditions, found, for example, in overworked muscles that are starved of oxygen. In many tissues, this is a cellular last resort for energy; most animal tissue cannot tolerate anaerobic conditions for an extended period of time.

Some organisms, such as yeast, convert NADH back to NAD^+ in a process called ethanol fermentation. In this process, the pyruvate is converted first to acetaldehyde and carbon dioxide, and then to ethanol.

Lactic acid fermentation and ethanol fermentation can occur in the absence of oxygen. This anaerobic fermentation allows many single-cell organisms to use glycolysis as their only energy source.

Anoxic regeneration of NAD^+ is only an effective means of energy production during short, intense exercise in vertebrates, for a period ranging from 10 seconds to 2 minutes during a maximal effort in humans. (At lower exercise intensities it can sustain muscle activity in diving animals, such as seals, whales and other aquatic vertebrates, for very much longer periods of time.) Under these conditions NAD^+ is replenished by NADH donating its electrons to pyruvate to form lactate. This produces 2 ATP molecules per glucose molecule, or about 5% of glucose's energy potential (38 ATP molecules in bacteria). But the speed at which ATP is produced in this manner is about 100 times that of oxidative phosphorylation. The pH in the cytoplasm quickly drops when hydrogen ions accumulate in the muscle, eventually inhibiting the enzymes involved in glycolysis.

The burning sensation in muscles during hard exercise can be attributed to the release of hydrogen ions during the shift to glucose fermentation from glucose oxidation to carbon dioxide and water, when aerobic metabolism can no longer keep pace with the energy demands of the muscles. These hydrogen ions form a part of lactic acid. The body falls back on this less efficient but faster method of producing ATP under low oxygen conditions. This is thought to have been the primary means of energy production in earlier organisms before oxygen reached high concentrations in the atmosphere between 2000 and 2500 million years ago, and thus would represent a more ancient form of energy production than the aerobic replenishment of NAD^+ in cells.

The liver in mammals gets rid of this excess lactate by transforming it back into pyruvate under aerobic conditions.

Fermentation of pyruvate to lactate is sometimes also called "anaerobic glycolysis", however, glycolysis ends with the production of pyruvate regardless of the presence or absence of oxygen.

In the above two examples of fermentation, NADH is oxidized by transferring two electrons to pyruvate. However, anaerobic bacteria use a wide variety of compounds as the terminal electron acceptors in cellular respiration: nitrogenous compounds, such as nitrates and nitrites; sulfur compounds, such as sulfates, sulfites, sulfur dioxide, and elemental sulfur; carbon dioxide; iron compounds; manganese compounds; cobalt compounds; and uranium compounds.

Aerobic Regeneration of NAD+ and Further Catabolism of Pyruvate

In aerobic eukaryotes, a complex mechanism has developed to use the oxygen in air as the final electron acceptor, in a process called oxidative phosphorylation. Aerobic prokaryotes, which lack mitochondria, use a variety of simpler mechanisms.

- **Firstly**, the NADH + H^+ generated by glycolysis has to be transferred to the mitochondrion to be oxidized, and thus to regenerate the NAD^+ necessary for glycolysis to continue. However the inner mitochondrial membrane is impermeable to NADH and NAD^+. Use is therefore made of two "shuttles" to transport the electrons from NADH across the mitochondrial membrane. They are the malate-aspartate shuttle and the glycerol phosphate shuttle. In the former the electrons from NADH are transferred to cytosolic oxaloacetate to form malate. The malate then traverses the inner mitochondrial membrane into the mitochondrial matrix, where it is reoxidized by NAD^+ forming intra-mitochondrial oxaloacetate and NADH. The oxaloacetate is then re-cycled to the cytosol via its conversion to aspartate which is readily transported out of the mitochondrion. In the glycerol phosphate shuttle electrons from cytosolic NADH are transferred to dihydroxyacetone to form glycerol-3-phosphate which readily traverses the outer mitochondrial membrane. Glycerol-3-phosphate is then reoxidized to dihydroxyacetone, donating its electrons to FAD instead of NAD^+. This reaction takes place on the inner mitochondrial membrane, allowing $FADH_2$ to donate its electrons directly to coenzyme Q (ubiquinone) which is part of the electron transport chain which ultimately transfers electrons to molecular oxygen O_2, with the formation of water, and the release of energy eventually captured in the form of ATP.
- The glycolytic end-product, pyruvate (plus NAD^+) is converted to acetyl-CoA, CO_2 and NADH + H^+ within the mitochondria in a process called pyruvate decarboxylation.
- The resulting acetyl-CoA enters the citric acid cycle (or Krebs Cycle), where the acetyl group of the acetyl-CoA is converted into carbon dioxide by two decarboxylation reactions with the formation of yet more intra-mitochondrial NADH + H^+.

- The intra-mitochondrial NADH + H^+ is oxidized to NAD^+ by the electron transport chain, using oxygen as the final electron acceptor to form water. The energy released during this process is used to create a hydrogen ion (or proton) gradient across the inner membrane of the mitochondrion.
- Finally, the proton gradient is used to produce about 2.5 ATP for every NADH + H^+ oxidized in a process called oxidative phosphorylation.

Conversion of Carbohydrates into Fatty Acids and Cholesterol

The pyruvate produced by glycolysis is an important intermediary in the conversion of carbohydrates into fatty acids and cholesterol. This occurs via the conversion of pyruvate into acetyl-CoA in the mitochondrion. However, this acetyl CoA needs to be transported into cytosol where the synthesis of fatty acids and cholesterol occurs. This cannot occur directly. To obtain cytosolic acetyl-CoA, citrate (produced by the condensation of acetyl CoA with oxaloacetate) is removed from the citric acid cycle and carried across the inner mitochondrial membrane into the cytosol.

There it is cleaved by ATP citrate lyase into acetyl-CoA and oxaloacetate. The oxaloacetate is returned to mitochondrion as malate (and then back into oxaloacetate to transfer more acetyl-CoA out of the mitochondrion). The cytosolic acetyl-CoA can be carboxylated by acetyl-CoA carboxylase into malonyl CoA, the first committed step in the synthesis of fatty acids, or it can be combined with acetoacetyl-CoA to form 3-hydroxy-3-methylglutaryl-CoA (HMG-CoA) which is the rate limiting step controlling the synthesis of cholesterol. Cholesterol can be used as is, as a structural component of cellular membranes, or it can be used to synthesize the steroid hormones, bile salts, and vitamin D.

Conversion of Pyruvate into Oxaloacetate for the Citric Acid Cycle

Pyruvate molecules produced by glycolysis are actively transported across the inner mitochondrial membrane, and into the matrix where they can either be oxidized and combined with coenzyme A to form CO_2, acetyl-CoA, and NADH, or they can be carboxylated (by pyruvate carboxylase) to form oxaloacetate. This latter reaction "fills up" the amount of oxaloacetate in the citric acid cycle, and is therefore an anaplerotic reaction (from the Greek meaning to "fill up"), increasing the cycle's capacity to metabolize acetyl-CoA when the tissue's energy needs (e.g. in heart and skeletal muscle) are suddenly increased by activity.

In the citric acid cycle all the intermediates (e.g. citrate, iso-citrate, alpha-ketoglutarate, succinate, fumarate, malate and oxaloacetate) are regenerated during each turn of the cycle. Adding more of any of these intermediates to the mitochondrion therefore means that that additional amount is retained within the cycle, increasing all the other intermediates as one is converted into the other.

Hence the addition of oxaloacetate greatly increases the amounts of all the citric acid intermediates, thereby increasing the cycle's capacity to metabolize acetyl CoA, converting its acetate component into CO_2 and water, with the release of enough energy to form 11 ATP and 1 GTP molecule for each additional molecule of acetyl CoA that combines with oxaloacetate in the cycle.

To cataplerotically remove oxaloacetate from the citric cycle, malate can be transported from the mitochondrion into the cytoplasm, decreasing the amount of oxaloacetate that can be regenerated. Furthermore, citric acid intermediates are constantly used to form a variety of substances such as the purines, pyrimidines and porphyrins.

16.1.6 Intermediates for other Pathways

This concentrates on the catabolic role of glycolysis with regard to converting potential chemical energy to usable chemical energy during the oxidation of glucose to pyruvate. Many of the metabolites in the glycolytic pathway are also used by anabolic pathways, and, as a consequence, flux through the pathway is critical to maintain a supply of carbon skeletons for biosynthesis.

The following metabolic pathways are all strongly reliant on glycolysis as a source of metabolites: and many more.

- Pentose phosphate pathway, which begins with the dehydrogenation of glucose-6-phosphate, the first intermediate to be produced by glycolysis, produces various pentose sugars, and NADPH for the synthesis of fatty acids and cholesterol.
- Glycogen synthesis also starts with glucose-6-phosphate at the beginning of the glycolytic pathway.
- Glycerol, for the formation of triglycerides and phospholipids, is produced from the glycolytic intermediate glyceraldehyde-3-phosphate.
- **Various Post-Glycolytic Pathways:**
  - Fatty acid synthesis.
  - Cholesterol synthesis.
- The Citric Acid Cycle which in Turn Leads to:
  - Amino acid synthesis.
  - Nucleotide synthesis.
  - Tetrapyrrole synthesis.

Although gluconeogenesis and glycolysis share many intermediates the one is not functionally a branch or tributary of the other. There are two regulatory steps in both pathways which, when active in the one pathway, are automatically inactive in the other. The two processes can therefore not be simultaneously active. Indeed, if both sets of reactions were highly active at the same time the net result would be the hydrolysis of four high energy phosphate bonds (two ATP and two GTP) per reaction cycle.

NAD^+ is the oxidizing agent in glycolysis, as it is in most other energy yielding metabolic reactions (e.g. beta-oxidation of fatty acids, and during the citric acid cycle). The NADH thus produced is primarily used to ultimately transfer electrons to O_2 to produce water, or, when O_2 is not available, to produce compounds such as lactate or ethanol. NADH is rarely used for synthetic processes, the notable exception being gluconeogenesis. During fatty acid and cholesterol synthesis the reducing agent is NADPH. This difference exemplifies a general principle that NADPH is consumed during biosynthetic reactions, whereas NADH is generated in energy-yielding reactions. The source of the NADPH is two-fold. When malate is oxidatively decarboxylated by "$NADP^+$-linked malic enzyme" pyruvate, CO_2 and NADPH are formed. NADPH is also formed by the pentose phosphate pathway which converts glucose into ribose, which can be used in synthesis of nucleotides and nucleic acids, or it can be catabolized to pyruvate.

16.1.7 Glycolysis in Disease

Diabetes

Cellular uptake of glucose occurs in response to insulin signals, and glucose is subsequently broken down through glycolysis, lowering blood sugar levels. However, insulin resistance or low insulin levels seen in diabetes result in hyperglycemia, where glucose levels in the blood rise and glucose is not properly taken up by cells. Hepatocytes further contribute to this hyperglycemia through gluconeogenesis. Glycolysis in hepatocytes controls hepatic glucose production, and when glucose is overproduced by the liver without having a means of being broken down by the body, hyperglycemia results.

Genetic Diseases

Glycolytic mutations are generally rare due to importance of the metabolic pathway; the majority of occurring mutations result in an inability of the cell to respire, and therefore cause the death of the cell at an early stage. However, some mutations (glycogen storage diseases and other inborn errors of carbohydrate metabolism) are seen with one notable example being pyruvate kinase deficiency, leading to chronic hemolytic anemia. In combined malonic and methylmalonic aciduria (CMAMMA) due to ACSF3 deficiency, glycolysis is reduced by -50%, which is caused by reduced lipoylation of mitochondrial enzymes such as the pyruvate dehydrogenase complex and α-ketoglutarate dehydrogenase complex.

Cancer

Malignant tumor cells perform glycolysis at a rate that is ten times faster than their noncancerous tissue counterparts. During their genesis, limited capillary support often results in hypoxia (decreased O2 supply) within the tumor cells. Thus, these cells rely on anaerobic metabolic processes such as glycolysis for ATP (adenosine triphosphate). Some tumor cells overexpress specific glycolytic enzymes which result in higher rates of glycolysis. Often these enzymes are Isoenzymes, of traditional glycolysis enzymes, that vary in their susceptibility to traditional feedback inhibition. The increase in glycolytic activity ultimately counteracts the effects of hypoxia by generating sufficient ATP from this anaerobic pathway. This phenomenon was first described in 1930 by Otto Warburg and is referred to as the Warburg effect. The Warburg hypothesis claims that cancer is primarily caused by dysfunctionality in mitochondrial metabolism, rather than because of the uncontrolled growth of cells. A number of theories have been advanced to explain the Warburg effect. One such theory suggests that the increased glycolysis is a normal protective process of the body and that malignant change could be primarily caused by energy metabolism.

This high glycolysis rate has important medical applications, as high aerobic glycolysis by malignant tumors is utilized clinically to diagnose and monitor treatment responses of cancers by imaging uptake of 2-^{18}F-2-deoxyglucose (FDG) (a radioactive modified hexokinase substrate) with positron emission tomography (PET).

There is ongoing research to affect mitochondrial metabolism and treat cancer by reducing glycolysis and thus starving cancerous cells in various new ways, including a ketogenic diet.

16.2 GLUCONEOGENESIS

Gluconeogenesis (GNG) is a metabolic pathway that results in the biosynthesis of glucose from certain non-carbohydrate carbon substrates. It is a ubiquitous process, present in plants, animals, fungi, bacteria, and other microorganisms. In vertebrates, gluconeogenesis occurs mainly in the liver and, to a lesser extent, in the cortex of the kidneys. It is one of two primary mechanisms – the other being degradation of glycogen (glycogenolysis) – used by humans and many other animals to maintain blood sugar levels, avoiding low levels (hypoglycemia). In ruminants, because dietary carbohydrates tend to be metabolized by rumen organisms, gluconeogenesis occurs regardless of fasting, low-carbohydrate diets, exercise, etc. In many other animals, the process occurs during periods of fasting, starvation, low-carbohydrate diets, or intense exercise.

In humans, substrates for gluconeogenesis may come from any non-carbohydrate sources that can be converted to pyruvate or intermediates of glycolysis. For the breakdown of proteins, these substrates include glucogenic amino acids (although not ketogenic amino acids); from breakdown of lipids (such as triglycerides), they include glycerol, odd-chain fatty acids (although not even-chain fatty acids); and from other parts of metabolism that includes lactate from the Cori cycle. Under conditions of prolonged fasting, acetone derived from ketone bodies can also serve as a substrate, providing a pathway from fatty acids to glucose. Although most gluconeogenesis occurs in the liver, the relative contribution of gluconeogenesis by the kidney is increased in diabetes and prolonged fasting.

The gluconeogenesis pathway is highly endergonic until it is coupled to the hydrolysis of ATP or GTP, effectively making the process exergonic. For example, the pathway leading from pyruvate to glucose-6-phosphate requires 4 molecules of ATP and 2 molecules of GTP to proceed spontaneously. These ATPs are supplied from fatty acid catabolism via beta oxidation.

16.2.1 Origins

Gluconeogenesis is considered one of the most ancient anabolic pathways and is likely to have been exhibited in the last universal common ancestor. Rafael F. Say and Georg Fuchs stated in 2010 that "all archaeal groups as well as the deeply branching bacterial lineages contain a bifunctional fructose 1,6-bisphosphate (FBP) aldolase/phosphatase with both FBP aldolase and FBP phosphatase activity. This enzyme is missing in most other Bacteria and in Eukaryota, and is heat-stabile even in mesophilic marine Crenarchaeota". It is proposed that fructose 1,6-bisphosphate aldolase/phosphatase was an ancestral gluconeogenic enzyme and had preceded glycolysis. But the chemical mechanisms between gluconeogenesis and glycolysis, whether it is anabolic or catabolic, are similar, suggesting they both originated at the same time. Fructose 1,6-bisphosphate is shown to be nonenzymatically synthesized continuously within a freezing solution.

The synthesis is accelerated in the presence of amino acids such as glycine and lysine implying that the first anabolic enzymes were amino acids. The prebiotic reactions in gluconeogenesis can also proceed nonenzymatically at dehydration-desiccation cycles. Such chemistry could have occurred in hydrothermal environments, including temperature gradients and cycling of freezing and thawing. Mineral surfaces might have played a role in the phosphorylation of metabolic intermediates from gluconeogenesis and have to been shown to produce tetrose, hexose phosphates, and pentose from formaldehyde, glyceraldehyde, and glycolaldehyde.

16.2.2 Precursors

In humans the main gluconeogenic precursors are lactate, glycerol (which is a part of the triglyceride molecule), alanine and glutamine. Altogether, they account for over 90% of the overall gluconeogenesis. Other glucogenic amino acids and all citric acid cycle intermediates (through conversion to oxaloacetate) can also function as substrates for gluconeogenesis. Generally, human consumption of gluconeogenic substrates in food does not result in increased gluconeogenesis.

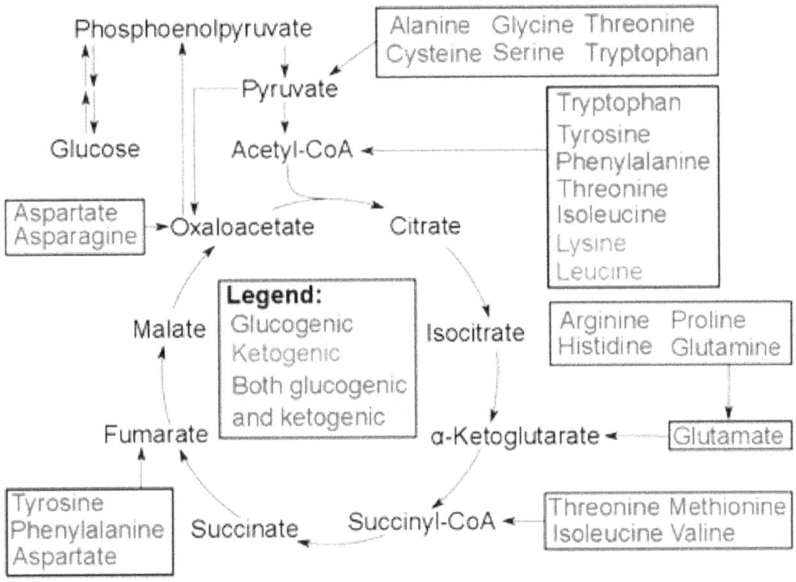

Fig 2: Catabolism of proteinogenic amino acids. Amino acids are classified according to the abilities of their products to enter gluconeogenesis:

- Glucogenic amino acids have this ability
- Ketogenic amino acids do not. These products may still be used for ketogenesis or lipid synthesis.
- Some amino acids are catabolized into both glucogenic and ketogenic products.

In ruminants, propionate is the principal gluconeogenic substrate. In nonruminants, including human beings, propionate arises from the β-oxidation of odd-chain and branched-chain fatty acids, and is a (relatively minor) substrate for gluconeogenesis.

Lactate is transported back to the liver where it is converted into pyruvate by the Cori cycle using the enzyme lactate dehydrogenase. Pyruvate, the first designated substrate of the gluconeogenic pathway, can then be used to generate glucose. Transamination or deamination of amino acids facilitates entering of their carbon skeleton into the cycle directly (as pyruvate or oxaloacetate), or indirectly via the citric acid cycle.

The contribution of Cori cycle lactate to overall glucose production increases with fasting duration. Specifically, after 12, 20, and 40 hours of fasting by human volunteers, the contribution of Cori cycle lactate to gluconeogenesis was 41%, 71%, and 92%, respectively.

Whether even-chain fatty acids can be converted into glucose in animals has been a longstanding question in biochemistry. Odd-chain fatty acids can be oxidized to yield acetyl-CoA and propionyl-CoA, the latter serving as a precursor to succinyl-CoA, which can be converted to oxaloacetate and enter into gluconeogenesis. In contrast, even-chain fatty acids are oxidized to yield only acetyl-CoA, whose entry into gluconeogenesis requires the presence of a glyoxylate cycle (also known as glyoxylate shunt) to produce four-carbon dicarboxylic acid precursors. The glyoxylate shunt comprises two enzymes, malate synthase and isocitrate lyase, and is present in fungi, plants, and bacteria.

Despite some reports of glyoxylate shunt enzymatic activities detected in animal tissues, genes encoding both enzymatic functions have only been found in nematodes, in which they exist as a single bi-functional enzyme. Genes coding for malate synthase alone (but not isocitrate lyase) have been identified in other animals including arthropods, echinoderms, and even some vertebrates. Mammals found to possess the malate synthase gene include monotremes (platypus) and marsupials (opossum), but not placental mammals.

The existence of the glyoxylate cycle in humans has not been established, and it is widely held that fatty acids cannot be converted to glucose in humans directly. Carbon-14 has been shown to end up in glucose when it is supplied in fatty acids, but this can be expected from the incorporation of labelled atoms derived from acetyl-CoA into citric acid cycle intermediates which are interchangeable with those derived from other physiological sources, such as glucogenic amino acids. In the absence of other glucogenic sources, the 2-carbon acetyl-CoA derived from the oxidation of fatty acids cannot produce a net yield of glucose via the citric acid cycle, since an equivalent two carbon atoms are released as carbon dioxide during the cycle. During ketosis, however, acetyl-CoA from fatty acids yields ketone bodies, including acetone, and up to ~60% of acetone may be oxidized in the liver to the pyruvate precursors acetol and methylglyoxal. Thus ketone bodies derived from fatty acids could account for up to 11% of gluconeogenesis during starvation. Catabolism of fatty acids also produces energy in the form of ATP that is necessary for the gluconeogenesis pathway.

16.2.3 Location

In mammals, gluconeogenesis has been believed to be restricted to the liver, the kidney, the intestine, and muscle, but recent evidence indicates gluconeogenesis occurring in astrocytes of the brain. These organs use somewhat different gluconeogenic precursors. The liver preferentially uses lactate, glycerol, and glucogenic amino acids (especially alanine) while the kidney preferentially uses lactate, glutamine and glycerol. Lactate from the Cori cycle is quantitatively the largest source of substrate for gluconeogenesis, especially for the kidney. The liver uses both glycogenolysis and gluconeogenesis to produce glucose, whereas the kidney only uses gluconeogenesis. After a meal, the liver shifts to glycogen synthesis, whereas the kidney increases gluconeogenesis. The intestine uses mostly glutamine and glycerol.

Propionate is the principal substrate for gluconeogenesis in the ruminant liver, and the ruminant liver may make increased use of gluconeogenic amino acids (e.g., alanine) when glucose demand is increased. The capacity of liver cells to use lactate for gluconeogenesis declines from the preruminant stage to the ruminant stage in calves and lambs. In sheep kidney tissue, very high rates of gluconeogenesis from propionate have been observed.

In all species, the formation of oxaloacetate from pyruvate and TCA cycle intermediates is restricted to the mitochondrion, and the enzymes that convert Phosphoenolpyruvic acid (PEP) to glucose-6-phosphate are found in the cytosol. The location of the enzyme that links these two parts of gluconeogenesis by converting oxaloacetate to PEP – PEP carboxykinase (PEPCK) – is variable by species: it can be found entirely within the mitochondria, entirely within the cytosol, or dispersed evenly between the two, as it is in humans.

Transport of PEP across the mitochondrial membrane is accomplished by dedicated transport proteins; however no such proteins exist for oxaloacetate. Therefore, in species that lack intra-mitochondrial PEPCK, oxaloacetate must be converted into malate or aspartate, exported from the mitochondrion, and converted back into oxaloacetate in order to allow gluconeogenesis to continue.

16.2.4 Pathway

Gluconeogenesis is a pathway consisting of a series of eleven enzyme-catalyzed reactions. The pathway will begin in either the liver or kidney, in the mitochondria or cytoplasm of those cells, this being dependent on the substrate being used. Many of the reactions are the reverse of steps found in glycolysis.

- Gluconeogenesis begins in the mitochondria with the formation of oxaloacetate by the carboxylation of pyruvate. This reaction also requires one molecule of ATP, and is catalyzed by pyruvate carboxylase. This enzyme is stimulated by high levels of acetyl-CoA (produced in β-oxidation in the liver) and inhibited by high levels of ADP and glucose.
- Oxaloacetate is reduced to malate using NADH, a step required for its transportation out of the mitochondria.

- Malate is oxidized to oxaloacetate using NAD^+ in the cytosol, where the remaining steps of gluconeogenesis take place.
- Oxaloacetate is decarboxylated and then phosphorylated to form phosphoenolpyruvate using the enzyme PEPCK. A molecule of GTP is hydrolyzed to GDP during this reaction.
- The next steps in the reaction are the same as reversed glycolysis. However, fructose 1,6-bisphosphatase converts fructose 1,6-bisphosphate to fructose 6-phosphate, using one water molecule and releasing one phosphate (in glycolysis, phosphofructokinase 1 converts F6P and ATP to F1,6BP and ADP). This is also the rate-limiting step of gluconeogenesis.
- Glucose-6-phosphate is formed from fructose 6-phosphate by phosphoglucoisomerase (the reverse of step 2 in glycolysis). Glucose-6-phosphate can be used in other metabolic pathways or dephosphorylated to free glucose. Whereas free glucose can easily diffuse in and out of the cell, the phosphorylated form (glucose-6-phosphate) is locked in the cell, a mechanism by which intracellular glucose levels are controlled by cells.
- The final step in gluconeogenesis, the formation of glucose, occurs in the lumen of the endoplasmic reticulum, where glucose-6-phosphate is hydrolyzed by glucose-6-phosphatase to produce glucose and release an inorganic phosphate. Like two steps prior, this step is not a simple reversal of glycolysis, in which hexokinase catalyzes the conversion of glucose and ATP into G6P and ADP. Glucose is shuttled into the cytoplasm by glucose transporters located in the endoplasmic reticulum's membrane.

16.2.5 Regulation

While most steps in gluconeogenesis are the reverse of those found in glycolysis, three regulated and strongly endergonic reactions are replaced with more kinetically favorable reactions. Hexokinase/glucokinase, phosphofructokinase, and pyruvate kinase enzymes of glycolysis are replaced with glucose-6-phosphatase, fructose-1,6-bisphosphatase, and PEP carboxykinase/pyruvate carboxylase.

These enzymes are typically regulated by similar molecules, but with opposite results. For example, acetyl CoA and citrate activate gluconeogenesis enzymes (pyruvate carboxylase and fructose-1,6-bisphosphatase, respectively), while at the same time inhibiting the glycolytic enzyme pyruvate kinase. This system of reciprocal control allow glycolysis and gluconeogenesis to inhibit each other and prevents a futile cycle of synthesizing glucose to only break it down. Pyruvate kinase can be also bypassed by 86 pathways not related to gluconeogenesis, for the purpose of forming pyruvate and subsequently lactate; some of these pathways use carbon atoms originated from glucose.

The majority of the enzymes responsible for gluconeogenesis are found in the cytosol; the exceptions are mitochondrial pyruvate carboxylase and, in animals, phosphoenolpyruvate carboxykinase. The latter exists as an isozyme located in both the mitochondrion and the cytosol. The rate of gluconeogenesis is ultimately controlled by the action of a key enzyme, fructose-1,6-bisphosphatase, which is also regulated through signal transduction by cAMP and its phosphorylation.

Global control of gluconeogenesis is mediated by glucagon (*released when blood glucose is low*); it triggers phosphorylation of enzymes and regulatory proteins by Protein Kinase A (a cyclic AMP regulated kinase) resulting in inhibition of glycolysis and stimulation of gluconeogenesis. Insulin counteracts glucagon by inhibiting gluconeogenesis. Type 2 diabetes is marked by excess glucagon and insulin resistance from the body. Insulin can no longer inhibit the gene expression of enzymes such as PEPCK which leads to increased levels of hyperglycemia in the body. The anti-diabetic drug metformin reduces blood glucose primarily through inhibition of gluconeogenesis, overcoming the failure of insulin to inhibit gluconeogenesis due to insulin resistance.

Studies have shown that the absence of hepatic glucose production has no major effect on the control of fasting plasma glucose concentration. Compensatory induction of gluconeogenesis occurs in the kidneys and intestine, driven by glucagon, glucocorticoids, and acidosis.

16.2.6 Insulin Resistance

In the liver, the FOX protein FOXO6 normally promotes gluconeogenesis in the fasted state, but insulin blocks FOXO6 upon feeding. In a condition of insulin resistance, insulin fails to block FOXO6 resulting in continued gluconeogenesis even upon feeding, resulting in high blood glucose (hyperglycemia).

Insulin resistance is a common feature of metabolic syndrome and type 2 diabetes. For this reason, gluconeogenesis is a target of therapy for type 2 diabetes, such as the antidiabetic drug metformin, which inhibits gluconeogenic glucose formation, and stimulates glucose uptake by cells.

CHAPTER 17

PYRUVATE DEHYDROGENASE AND THE CITRIC ACID CYCLE

17.1 PYRUVATE DEHYDROGENASE

Pyruvate dehydrogenase is an enzyme that catalyzes the reaction of pyruvate and a lipoamide to give the acetylated dihydrolipoamide and carbon dioxide. The conversion requires the coenzyme thiamine pyrophosphate.

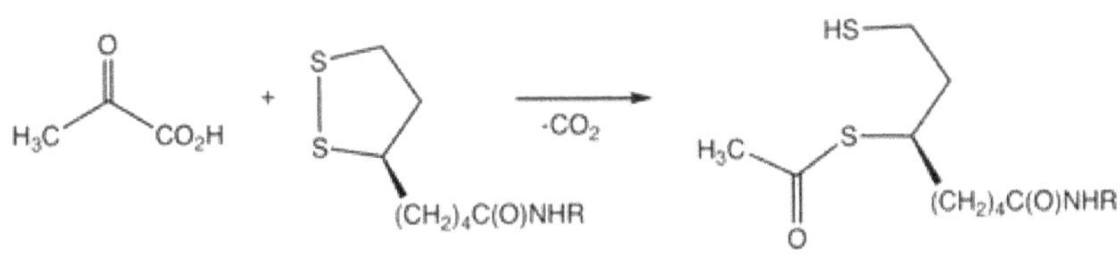

Pyruvate dehydrogenase is usually encountered as a component, referred to as E1, of the pyruvate dehydrogenase complex (PDC). PDC consists of other enzymes, referred to as E2 and E3. Collectively E1-E3 transform pyruvate, NAD^+, coenzyme A into acetyl-CoA, CO_2, and NADH. The conversion is crucial because acetyl-CoA may then be used in the citric acid cycle to carry out cellular respiration. To distinguish between this enzyme and the PDC, it is systematically called pyruvate dehydrogenase (acetyl-transferring).

17.1.1 Mechanism

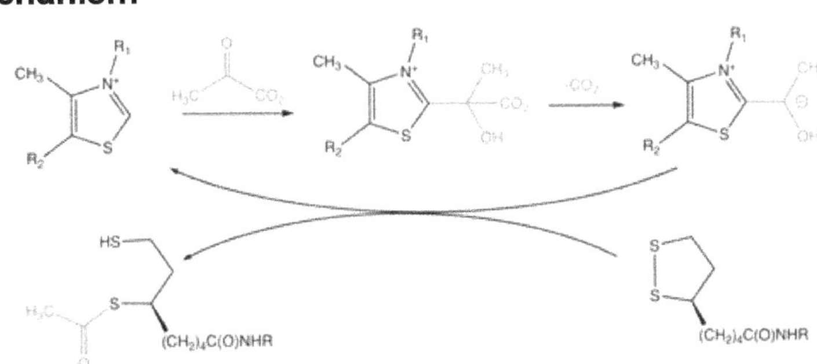

Fig 1: Simplified Mechanism for Pyruvate Dehydrogenase Reaction. The TPP Coenzyme is Shown with Abbreviated Substituents

The thiamine pyrophosphate (TPP) converts to an ylide by deprotonation. The ylide attack the ketone group of pyruvate. The resulting adduct decarboxylates. The resulting 1,3-dipole reductively acetylates lipoamide-E2.

In terms of details, biochemical and structural data for E1 revealed a mechanism of activation of TPP coenzyme by forming the conserved hydrogen bond with glutamate residue (Glu59 in human E1) and by imposing a V-conformation that brings the N4' atom of the aminopyrimidine to intramolecular hydrogen bonding with the thiazolium C2 atom. This unique combination of contacts and conformations of TPP leads to formation of the reactive C2-carbanion, eventually. After the cofactor TPP decarboxylates pyruvate, the acetyl portion becomes a hydroxyethyl derivative covalently attached to TPP.

17.1.2 Structure

E1 is a multimeric protein. Mammalian E1s, including human E1, are tetrameric, composed of two α- and two β- subunits. Some bacterial E1s, including E1 from *Escherichia coli*, are composed of two similar subunits, each being as large as the sum of molecular masses of α- and β- subunits.

17.1.3 Regulation

Phosphorylation of E1 by pyruvate dehydrogenase kinase (PDK) inactivates E1 and subsequently the entire complex. PDK is inhibited by dichloroacetic acid and pyruvate, resulting in a higher quantity of active, unphosphorylated PDH. Phosphorylation is reversed by pyruvate dehydrogenase phosphatase, which is stimulated by insulin, PEP, and AMP, but competitively inhibited by ATP, NADH, and Acetyl-CoA.

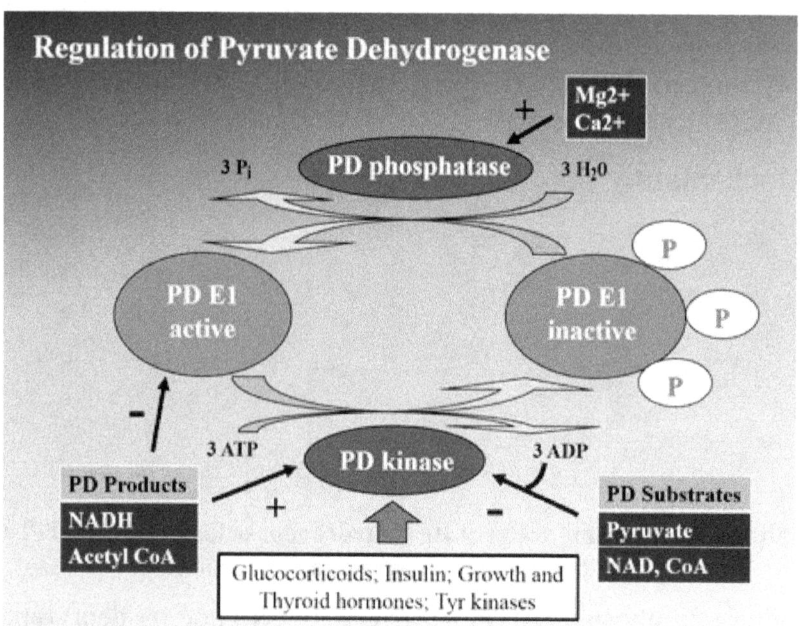

17.1.4 Active Site

E1 has two catalytic sites, each providing thiamine pyrophosphate (TPP) and magnesium ion as cofactors. The α-subunit binds magnesium ion and pyrophosphate fragment while the β-subunit binds pyrimidine fragment of TPP, forming together a catalytic site at the interface of subunits.

The active site for pyruvate dehydrogenase (image created from PDB: 1NI4) holds TPP through metal ligation to a magnesium ion (purple sphere) and through hydrogen bonding to amino acids. While over 20 amino acids can be found in the active site, amino acids Tyr 89, Arg 90, Gly 136, Val 138, Asp 167, Gly 168, Ala 169, Asn, 196, and His 263 actually participate in hydrogen bonding to hold TPP and pyruvate (not shown here) in the active site. The amino acids are shown as wires, and the TPP is in ball and stick form. The active site also aids in the transfer of the acyl on the TPP to a lipoamide waiting on E2.

17.1.5 Pathology

Pyruvate dehydrogenase is targeted by an autoantigen known as anti-mitochondrial antibodies (AMA), which results in progressive destruction of the small bile ducts of the liver, leading to primary biliary cirrhosis. These antibodies appear to recognize oxidized protein that has resulted from inflammatory immune responses. Some of these inflammatory responses could be related to gluten sensitivity as over 50% of the acute liver failure patients in one study exhibited a nonmitochondrial autoantibody against tissue transglutaminase. Other mitochondrial autoantigens include oxoglutarate dehydrogenase and branched-chain alpha-keto acid dehydrogenase complex, which are antigens recognized by anti-mitochondrial antibodies.

Increased pyruvate dehydrogenase (PDH) activity can cause oncogene-induced cellular senescence, as well as promoting aging. Decreased activity of mitochondrial PDH with age has been shown in the heart as well as in certain regions of the brain (the striatum and brainstem).

Pyruvate dehydrogenase (PDH) deficiency is a congenital degenerative metabolic disease resulting from a mutation of the pyruvate dehydrogenase complex (PDC) located on the X chromosome. While defects have been identified in all 3 enzymes of the complex, the E1-α subunit is predominantly the culprit. Malfunction of the citric acid cycle due to PDH deficiency deprives the body of energy and leads to an abnormal buildup of lactate. PDH deficiency is a common cause of lactic acidosis in newborns and often presents with severe lethargy, poor feeding, tachypnea, and cases of death have occurred.

17.1.6 Related Enzymes

In bacteria, a form of pyruvate dehydrogenase (also called pyruvate oxidase, EC 1.2.2.2) exists that links the oxidation of pyruvate into acetate and carbon dioxide to the reduction of ferrocytochrome. In *E. coli* this enzyme is encoded by the *pox B* gene and the protein has a flavin cofactor. This enzyme increases the efficiency of growth of *E. coli* under aerobic conditions.

17.2 THE CITRIC ACID CYCLE

The citric acid cycle—also known as the Krebs cycle, Szent–Györgyi–Krebs cycle, or TCA cycle (tricarboxylic acid cycle)—is a series of biochemical reactions to release the energy stored in nutrients through the oxidation of acetyl-CoA derived from carbohydrates, fats, proteins, and alcohol. The chemical energy released is available in the form of ATP. The Krebs cycle is used by organisms that respire (as opposed to organisms that ferment) to generate energy, either by anaerobic respiration or aerobic respiration. In addition, the cycle provides precursors of certain amino acids, as well as the reducing agent NADH, that are used in numerous other reactions. Its central importance to many biochemical pathways suggests that it was one of the earliest components of metabolism. Even though it is branded as a "cycle", it is not necessary for metabolites to follow only one specific route; at least three alternative segments of the citric acid cycle have been recognized.

The name of this metabolic pathway is derived from the citric acid (a tricarboxylic acid, often called citrate, as the ionized form predominates at biological pH) that is consumed and then regenerated by this sequence of reactions to complete the cycle. The cycle consumes acetate (in the form of acetyl-CoA) and water, reduces NAD^+ to NADH, releasing carbon dioxide. The NADH generated by the citric acid cycle is fed into the oxidative phosphorylation (electron transport) pathway. The net result of these two closely linked pathways is the oxidation of nutrients to produce usable chemical energy in the form of ATP.

In eukaryotic cells, the citric acid cycle occurs in the matrix of the mitochondrion. In prokaryotic cells, such as bacteria, which lack mitochondria, the citric acid cycle reaction sequence is performed in the cytosol with the proton gradient for ATP production being across the cell's surface (plasma membrane) rather than the inner membrane of the mitochondrion.

For each pyruvate molecule (from glycolysis), the overall yield of energy-containing compounds from the citric acid cycle is three NADH, one $FADH_2$, and one GTP.

17.2.1 Discovery

Several of the components and reactions of the citric acid cycle were established in the 1930s by the research of Albert Szent-Györgyi, who received the Nobel Prize in Physiology or Medicine in 1937 specifically for his discoveries pertaining to fumaric acid, a component of the cycle. He made this discovery by studying pigeon breast muscle. Because this tissue maintains its oxidative capacity well after breaking down in the Latapie mincer and releasing in aqueous solutions, breast muscle of the pigeon was very well qualified for the study of oxidative reactions.

The citric acid cycle itself was finally identified in 1937 by Hans Adolf Krebs and William Arthur Johnson while at the University of Sheffield, for which the former received the Nobel Prize for Physiology or Medicine in 1953, and for whom the cycle is sometimes named the "Krebs cycle".

17.2.2 Overview

The citric acid cycle is a metabolic pathway that connects carbohydrate, fat, and protein metabolism. The reactions of the cycle are carried out by eight enzymes that completely oxidize acetate (a two carbon molecule), in the form of acetyl-CoA, into two molecules each of carbon dioxide and water. Through catabolism of sugars, fats, and proteins, the two-carbon organic product acetyl-CoA is produced which enters the citric acid cycle. The reactions of the cycle also convert three equivalents of nicotinamide adenine dinucleotide (NAD^+) into three equivalents of reduced NAD (NADH), one equivalent of flavin adenine dinucleotide (FAD) into one equivalent of $FADH_2$, and one equivalent each of guanosine diphosphate (GDP) and inorganic phosphate (P_i) into one equivalent of guanosine triphosphate (GTP). The NADH and $FADH_2$ generated by the citric acid cycle are, in turn, used by the oxidative phosphorylation pathway to generate energy-rich ATP. One of the primary sources of acetyl-CoA is from the breakdown of sugars by glycolysis which yield pyruvate that in turn is decarboxylated by the pyruvate dehydrogenase complex generating acetyl-CoA according to the following reaction scheme:

$$\underset{\text{pyruvate}}{CH_3C(=O)C(=O)O^-} + HSCoA + NAD^+ \rightarrow \underset{\text{acetyl-CoA}}{CH_3C(=O)SCoA} + NADH + CO_2$$

The product of this reaction, acetyl-CoA, is the starting point for the citric acid cycle. Acetyl-CoA may also be obtained from the oxidation of fatty acids. Below is a schematic outline of the cycle:

- The citric acid cycle begins with the transfer of a two-carbon acetyl group from acetyl-CoA to the four-carbon acceptor compound (oxaloacetate) to form a six-carbon compound (citrate).
- The citrate then goes through a series of chemical transformations, losing two carboxyl groups as CO_2. The carbons lost as CO_2 originate from what was oxaloacetate, not directly from acetyl-CoA. The carbons donated by acetyl-CoA become part of the oxaloacetate carbon backbone after the first turn of the citric acid cycle. Loss of the acetyl-CoA-donated carbons as CO_2 requires several turns of the citric acid cycle. However, because of the role of the citric acid cycle in anabolism, they might not be lost, since many citric acid cycle intermediates are also used as precursors for the biosynthesis of other molecules.
- Most of the electrons made available by the oxidative steps of the cycle

are transferred to NAD^+, forming NADH. For each acetyl group that enters the citric acid cycle, three molecules of NADH are produced. The citric acid cycle includes a series of redox reactions in mitochondria.

- In addition, electrons from the succinate oxidation step are transferred first to the FAD cofactor of succinate dehydrogenase, reducing it to $FADH_2$, and eventually to ubiquinone (Q) in the mitochondrial membrane, reducing it to ubiquinol (QH_2) which is a substrate of the electron transfer chain at the level of Complex III.
- For every NADH and $FADH_2$ that are produced in the citric acid cycle, 2.5 and 1.5 ATP molecules are generated in oxidative phosphorylation, respectively.
- At the end of each cycle, the four-carbon oxaloacetate has been regenerated, and the cycle continues.

17.2.3 Evolution

It is believed that components of the citric acid cycle were derived from anaerobic bacteria, and that the TCA cycle itself may have evolved more than once. It may even predate biosis: the substrates appear to undergo most of the reactions spontaneously in the presence of persulfate radicals. Theoretically, several alternatives to the TCA cycle exist; however, the TCA cycle appears to be the most efficient. If several TCA alternatives had evolved independently, they all appear to have converged to the TCA cycle.

17.2.4 Steps

There are ten basic steps in the citric acid cycle, as outlined below. The cycle is continuously supplied with new carbon in the form of acetyl-CoA, entering at step 0 in the table.

	Reaction Type	Substrates	Enzyme	Products	Comment
0 / 10	Aldol condensation	Oxaloacetate + Acetyl CoA + H_2O	Citrate synthase	Citrate + CoA-SH	irreversible, extends the 4C oxaloacetate to a 6C molecule
1	Dehydration	Citrate	Aconitase	*cis*-Aconitate + H_2O	reversible isomerisation
2	Hydration	*cis*-Aconitate + H_2O		Isocitrate	
3	Oxidation	Isocitrate + NAD^+	Isocitrate dehydrogenase	Oxalosuccinate + NADH + H^+	generates NADH (equivalent of 2.5 ATP)
4	Decarboxylation	Oxalosuccinate		α-Ketoglutarate + CO_2	rate-limiting, irreversible stage, generates a 5C molecule

	Reaction Type	Substrates	Enzyme	Products	Comment
5	Oxidative decarboxylation	α-Ketoglutarate + NAD^+ + CoA-SH	α-Ketoglutarate dehydrogenase, Thiamine pyrophosphate, Lipoic acid, Mg++,transsuccinytase	Succinyl-CoA + NADH + H^+ + CO_2	irreversible stage, generates NADH (equivalent of 2.5 ATP), regenerates the 4C chain (CoA excluded)
6	substrate-level phosphorylation	Succinyl-CoA + GDP + P_i	Succinyl-CoA synthetase	Succinate + CoA-SH + GTP	or ADP→ATP instead of GDP→GTP, generates 1 ATP or equivalent. Condensation reaction of GDP + P_i and hydrolysis of succinyl-CoA involve the H_2O needed for balanced equation.
7	Oxidation	Succinate + ubiquinone (Q)	Succinate dehydrogenase	Fumarate + ubiquinol (QH_2)	uses FAD as a prosthetic group (FAD→$FADH_2$ in the first step of the reaction) in the enzyme. These two electrons are later transferred to QH_2 during Complex II of the ETC, where they generate the equivalent of 1.5 ATP
8	Hydration	Fumarate + H_2O	Fumarase	L-Malate	Hydration of C-C double bond
9	Oxidation	L-Malate + NAD^+	Malate dehydrogenase	Oxaloacetate + NADH + H^+	reversible (in fact, equilibrium favors malate), generates NADH (equivalent of 2.5 ATP)
10 / 0	Aldol condensation	Oxaloacetate + Acetyl CoA + H_2O	Citrate synthase	Citrate + CoA-SH	This is the same as step 0 and restarts the cycle. The reaction is irreversible and extends the 4C oxaloacetate to a 6C molecule

Two carbon atoms are oxidized to CO_2, the energy from these reactions is transferred to other metabolic processes through GTP (or ATP), and as electrons in NADH and QH_2. The NADH generated in the citric acid cycle may later be oxidized (donate its electrons) to drive ATP synthesis in a type of process called oxidative phosphorylation.

$FADH_2$ is covalently attached to succinate dehydrogenase, an enzyme which functions both in the citric acid cycle and the mitochondrial electron transport chain in oxidative phosphorylation. $FADH_2$, therefore, facilitates transfer of electrons to coenzyme Q, which is the final electron acceptor of the reaction catalyzed by the succinate:ubiquinone oxidoreductase complex, also acting as an intermediate in the electron transport chain.

Mitochondria in animals, including humans, possess two succinyl-CoA synthetases: one that produces GTP from GDP, and another that produces ATP from ADP. Plants have the type that produces ATP (ADP-forming succinyl-CoA synthetase). Several of the enzymes in the cycle may be loosely associated in a multienzyme protein complex within the mitochondrial matrix.

The GTP that is formed by GDP-forming succinyl-CoA synthetase may be utilized by nucleoside-diphosphate kinase to form ATP (the catalyzed reaction is GTP + ADP → GDP + ATP).

17.2.5 Products

Products of the first turn of the cycle are one GTP (or ATP), three NADH, one $FADH_2$ and two CO_2.

Because two acetyl-CoA molecules are produced from each glucose molecule, two cycles are required per glucose molecule. Therefore, at the end of two cycles, the products are: two GTP, six NADH, two $FADH_2$, and four CO_2.

Description	Reactants	Products
The sum of all reactions in the citric acid cycle is:	Acetyl-CoA + 3 NAD^+ + FAD + GDP + P_i + 2 H_2O	→ CoA-SH + 3 NADH + $FADH_2$ + 3 H^+ + GTP + 2 CO_2
Combining the reactions occurring during the pyruvate oxidation with those occurring during the citric acid cycle, the following overall pyruvate oxidation reaction is obtained:	Pyruvate ion + 4 NAD^+ + FAD + GDP + P_i + 2 H_2O	→ 4 NADH + $FADH_2$ + 4 H^+ + GTP + 3 CO_2
Combining the above reaction with the ones occurring in the course of glycolysis, the following overall glucose oxidation reaction (excluding reactions in the respiratory chain) is obtained:	Glucose + 10 NAD^+ + 2 FAD + 2 ADP + 2 GDP + 4 P_i + 2 H_2O	→ 10 NADH + 2 $FADH_2$ + 10 H^+ + 2 ATP + 2 GTP + 6 CO_2

The above reactions are balanced if P_i represents the $H_2PO_4^-$ ion, ADP and GDP the ADP^{2-} and GDP^{2-} ions, respectively, and ATP and GTP the ATP^{3-} and GTP^{3-} ions, respectively.

The total number of ATP molecules obtained after complete oxidation of one glucose in glycolysis, citric acid cycle, and oxidative phosphorylation is estimated to be between 30 and 38.

17.2.6 Efficiency

The theoretical maximum yield of ATP through oxidation of one molecule of glucose in glycolysis, citric acid cycle, and oxidative phosphorylation is 38 (assuming 3 molar equivalents of ATP per equivalent NADH and 2 ATP per $FADH_2$). In eukaryotes, two equivalents of NADH and two equivalents of ATP are generated in glycolysis, which takes place in the cytoplasm. If transported using the glycerol phosphate shuttle rather than the malate–aspartate shuttle, transport of two of these equivalents of NADH into the mitochondria effectively consumes two equivalents of ATP, thus reducing the net production of ATP to 36. Furthermore, inefficiencies in oxidative phosphorylation due to leakage of protons across the mitochondrial membrane and slippage of the ATP synthase/proton pump commonly reduces the ATP yield from NADH and $FADH_2$ to less than the theoretical maximum yield. The observed yields are, therefore, closer to ~2.5 ATP per NADH and ~1.5 ATP per $FADH_2$, further reducing the total net production of ATP to approximately 30. An assessment of the total ATP yield with newly revised proton-to-ATP ratios provides an estimate of 29.85 ATP per glucose molecule.

17.2.7 Variation

While the citric acid cycle is in general highly conserved, there is significant variability in the enzymes found in different taxa (note that the diagrams on this page are specific to the mammalian pathway variant).

Some differences exist between eukaryotes and prokaryotes. The conversion of D-*threo*-isocitrate to 2-oxoglutarate is catalyzed in eukaryotes by the NAD^+-dependent EC 1.1.1.41, while prokaryotes employ the $NADP^+$-dependent EC 1.1.1.42. Similarly, the conversion of (*S*)-malate to oxaloacetate is catalyzed in eukaryotes by the NAD^+-dependent EC 1.1.1.37, while most prokaryotes utilize a quinone-dependent enzyme, EC 1.1.5.4.

A step with significant variability is the conversion of succinyl-CoA to succinate. Most organisms utilize EC 6.2.1.5, succinate–CoA ligase (ADP-forming) (despite its name, the enzyme operates in the pathway in the direction of ATP formation). In mammals a GTP-forming enzyme, succinate–CoA ligase (GDP-forming) (EC 6.2.1.4) also operates. The level of utilization of each isoform is tissue dependent. In some acetate-producing bacteria, such as *Acetobacter aceti*, an entirely different enzyme catalyzes this conversion – EC 2.8.3.18, succinyl-CoA:acetate CoA-transferase. This specialized enzyme links the TCA cycle with acetate metabolism in these organisms. Some bacteria, such as *Helicobacter pylori*, employ yet another enzyme for this conversion – succinyl-CoA:acetoacetate CoA-transferase (EC 2.8.3.5).

Some variability also exists at the previous step – the conversion of 2-oxoglutarate to succinyl-CoA. While most organisms utilize the ubiquitous NAD^+-dependent 2-oxoglutarate dehydrogenase, some bacteria utilize a ferredoxin-dependent 2-oxoglutarate synthase (EC 1.2.7.3). Other organisms, including obligately autotrophic and methanotrophic bacteria and archaea, bypass succinyl-CoA entirely, and convert 2-oxoglutarate to succinate via succinate semialdehyde, using EC 4.1.1.71, 2-oxoglutarate decarboxylase, and EC 1.2.1.79, succinate-semialdehyde dehydrogenase.

In cancer, there are substantial metabolic derangements that occur to ensure the proliferation of tumor cells, and consequently metabolites can accumulate which serve to facilitate tumorigenesis, dubbed oncometabolites.

Among the best characterized oncometabolites is 2-hydroxyglutarate which is produced through a heterozygous gain-of-function mutation (specifically a neomorphic one) in isocitrate dehydrogenase (IDH) (which under normal circumstances catalyzes the oxidation of isocitrate to oxalosuccinate, which then spontaneously decarboxylates to alpha-ketoglutarate, as discussed above; in this case an additional reduction step occurs after the formation of alpha-ketoglutarate via NADPH to yield 2-hydroxyglutarate), and hence IDH is considered an oncogene. Under physiological conditions, 2-hydroxyglutarate is a minor product of several metabolic pathways as an error but readily converted to alpha-ketoglutarate via hydroxyglutarate dehydrogenase enzymes (L2HGDH and D2HGDH) but does not have a known physiologic role in mammalian cells; of note, in cancer, 2-hydroxyglutarate is likely a terminal metabolite as isotope labelling experiments of colorectal cancer cell lines show that its conversion back to alpha-ketoglutarate is too low to measure.

In cancer, 2-hydroxyglutarate serves as a competitive inhibitor for a number of enzymes that facilitate reactions via alpha-ketoglutarate in alpha-ketoglutarate-dependent dioxygenases. This mutation results in several important changes to the metabolism of the cell. For one thing, because there is an extra NADPH-catalyzed reduction, this can contribute to depletion of cellular stores of NADPH and also reduce levels of alpha-ketoglutarate available to the cell. In particular, the depletion of NADPH is problematic because NADPH is highly compartmentalized and cannot freely diffuse between the organelles in the cell. It is produced largely via the pentose phosphate pathway in the cytoplasm.

The depletion of NADPH results in increased oxidative stress within the cell as it is a required cofactor in the production of GSH, and this oxidative stress can result in DNA damage. There are also changes on the genetic and epigenetic level through the function of histone lysine demethylases (KDMs) and ten-eleven translocation (TET) enzymes; ordinarily TETs hydroxylate 5-methylcytosines to prime them for demethylation.

However, in the absence of alpha-ketoglutarate this cannot be done and there is hence hypermethylation of the cell's DNA, serving to promote epithelial-mesenchymal transition (EMT) and inhibit cellular differentiation. A similar phenomenon is observed for the Jumonji C family of KDMs which require a hydroxylation to perform demethylation at the epsilon-amino methyl group. Additionally, the inability of prolyl hydroxylases to catalyze reactions results in stabilization of hypoxia-inducible factor alpha, which is necessary to promote degradation of the latter (as under conditions of low oxygen there will not be adequate substrate for hydroxylation). This results in a pseudohypoxic phenotype in the cancer cell that promotes angiogenesis, metabolic reprogramming, cell growth, and migration.

17.2.8 Regulation

Allosteric Regulation by Metabolites

The regulation of the citric acid cycle is largely determined by product inhibition and substrate availability. If the cycle were permitted to run unchecked, large amounts of metabolic energy could be wasted in overproduction of reduced coenzyme such as NADH and ATP. The major eventual substrate of the cycle is ADP which gets converted to ATP. A reduced amount of ADP causes accumulation of precursor NADH which in turn can inhibit a number of enzymes. NADH, a product of all dehydrogenases in the citric acid cycle with the exception of succinate dehydrogenase, inhibits pyruvate dehydrogenase, isocitrate dehydrogenase, α-ketoglutarate dehydrogenase, and also citrate synthase. Acetyl-coA inhibits pyruvate dehydrogenase, while succinyl-CoA inhibits alpha-ketoglutarate dehydrogenase and citrate synthase. When tested in vitro with TCA enzymes, ATP inhibits citrate synthase and α-ketoglutarate dehydrogenase; however, ATP levels do not change more than 10% in vivo between rest and vigorous exercise. There is no known allosteric mechanism that can account for large changes in reaction rate from an allosteric effector whose concentration changes less than 10%.

Citrate

Citrate is used for feedback inhibition, as it inhibits phosphofructokinase, an enzyme involved in glycolysis that catalyses formation of fructose 1,6-bisphosphate, a precursor of pyruvate. This prevents a constant high rate of flux when there is an accumulation of citrate and a decrease in substrate for the enzyme.

Regulation by Calcium

Calcium is also used as a regulator in the citric acid cycle. Calcium levels in the mitochondrial matrix can reach up to the tens of micromolar levels during cellular activation. It activates pyruvate dehydrogenase phosphatase which in turn activates the pyruvate dehydrogenase complex. Calcium also activates isocitrate dehydrogenase and α-ketoglutarate dehydrogenase. This increases the reaction rate of many of the steps in the cycle, and therefore increases flux throughout the pathway.

Transcriptional Regulation

There is a link between intermediates of the citric acid cycle and the regulation of hypoxia-inducible factors (HIF). HIF plays a role in the regulation of oxygen homeostasis, and is a transcription factor that targets angiogenesis, vascular remodeling, glucose utilization, iron transport and apoptosis. HIF is synthesized constitutively, and hydroxylation of at least one of two critical proline residues mediates their interaction with the von Hippel Lindau E3 ubiquitin ligase complex, which targets them for rapid degradation. This reaction is catalysed by prolyl 4-hydroxylases. Fumarate and succinate have been identified as potent inhibitors of prolyl hydroxylases, thus leading to the stabilisation of HIF.

17.2.9 Major Metabolic Pathways Converging on the Citric Acid Cycle

Several catabolic pathways converge on the citric acid cycle. Most of these reactions add intermediates to the citric acid cycle, and are therefore known as anaplerotic reactions, from the Greek meaning to "fill up". These increase the amount of acetyl CoA that the cycle is able to carry, increasing the mitochondrion's capability to carry out respiration if this is otherwise a limiting factor. Processes that remove intermediates from the cycle are termed "cataplerotic" reactions.

Pyruvate molecules produced by glycolysis are actively transported across the inner mitochondrial membrane, and into the matrix. Here they can be oxidized and combined with coenzyme A to form CO_2, *acetyl-CoA*, and NADH, as in the normal cycle.

However, it is also possible for pyruvate to be carboxylated by pyruvate carboxylase to form *oxaloacetate*. This latter reaction "fills up" the amount of *oxaloacetate* in the citric acid cycle, and is therefore an anaplerotic reaction, increasing the cycle's capacity to metabolize *acetyl-CoA* when the tissue's energy needs (e.g. in muscle) are suddenly increased by activity.

In the citric acid cycle all the intermediates (e.g. *citrate, iso-citrate, alpha-ketoglutarate, succinate, fumarate, malate*, and *oxaloacetate*) are regenerated during each turn of the cycle. Adding more of any of these intermediates to the mitochondrion therefore means that that additional amount is retained within the cycle, increasing all the other intermediates as one is converted into the other. Hence the addition of any one of them to the cycle has an anaplerotic effect, and its removal has a cataplerotic effect. These anaplerotic and cataplerotic reactions will, during the course of the cycle, increase or decrease the amount of *oxaloacetate* available to combine with *acetyl-CoA* to form *citric acid*. This in turn increases or decreases the rate of ATP production by the mitochondrion, and thus the availability of ATP to the cell.

Acetyl-CoA, on the other hand, derived from pyruvate oxidation, or from the beta-oxidation of fatty acids, is the only fuel to enter the citric acid cycle. With each turn of the cycle one molecule of *acetyl-CoA* is consumed for every molecule of *oxaloacetate* present in the mitochondrial matrix, and is never regenerated. It is the oxidation of the acetate portion of *acetyl-CoA* that produces CO_2 and water, with the energy thus released captured in the form of ATP. The three steps of beta-oxidation resemble the steps that occur in the production of oxaloacetate from succinate in the TCA cycle. Acyl-CoA is oxidized to trans-Enoyl-CoA while FAD is reduced to $FADH_2$, which is similar to the oxidation of succinate to fumarate. Following, trans-enoyl-CoA is hydrated across the double bond to beta-hydroxyacyl-CoA, just like fumarate is hydrated to malate. Lastly, beta-hydroxyacyl-CoA is oxidized to beta-ketoacyl-CoA while NAD+ is reduced to NADH, which follows the same process as the oxidation of malate to oxaloacetate.

In the liver, the carboxylation of cytosolic pyruvate into intra-mitochondrial *oxaloacetate* is an early step in the gluconeogenic pathway which converts lactate and de-aminated alanine into glucose, under the influence of high levels of glucagon and/or epinephrine in the blood. Here the addition of *oxaloacetate* to the mitochondrion does not have a net anaplerotic effect, as another citric acid cycle intermediate (*malate*) is immediately removed from the mitochondrion to be converted into cytosolic oxaloacetate, which is ultimately converted into glucose, in a process that is almost the reverse of glycolysis.

In protein catabolism, proteins are broken down by proteases into their constituent amino acids. Their carbon skeletons (i.e. the de-aminated amino acids) may either enter the citric acid cycle as intermediates (e.g. *alpha-ketoglutarate* derived from glutamate or glutamine), having an anaplerotic effect on the cycle, or, in the case of leucine, isoleucine, lysine, phenylalanine, tryptophan, and tyrosine, they are converted into *acetyl-CoA* which can be burned to CO_2 and water, or used to form ketone bodies, which too can only be burned in tissues other than the liver where they are formed, or excreted via the urine or breath.

These latter amino acids are therefore termed "ketogenic" amino acids, whereas those that enter the citric acid cycle as intermediates can only be cataplerotically removed by entering the gluconeogenic pathway via *malate* which is transported out of the mitochondrion to be converted into cytosolic oxaloacetate and ultimately into glucose.

These are the so-called "glucogenic" amino acids. De-aminated alanine, cysteine, glycine, serine, and threonine are converted to pyruvate and can consequently either enter the citric acid cycle as *oxaloacetate* (an anaplerotic reaction) or as *acetyl-CoA* to be disposed of as CO_2 and water.

In fat catabolism, triglycerides are hydrolyzed to break them into fatty acids and glycerol. In the liver the glycerol can be converted into glucose via dihydroxyacetone phosphate and glyceraldehyde-3-phosphate by way of gluconeogenesis. In skeletal muscle, glycerol is used in glycolysis by converting glycerol into glycerol-3-phosphate, then into dihydroxyacetone phosphate (DHAP), then into glyceraldehyde-3-phosphate.

In many tissues, especially heart and skeletal muscle tissue, fatty acids are broken down through a process known as beta oxidation, which results in the production of mitochondrial *acetyl-CoA*, which can be used in the citric acid cycle. Beta oxidation of fatty acids with an odd number of methylene bridges produces propionyl-CoA, which is then converted into *succinyl-CoA* and fed into the citric acid cycle as an anaplerotic intermediate.

The total energy gained from the complete breakdown of one (six-carbon) molecule of glucose by glycolysis, the formation of 2 *acetyl-CoA* molecules, their catabolism in the citric acid cycle, and oxidative phosphorylation equals about 30 ATP molecules, in eukaryotes. The number of ATP molecules derived from the beta oxidation of a 6 carbon segment of a fatty acid chain, and the subsequent oxidation of the resulting 3 molecules of *acetyl-CoA* is 40.

17.2.10 Citric Acid Cycle Intermediates Serve as Substrates for Biosynthetic Processes

Several of the citric acid cycle intermediates are used for the synthesis of important compounds, which will have significant cataplerotic effects on the cycle. *Acetyl-CoA* cannot be transported out of the mitochondrion. To obtain cytosolic acetyl-CoA, *citrate* is removed from the citric acid cycle and carried across the inner mitochondrial membrane into the cytosol. There it is cleaved by ATP citrate lyase into acetyl-CoA and oxaloacetate.

The oxaloacetate is returned to mitochondrion as *malate* (and then converted back into *oxaloacetate* to transfer more *acetyl-CoA* out of the mitochondrion). The cytosolic acetyl-CoA is used for fatty acid synthesis and the production of cholesterol. Cholesterol can, in turn, be used to synthesize the steroid hormones, bile salts, and vitamin D.

The carbon skeletons of many non-essential amino acids are made from citric acid cycle intermediates. To turn them into amino acids the alpha keto-acids formed from the citric acid cycle intermediates have to acquire their amino groups from glutamate in a transamination reaction, in which pyridoxal phosphate is a cofactor. In this reaction the glutamate is converted into *alpha-ketoglutarate*, which is a citric acid cycle intermediate. The intermediates that can provide the carbon skeletons for amino acid synthesis are *oxaloacetate* which forms aspartate and asparagine; and *alpha-ketoglutarate* which forms glutamine, proline, and arginine.

Of these amino acids, aspartate and glutamine are used, together with carbon and nitrogen atoms from other sources, to form the purines that are used as the bases in DNA and RNA, as well as in ATP, AMP, GTP, NAD, FAD and CoA.

The pyrimidines are partly assembled from aspartate (derived from *oxaloacetate*). The pyrimidines, thymine, cytosine and uracil, form the complementary bases to the purine bases in DNA and RNA, and are also components of CTP, UMP, UDP and UTP. The majority of the carbon atoms in the porphyrins come from the citric acid cycle intermediate, *succinyl-CoA*. These molecules are an important component of the hemoproteins, such as hemoglobin, myoglobin and various cytochromes.

During gluconeogenesis mitochondrial *oxaloacetate* is reduced to *malate* which is then transported out of the mitochondrion, to be oxidized back to oxaloacetate in the cytosol. Cytosolic oxaloacetate is then decarboxylated to phosphoenolpyruvate by phosphoenolpyruvate carboxykinase, which is the rate limiting step in the conversion of nearly all the gluconeogenic precursors (such as the glucogenic amino acids and lactate) into glucose by the liver and kidney.

17.2.11 Glucose Feeds the TCA Cycle via Circulating Lactate

The metabolic role of lactate is well recognized as a fuel for tissues, mitochondrial cytopathies such as DPH Cytopathy, and the scientific field of oncology (tumors). In the classical Cori cycle, muscles produce lactate which is then taken up by the liver for gluconeogenesis. New studies suggest that lactate can be used as a source of carbon for the TCA cycle.

CHAPTER 18
OXIDATIVE PHOSPHORYLATION

18.1 AN INTRODUCTION

Oxidative phosphorylation or electron transport-linked phosphorylation or terminal oxidation is the metabolic pathway in which cells use enzymes to oxidize nutrients, thereby releasing chemical energy in order to produce adenosine triphosphate (ATP). In eukaryotes, this takes place inside mitochondria. Almost all aerobic organisms carry out oxidative phosphorylation. This pathway is so pervasive because it releases more energy than alternative fermentation processes such as anaerobic glycolysis.

The energy stored in the chemical bonds of glucose is released by the cell in the citric acid cycle, producing carbon dioxide and the energetic electron donors NADH and FADH. Oxidative phosphorylation uses these molecules and O_2 to produce ATP, which is used throughout the cell whenever energy is needed. During oxidative phosphorylation, electrons are transferred from the electron donors to a series of electron acceptors in a series of redox reactions ending in oxygen, whose reaction releases half of the total energy.

In eukaryotes, these redox reactions are catalyzed by a series of protein complexes within the inner membrane of the cell's mitochondria, whereas, in prokaryotes, these proteins are located in the cell's outer membrane. These linked sets of proteins are called the electron transport chain. In eukaryotes, five main protein complexes are involved, whereas in prokaryotes many different enzymes are present, using a variety of electron donors and acceptors.

The energy transferred by electrons flowing through this electron transport chain is used to transport protons across the inner mitochondrial membrane, in a process called *electron transport*. This generates potential energy in the form of a pH gradient and the resulting electrical potential across this membrane. This store of energy is tapped when protons flow back across the membrane and down the potential energy gradient, through a large enzyme called ATP synthase in a process called chemiosmosis. The ATP synthase uses the energy to transform adenosine diphosphate (ADP) into adenosine triphosphate, in a phosphorylation reaction. The reaction is driven by the proton flow, which forces the rotation of a part of the enzyme. The ATP synthase is a rotary mechanical motor.

Although oxidative phosphorylation is a vital part of metabolism, it produces reactive oxygen species such as superoxide and hydrogen peroxide, which lead to propagation of free radicals, damaging cells and contributing to disease and, possibly, aging and senescence. The enzymes carrying out this metabolic pathway are also the target of many drugs and poisons that inhibit their activities.

18.2 HISTORY

The field of oxidative phosphorylation began with the report in 1906 by Arthur Harden of a vital role for phosphate in cellular fermentation, but initially only sugar phosphates were known to be involved. However, in the early 1940s, the link between the oxidation of sugars and the generation of ATP was firmly established by Herman Kalckar, confirming the central role of ATP in energy transfer that had been proposed by Fritz Albert Lipmann in 1941. Later, in 1949, Morris Friedkin and Albert L. Lehninger proved that the coenzyme NADH linked metabolic pathways such as the citric acid cycle and the synthesis of ATP. The term *oxidative phosphorylation* was coined by Volodymyr Belitser [uk] in 1939.

For another twenty years, the mechanism by which ATP is generated remained mysterious, with scientists searching for an elusive "high-energy intermediate" that would link oxidation and phosphorylation reactions. This puzzle was solved by Peter D. Mitchell with the publication of the chemiosmotic theory in 1961. At first, this proposal was highly controversial, but it was slowly accepted and Mitchell was awarded a Nobel prize in 1978. Subsequent research concentrated on purifying and characterizing the enzymes involved, with major contributions being made by David E. Green on the complexes of the electron-transport chain, as well as Efraim Racker on the ATP synthase. A critical step towards solving the mechanism of the ATP synthase was provided by Paul D. Boyer, by his development in 1973 of the "binding change" mechanism, followed by his radical proposal of rotational catalysis in 1982. More recent work has included structural studies on the enzymes involved in oxidative phosphorylation by John E. Walker, with Walker and Boyer being awarded a Nobel Prize in 1997.

18.3 CHEMIOSMOSIS

Oxidative phosphorylation works by using energy-releasing chemical reactions to drive energy-requiring reactions. The two sets of reactions are said to be *coupled.* This means one cannot occur without the other. The chain of redox reactions driving the flow of electrons through the electron transport chain, from electron donors such as NADH to electron acceptors such as oxygen and hydrogen (protons), is an exergonic process – it releases energy, whereas the synthesis of ATP is an endergonic process, which requires an input of energy. Both the electron transport chain and the ATP synthase are embedded in a membrane, and energy is transferred from the electron transport chain to the ATP synthase by movements of protons across this membrane, in a process called *chemiosmosis*. A current of protons is driven from the negative N-side of the membrane to the positive P-side through the proton-pumping enzymes of the electron transport chain. The movement of protons creates an electrochemical gradient across the membrane, is called the proton-motive force. It has two components: a difference in proton concentration (a H^+ gradient, ΔpH) and a difference in electric potential, with the N-side having a negative charge.

ATP synthase releases this stored energy by completing the circuit and allowing protons to flow down the electrochemical gradient, back to the N-side of the membrane. The electrochemical gradient drives the rotation of part of the enzyme's structure and couples this motion to the synthesis of ATP.

The two components of the proton-motive force are thermodynamically equivalent: In mitochondria, the largest part of energy is provided by the potential; in alkaliphile bacteria the electrical energy even has to compensate for a counteracting inverse pH difference. Inversely, chloroplasts operate mainly on ΔpH. However, they also require a small membrane potential for the kinetics of ATP synthesis. In the case of the fusobacterium *Propionigenium modestum* it drives the counter-rotation of subunits a and c of the F_0 motor of ATP synthase.

The amount of energy released by oxidative phosphorylation is high, compared with the amount produced by anaerobic fermentation. Glycolysis produces only 2 ATP molecules, but somewhere between 30 and 36 ATPs are produced by the oxidative phosphorylation of the 10 NADH and 2 succinate molecules made by converting one molecule of glucose to carbon dioxide and water, while each cycle of beta oxidation of a fatty acid yields about 14 ATPs.

These ATP yields are theoretical maximum values; in practice, some protons leak across the membrane, lowering the yield of ATP.

18.4 ELECTRON AND PROTON TRANSFER MOLECULES

The electron transport chain carries both protons and electrons, passing electrons from donors to acceptors, and transporting protons across a membrane.

These processes use both soluble and protein-bound transfer molecules. In the mitochondria, electrons are transferred within the intermembrane space by the water-soluble electron transfer protein cytochrome c.

This carries only electrons, and these are transferred by the reduction and oxidation of an iron atom that the protein holds within a heme group in its structure. Cytochrome c is also found in some bacteria, where it is located within the periplasmic space.

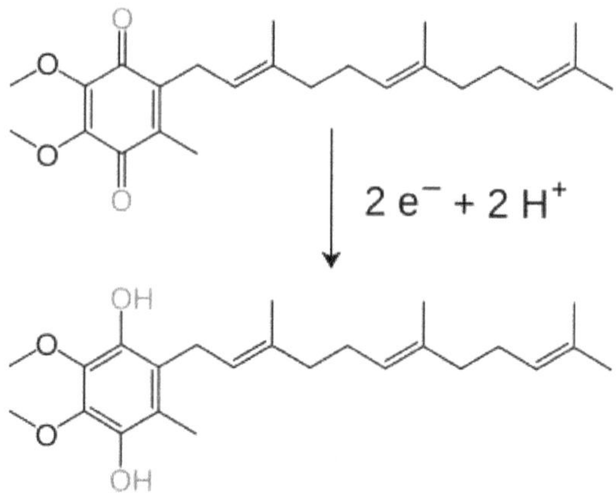

Fig 1: Reduction of Coenzyme Q from its Ubiquinone form (Q) to the Reduced Ubiquinol form (QH_2)

Within the inner mitochondrial membrane, the lipid-soluble electron carrier coenzyme Q10 (Q) carries both electrons and protons by a redox cycle. This small benzoquinone molecule is very hydrophobic, so it diffuses freely within the membrane. When Q accepts two electrons and two protons, it becomes reduced to the *ubiquinol* form (QH_2); when QH_2 releases two electrons and two protons, it becomes oxidized back to the *ubiquinone* (Q) form. As a result, if two enzymes are arranged so that Q is reduced on one side of the membrane and QH_2 oxidized on the other, ubiquinone will couple these reactions and shuttle protons across the membrane. Some bacterial electron transport chains use different quinones, such as menaquinone, in addition to ubiquinone.

Within proteins, electrons are transferred between flavin cofactors, iron–sulfur clusters and cytochromes. There are several types of iron–sulfur cluster. The simplest kind found in the electron transfer chain consists of two iron atoms joined by two atoms of inorganic sulfur; these are called [2Fe–2S] clusters. The second kind, called [4Fe–4S], contains a cube of four iron atoms and four sulfur atoms. Each iron atom in these clusters is coordinated by an additional amino acid, usually by the sulfur atom of cysteine. Metal ion cofactors undergo redox reactions without binding or releasing protons, so in the electron transport chain they serve solely to transport electrons through proteins. Electrons move quite long distances through proteins by hopping along chains of these cofactors. This occurs by quantum tunnelling, which is rapid over distances of less than 1.4×10^{-9} m.

18.5 EUKARYOTIC ELECTRON TRANSPORT CHAINS

Many catabolic biochemical processes, such as glycolysis, the citric acid cycle, and beta oxidation, produce the reduced coenzyme NADH. This coenzyme contains electrons that have a high transfer potential; in other words, they will release a large amount of energy upon oxidation. However, the cell does not release this energy all at once, as this would be an uncontrollable reaction. Instead, the electrons are removed from NADH and passed to oxygen through a series of enzymes that each release a small amount of the energy. This set of enzymes, consisting of complexes I through IV, is called the electron transport chain and is found in the inner membrane of the mitochondrion. Succinate is also oxidized by the electron transport chain, but feeds into the pathway at a different point.

In eukaryotes, the enzymes in this electron transport system use the energy released from O_2 by NADH to pump protons across the inner membrane of the mitochondrion. This causes protons to build up in the intermembrane space, and generates an electrochemical gradient across the membrane. The energy stored in this potential is then used by ATP synthase to produce ATP. Oxidative phosphorylation in the eukaryotic mitochondrion is the best-understood example of this process. The mitochondrion is present in almost all eukaryotes, with the exception of anaerobic protozoa such as *Trichomonas vaginalis* that instead reduce protons to hydrogen in a remnant mitochondrion called a hydrogenosome.

Table 1: Typical Respiratory Enzymes and Substrates in Eukaryotes

Respiratory Enzyme	Redox Pair	Midpoint Potential (Volts)
NADH dehydrogenase	NAD^+ / NADH	−0.32
Succinate dehydrogenase	FMN or FAD / $FMNH_2$ or $FADH_2$	−0.20
Cytochrome bc_1 complex	Coenzyme $Q10_{ox}$ / Coenzyme $Q10_{red}$	+0.06
Cytochrome bc_1 complex	Cytochrome b_{ox} / Cytochrome b_{red}	+0.12
Complex IV	Cytochrome c_{ox} / Cytochrome c_{red}	+0.22
Complex IV	Cytochrome a_{ox} / Cytochrome a_{red}	+0.29
Complex IV	O_2 / HO^-	+0.82
Conditions: pH = 7		

18.5.1 NADH-Coenzyme Q Oxidoreductase (Complex I)

NADH-coenzyme Q oxidoreductase, also known as *NADH dehydrogenase* or *complex I*, is the first protein in the electron transport chain. Complex I is a giant enzyme with the mammalian complex I having 46 subunits and a molecular mass of about 1,000 kilodaltons (kDa). The structure is known in detail only from a bacterium; in most organisms the complex resembles a boot with a large "ball" poking out from the membrane into the mitochondrion. The genes that encode the individual proteins are contained in both the cell nucleus and the mitochondrial genome, as is the case for many enzymes present in the mitochondrion.

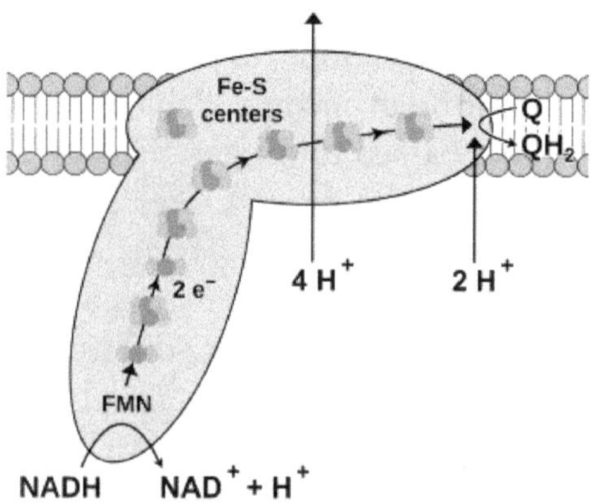

Fig 2: Complex I or NADH-Q Oxidoreductase

The reaction that is catalyzed by this enzyme is the two electron oxidation of NADH by coenzyme Q10 or *ubiquinone* (represented as Q in the equation below), a lipid-soluble quinone that is found in the mitochondrion membrane:

$$NADH + Q + 5H^{+}{}_{matrix} \rightarrow NAD^{+} + QH_{2} + 4H^{+}{}_{intermembrane}$$

The start of the reaction, and indeed of the entire electron chain, is the binding of a NADH molecule to complex I and the donation of two electrons. The electrons enter complex I via a prosthetic group attached to the complex, flavin mononucleotide (FMN). The addition of electrons to FMN converts it to its reduced form, $FMNH_2$. The electrons are then transferred through a series of iron–sulfur clusters: the second kind of prosthetic group present in the complex. There are both [2Fe–2S] and [4Fe–4S] iron–sulfur clusters in complex I.

As the electrons pass through this complex, four protons are pumped from the matrix into the intermembrane space. Exactly how this occurs is unclear, but it seems to involve conformational changes in complex I that cause the protein to bind protons on the N-side of the membrane and release them on the P-side of the membrane. Finally, the electrons are transferred from the chain of iron–sulfur clusters to a ubiquinone molecule in the membrane. Reduction of ubiquinone also contributes to the generation of a proton gradient, as two protons are taken up from the matrix as it is reduced to ubiquinol (QH_2).

18.5.2 Succinate-Q Oxidoreductase (Complex II)

Succinate-Q oxidoreductase, also known as *complex II* or *succinate dehydrogenase*, is a second entry point to the electron transport chain. It is unusual because it is the only enzyme that is part of both the citric acid cycle and the electron transport chain. Complex II consists of four protein subunits and contains a bound flavin adenine dinucleotide (FAD) cofactor, iron–sulfur clusters, and a heme group that does not participate in electron transfer to coenzyme Q, but is believed to be important in decreasing production of reactive oxygen species. It oxidizes succinate to fumarate and reduces ubiquinone.

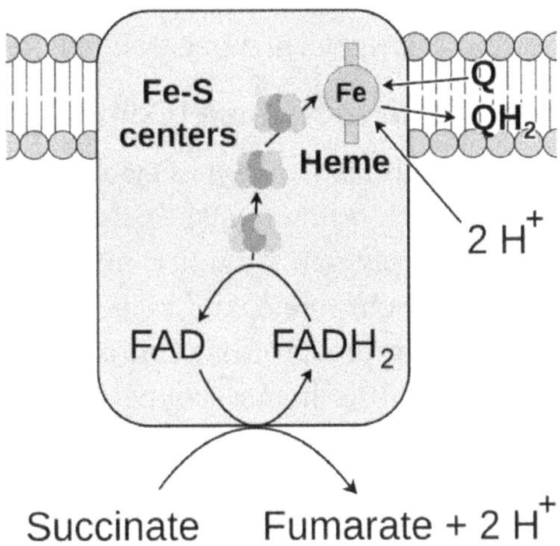

Fig 3: Complex II: Succinate-Q Oxidoreductase.

As this reaction releases less energy than the oxidation of NADH, complex II does not transport protons across the membrane and does not contribute to the proton gradient.

$$Succinate + Q \rightarrow Fumarate + QH_2$$

In some eukaryotes, such as the parasitic worm *Ascaris suum*, an enzyme similar to complex II, fumarate reductase (menaquinol:fumarate oxidoreductase, or QFR), operates in reverse to oxidize ubiquinol and reduce fumarate. This allows the worm to survive in the anaerobic environment of the large intestine, carrying out anaerobic oxidative phosphorylation with fumarate as the electron acceptor. Another unconventional function of complex II is seen in the malaria parasite *Plasmodium falciparum*. Here, the reversed action of complex II as an oxidase is important in regenerating ubiquinol, which the parasite uses in an unusual form of pyrimidine biosynthesis.

18.5.3 Electron Transfer Flavoprotein-Q Oxidoreductase

Electron transfer flavoprotein-ubiquinone oxidoreductase (ETF-Q oxidoreductase), also known as *electron transferring-flavoprotein dehydrogenase*, is a third entry point to the electron transport chain. It is an enzyme that accepts electrons from electron-transferring flavoprotein in the mitochondrial matrix, and uses these electrons to reduce ubiquinone. This enzyme contains a flavin and a [4Fe–4S] cluster, but, unlike the other respiratory complexes, it attaches to the surface of the membrane and does not cross the lipid bilayer.

$$ETF_{red} + Q \rightarrow ETF_{ox} + QH_2$$

In mammals, this metabolic pathway is important in beta oxidation of fatty acids and catabolism of amino acids and choline, as it accepts electrons from multiple acetyl-CoA dehydrogenases. In plants, ETF-Q oxidoreductase is also important in the metabolic responses that allow survival in extended periods of darkness.

18.5.4 Q-Cytochrome C Oxidoreductase (Complex III)

Q-cytochrome c oxidoreductase is also known as *cytochrome c reductase, cytochrome* bc_1 *complex*, or simply *complex III*. In mammals, this enzyme is a dimer, with each subunit complex containing 11 protein subunits, an [2Fe-2S] iron–sulfur cluster and three cytochromes: one cytochrome c_1 and two b cytochromes. A cytochrome is a kind of electron-transferring protein that contains at least one heme group. The iron atoms inside complex III's heme groups alternate between a reduced ferrous (+2) and oxidized ferric (+3) state as the electrons are transferred through the protein.

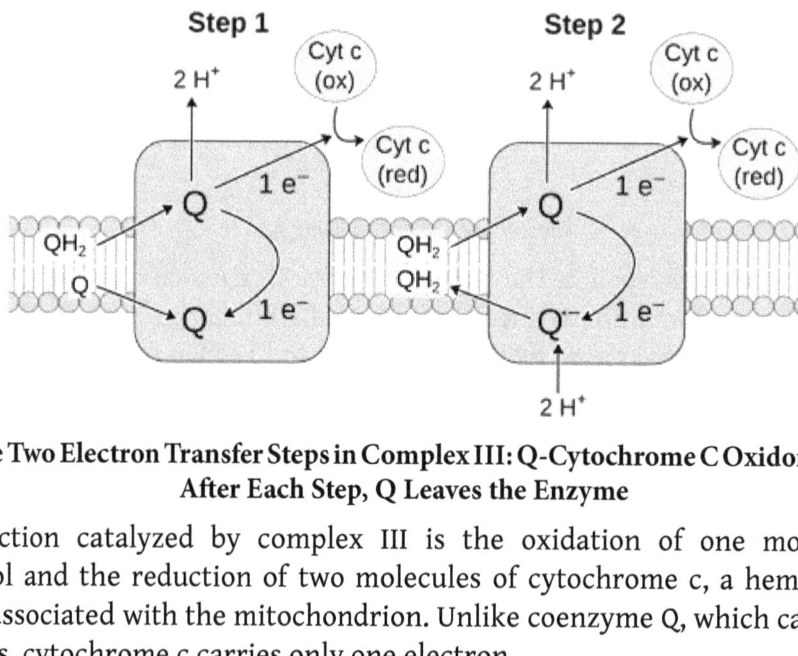

Fig 4: The Two Electron Transfer Steps in Complex III: Q-Cytochrome C Oxidoreductase. After Each Step, Q Leaves the Enzyme

The reaction catalyzed by complex III is the oxidation of one molecule of ubiquinol and the reduction of two molecules of cytochrome c, a heme protein loosely associated with the mitochondrion. Unlike coenzyme Q, which carries two electrons, cytochrome c carries only one electron.

$$QH_2 + 2Cyt\,c_{ox} + 2H^+{}_{matrix} \rightarrow Q + 2Cyt\,c_{red} + 4H^+{}_{intermembrane}$$

As only one of the electrons can be transferred from the QH_2 donor to a cytochrome c acceptor at a time, the reaction mechanism of complex III is more elaborate than those of the other respiratory complexes, and occurs in two steps called the Q cycle. In the first step, the enzyme binds three substrates, first, QH_2, which is then oxidized, with one electron being passed to the second substrate, cytochrome c. The two protons released from QH_2 pass into the intermembrane space. The third substrate is Q, which accepts the second electron from the QH_2 and is reduced to

$Q^{\cdot-}$, which is the ubisemiquinone free radical. The first two substrates are released, but this ubisemiquinone intermediate remains bound. In the second step, a second molecule of QH_2 is bound and again passes its first electron to a cytochrome c acceptor. The second electron is passed to the bound ubisemiquinone, reducing it to QH_2 as it gains two protons from the mitochondrial matrix. This QH_2 is then released from the enzyme.

As coenzyme Q is reduced to ubiquinol on the inner side of the membrane and oxidized to ubiquinone on the other, a net transfer of protons across the membrane occurs, adding to the proton gradient. The rather complex two-step mechanism by which this occurs is important, as it increases the efficiency of proton transfer. If, instead of the Q cycle, one molecule of QH_2 were used to directly reduce two molecules of cytochrome c, the efficiency would be halved, with only one proton transferred per cytochrome c reduced.

18.5.5 Cytochrome C Oxidase (Complex IV)

Cytochrome c oxidase, also known as *complex IV*, is the final protein complex in the electron transport chain. The mammalian enzyme has an extremely complicated structure and contains 13 subunits, two heme groups, as well as multiple metal ion cofactors – in all, three atoms of copper, one of magnesium and one of zinc.

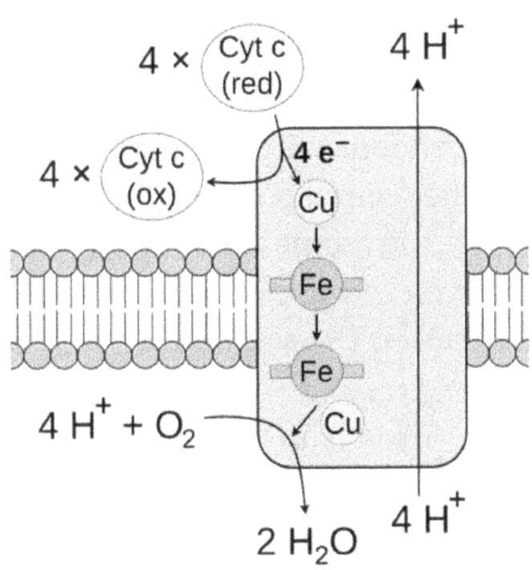

Fig 5: Complex IV: Cytochrome C Oxidase.

This enzyme mediates the final reaction in the electron transport chain and transfers electrons to oxygen and hydrogen (protons), while pumping protons across the membrane.

The final electron acceptor oxygen is reduced to water in this step. Both the direct pumping of protons and the consumption of matrix protons in the reduction of oxygen contribute to the proton gradient. The reaction catalyzed is the oxidation of cytochrome c and the reduction of oxygen:

$$4Cyt\,c_{red} + O_2 + 8H^+_{matrix} \rightarrow 4Cyt\,c_{ox} + 2H_2O + 4H^+_{intermembrane}$$

18.5.6 Alternative Reductases and Oxidases

Many eukaryotic organisms have electron transport chains that differ from the much-studied mammalian enzymes described above. For example, plants have alternative NADH oxidases, which oxidize NADH in the cytosol rather than in the mitochondrial matrix, and pass these electrons to the ubiquinone pool. These enzymes do not transport protons, and, therefore, reduce ubiquinone without altering the electrochemical gradient across the inner membrane.

Another example of a divergent electron transport chain is the *alternative oxidase*, which is found in plants, as well as some fungi, protists, and possibly some animals. This enzyme transfers electrons directly from ubiquinol to oxygen.

The electron transport pathways produced by these alternative NADH and ubiquinone oxidases have lower ATP yields than the full pathway. The advantages produced by a shortened pathway are not entirely clear. However, the alternative oxidase is produced in response to stresses such as cold, reactive oxygen species, and infection by pathogens, as well as other factors that inhibit the full electron transport chain. Alternative pathways might, therefore, enhance an organism's resistance to injury, by reducing oxidative stress.

18.5.7 Organization of Complexes

The original model for how the respiratory chain complexes are organized was that they diffuse freely and independently in the mitochondrial membrane. However, recent data suggest that the complexes might form higher-order structures called supercomplexes or "respirasomes". In this model, the various complexes exist as organized sets of interacting enzymes. These associations might allow channeling of substrates between the various enzyme complexes, increasing the rate and efficiency of electron transfer. Within such mammalian supercomplexes, some components would be present in higher amounts than others, with some data suggesting a ratio between complexes I/II/III/IV and the ATP synthase of approximately 1:1:3:7:4. However, the debate over this supercomplex hypothesis is not completely resolved, as some data do not appear to fit with this model.

18.6 PROKARYOTIC ELECTRON TRANSPORT CHAINS

In contrast to the general similarity in structure and function of the electron transport chains in eukaryotes, bacteria and archaea possess a large variety of electron-transfer enzymes. These use an equally wide set of chemicals as substrates. In common with eukaryotes, prokaryotic electron transport uses the energy released from the oxidation of a substrate to pump ions across a membrane and generate an electrochemical gradient. In the bacteria, oxidative phosphorylation in *Escherichia coli* is understood in most detail, while archaeal systems are at present poorly understood.

The main difference between eukaryotic and prokaryotic oxidative phosphorylation is that bacteria and archaea use many different substances to donate or accept electrons. This allows prokaryotes to grow under a wide variety of environmental conditions. In *E. coli*, for example, oxidative phosphorylation can be driven by a large number of pairs of reducing agents and oxidizing agents, which are listed below. The midpoint potential of a chemical measures how much energy is released when it is oxidized or reduced, with reducing agents having negative potentials and oxidizing agents positive potentials.

Table 2: Respiratory Enzymes and Substrates in *E. Coli*

Respiratory Enzyme	Redox Pair	Midpoint potential (Volts)
Formate dehydrogenase	Bicarbonate / Formate	-0.43
Hydrogenase	Proton / Hydrogen	-0.42
NADH dehydrogenase	NAD^+ / NADH	-0.32
Glycerol-3-phosphate dehydrogenase	DHAP / Gly-3-P	-0.19
Pyruvate oxidase	Acetate + Carbon dioxide / Pyruvate	-
Lactate dehydrogenase	Pyruvate / Lactate	-0.19
D-amino acid dehydrogenase	2-oxoacid + ammonia / D-amino acid	-
Succinate dehydrogenase	Fumarate / Succinate	+0.03
Ubiquinol oxidase	Oxygen / Water	+0.82
Nitrate reductase	Nitrate / Nitrite	+0.42
Nitrite reductase	Nitrite / Ammonia	+0.36
Dimethyl sulfoxide reductase	DMSO / DMS	+0.16
Trimethylamine *N*-oxide reductase	TMAO / TMA	+0.13
Fumarate reductase	Fumarate / Succinate	+0.03

As shown above, *E. coli* can grow with reducing agents such as formate, hydrogen, or lactate as electron donors, and nitrate, DMSO, or oxygen as acceptors. The larger the difference in midpoint potential between an oxidizing and reducing agent, the more energy is released when they react. Out of these compounds, the succinate/fumarate pair is unusual, as its midpoint potential is close to zero. Succinate can therefore be oxidized to fumarate if a strong oxidizing agent such as oxygen is available, or fumarate can be reduced to succinate using a strong reducing agent such as formate. These alternative reactions are catalyzed by succinate dehydrogenase and fumarate reductase, respectively.

Some prokaryotes use redox pairs that have only a small difference in midpoint potential. For example, nitrifying bacteria such as *Nitrobacter* oxidize nitrite to nitrate, donating the electrons to oxygen. The small amount of energy released in this reaction is enough to pump protons and generate ATP, but not enough to produce NADH or NADPH directly for use in anabolism. This problem is solved by using a nitrite oxidoreductase to produce enough proton-motive force to run part of the electron transport chain in reverse, causing complex I to generate NADH.

Prokaryotes control their use of these electron donors and acceptors by varying which enzymes are produced, in response to environmental conditions. This flexibility is possible because different oxidases and reductases use the same ubiquinone pool. This allows many combinations of enzymes to function together, linked by the common ubiquinol intermediate. These respiratory chains therefore have a modular design, with easily interchangeable sets of enzyme systems.

In addition to this metabolic diversity, prokaryotes also possess a range of isozymes – different enzymes that catalyze the same reaction. For example, in *E. coli*, there are two different types of ubiquinol oxidase using oxygen as an electron acceptor. Under highly aerobic conditions, the cell uses an oxidase with a low affinity for oxygen that can transport two protons per electron. However, if levels of oxygen fall, they switch to an oxidase that transfers only one proton per electron, but has a high affinity for oxygen.

18.7 ATP SYNTHASE (COMPLEX V)

ATP synthase, also called *complex V*, is the final enzyme in the oxidative phosphorylation pathway. This enzyme is found in all forms of life and functions in the same way in both prokaryotes and eukaryotes. The enzyme uses the energy stored in a proton gradient across a membrane to drive the synthesis of ATP from ADP and phosphate (P_i). Estimates of the number of protons required to synthesize one ATP have ranged from three to four, with some suggesting cells can vary this ratio, to suit different conditions.

$$ADP + P_i + 4H^+_{intermembrane} \;\square\; ATP + H_2O + 4H^+_{matrix}$$

This phosphorylation reaction is an equilibrium, which can be shifted by altering the proton-motive force. In the absence of a proton-motive force, the ATP synthase reaction will run from right to left, hydrolyzing ATP and pumping protons out of the matrix across the membrane. However, when the proton-motive force is high, the reaction is forced to run in the opposite direction; it proceeds from left to right, allowing protons to flow down their concentration gradient and turning ADP into ATP. Indeed, in the closely related vacuolar type H+-ATPases, the hydrolysis reaction is used to acidify cellular compartments, by pumping protons and hydrolysing ATP.

ATP synthase is a massive protein complex with a mushroom-like shape. The mammalian enzyme complex contains 16 subunits and has a mass of approximately 600 kilodaltons. The portion embedded within the membrane is called F_0 and contains a ring of c subunits and the proton channel. The stalk and the ball-shaped headpiece is called F_1 and is the site of ATP synthesis. The ball-shaped complex at the end of the F_1 portion contains six proteins of two different kinds (three α subunits and three β subunits), whereas the "stalk" consists of one protein: the γ subunit, with the tip of the stalk extending into the ball of α and β subunits. Both the α and β subunits bind nucleotides, but only the β subunits catalyze the ATP synthesis reaction.

Reaching along the side of the F_1 portion and back into the membrane is a long rod-like subunit that anchors the α and β subunits into the base of the enzyme.

As protons cross the membrane through the channel in the base of ATP synthase, the F_0 proton-driven motor rotates. Rotation might be caused by changes in the ionization of amino acids in the ring of c subunits causing electrostatic interactions that propel the ring of c subunits past the proton channel.

This rotating ring in turn drives the rotation of the central axle (the γ subunit stalk) within the α and β subunits. The α and β subunits are prevented from rotating themselves by the side-arm, which acts as a stator. This movement of the tip of the γ subunit within the ball of α and β subunits provides the energy for the active sites in the β subunits to undergo a cycle of movements that produces and then releases ATP.

This ATP synthesis reaction is called the *binding change mechanism* and involves the active site of a β subunit cycling between three states. In the "open" state, ADP and phosphate enter the active site. The protein then closes up around the molecules and binds them loosely – the "loose" state.

The enzyme then changes shape again and forces these molecules together, with the active site in the resulting "tight" state (shown in pink) binding the newly produced ATP molecule with very high affinity. Finally, the active site cycles back to the open state, releasing ATP and binding more ADP and phosphate, ready for the next cycle.

In some bacteria and archaea, ATP synthesis is driven by the movement of sodium ions through the cell membrane, rather than the movement of protons. Archaea such as *Methanococcus* also contain the A_1A_0 synthase, a form of the enzyme that contains additional proteins with little similarity in sequence to other bacterial and eukaryotic ATP synthase subunits. It is possible that, in some species, the A_1A_0 form of the enzyme is a specialized sodium-driven ATP synthase, but this might not be true in all cases.

18.8 OXIDATIVE PHOSPHORYLATION—ENERGETICS

The transport of electrons from redox pair NAD^+/ NADH to the final redox pair 1/2 O_2/ H_2O can be summarized as

$$1/2\ O_2 + NADH + H^+ \rightarrow H_2O + NAD^+$$

The potential difference between these two redox pairs is 1.14 volt, which is equivalent to -52 kcal/mol or -2600 kJ per 6 mol of O_2.

When one NADH is oxidized through the electron transfer chain, three ATPs are produced, which is equivalent to 7.3 kcal/mol x 3 = 21.9 kcal/mol.

The conservation of the energy can be calculated by the following formula

$$\text{Efficiency} = (21.9 \times 100\%) / 52 = 42\%$$

So we can conclude that when NADH is oxidized, about 42% of energy is conserved in the form of three ATPs and the remaining (58%) energy is lost as heat (unless the chemical energy of ATP under physiological conditions was underestimated).

18.9 REACTIVE OXYGEN SPECIES

Molecular oxygen is a good terminal electron acceptor because it is a strong oxidizing agent. The reduction of oxygen does involve potentially harmful intermediates. Although the transfer of four electrons and four protons reduces oxygen to water, which is harmless, transfer of one or two electrons produces superoxide or peroxide anions, which are dangerously reactive.

$$O2 \rightarrow^{e^-} \underset{Superoxide}{O2^{\pm}} \rightarrow^{e^-} \underset{Peroxide}{O2^{2-}}$$

These reactive oxygen species and their reaction products, such as the hydroxyl radical, are very harmful to cells, as they oxidize proteins and cause mutations in DNA. This cellular damage may contribute to disease and is proposed as one cause of aging.

The cytochrome c oxidase complex is highly efficient at reducing oxygen to water, and it releases very few partly reduced intermediates; however small amounts of superoxide anion and peroxide are produced by the electron transport chain. Particularly important is the reduction of coenzyme Q in complex III, as a highly reactive ubisemiquinone free radical is formed as an intermediate in the Q cycle. This unstable species can lead to electron "leakage" when electrons transfer directly to oxygen, forming superoxide. As the production of reactive oxygen species by these proton-pumping complexes is greatest at high membrane potentials, it has been proposed that mitochondria regulate their activity to maintain the membrane potential within a narrow range that balances ATP production against oxidant generation. For instance, oxidants can activate uncoupling proteins that reduce membrane potential.

To counteract these reactive oxygen species, cells contain numerous antioxidant systems, including antioxidant vitamins such as vitamin C and vitamin E, and antioxidant enzymes such as superoxide dismutase, catalase, and peroxidases, which detoxify the reactive species, limiting damage to the cell.

18.10 OXIDATIVE PHOSPHORYLATION IN HYPOXIC/ANOXIC CONDITIONS

As oxygen is fundamental for oxidative phosphorylation, a shortage in O_2 level can alter ATP production rates.

Under anoxic conditions, ATP-synthase will commit 'cellular treason' and run in reverse, forcing protons from the matrix back into the inner membrane space, using up ATP in the process.

The proton motive force and ATP production can be maintained by intracellular acidosis. Cytosolic protons that have accumulated with ATP hydrolysis and lactic acidosis can freely diffuse across the mitochondrial outer-membrane and acidify the inter-membrane space, hence directly contributing to the proton motive force and ATP production.

18.11 INHIBITORS

There are several well-known drugs and toxins that inhibit oxidative phosphorylation. Although any one of these toxins inhibits only one enzyme in the electron transport chain, inhibition of any step in this process will halt the rest of the process.

For example, if oligomycin inhibits ATP synthase, protons cannot pass back into the mitochondrion. As a result, the proton pumps are unable to operate, as the gradient becomes too strong for them to overcome. NADH is then no longer oxidized and the citric acid cycle ceases to operate because the concentration of NAD^+ falls below the concentration that these enzymes can use.

Many site-specific inhibitors of the electron transport chain have contributed to the present knowledge of mitochondrial respiration. Synthesis of ATP is also dependent on the electron transport chain, so all site-specific inhibitors also inhibit ATP formation.

The fish poison rotenone, the barbiturate drug amytal, and the antibiotic piericidin A inhibit NADH and coenzyme Q.

Carbon monoxide, cyanide, hydrogen sulphide and azide effectively inhibit cytochrome oxidase. Carbon monoxide reacts with the reduced form of the cytochrome while cyanide and azide react with the oxidised form. An antibiotic, antimycin A, and British anti-Lewisite, an antidote used against chemical weapons, are the two important inhibitors of the site between cytochrome B and C1.

Compounds	Use	Site of Action	Effect on Oxidative Phosphorylation
Cyanide Carbon monoxide Azide Hydrogen sulfide	Poisons	Complex IV	Inhibit the electron transport chain by binding more strongly than oxygen to the Fe–Cu center in cytochrome c oxidase, preventing the reduction of oxygen.
Oligomycin	Antibiotic	Complex V	Inhibits ATP synthase by blocking the flow of protons through the F_o subunit.
CCCP 2,4-Dinitrophenol	Poisons, weight-loss	Inner membrane	Ionophores that disrupt the proton gradient by carrying protons across a membrane. This ionophore uncouples proton pumping from ATP synthesis because it carries protons across the inner mitochondrial membrane.
Rotenone	Pesticide	Complex I	Prevents the transfer of electrons from complex I to ubiquinone by blocking the ubiquinone-binding site.
Malonate and oxaloacetate	Poisons	Complex II	Competitive inhibitors of succinate dehydrogenase (complex II).
Antimycin A	Piscicide	Complex III	Binds to the Qi site of cytochrome c reductase, thereby inhibiting the oxidation of ubiquinol.

Not all inhibitors of oxidative phosphorylation are toxins. In brown adipose tissue, regulated proton channels called uncoupling proteins can uncouple respiration from ATP synthesis. This rapid respiration produces heat, and is particularly important as a way of maintaining body temperature for hibernating animals, although these proteins may also have a more general function in cells' responses to stress.

CHAPTER 19

PHOTOTROPHY AND THE LIGHT REACTIONS OF PHOTOSYNTHESIS

Phototrophy, the process by which organisms harness light energy to fuel their metabolic activities, is essential to life on Earth. The most well-known form of phototrophy is photosynthesis, a process carried out by plants, algae, and certain bacteria that uses light to convert carbon dioxide and water into energy-rich organic compounds. The light reactions of photosynthesis, also known as the "light-dependent reactions," play a central role in capturing and converting solar energy into chemical energy. This energy drives the synthesis of carbohydrates in the subsequent steps of photosynthesis.

19.1 UNDERSTANDING PHOTOTROPHY AND PHOTOSYNTHESIS

Phototrophy is a mode of energy acquisition that enables organisms to utilize light as a primary energy source. Organisms capable of phototrophy are called **phototrophs** and include plants, algae, cyanobacteria, and some photosynthetic bacteria. Phototrophs can be divided into two main types based on how they use the energy they obtain:

- **Photoautotrophs**: Organisms that use light energy to convert carbon dioxide into organic compounds through photosynthesis.
- **Photoheterotrophs**: Organisms that use light energy but rely on organic carbon sources rather than carbon dioxide.

Photosynthesis, the process by which photoautotrophs convert light energy into chemical energy, is the most crucial phototrophic process for life on Earth. It sustains nearly all ecosystems, directly or indirectly, by producing oxygen as a byproduct and generating the organic compounds that form the base of the food web. The process is generally divided into two stages:

- **Light Reactions**: The energy-capturing phase that converts light into chemical energy in the form of ATP and NADPH.
- **Calvin Cycle** (or Dark Reactions): The carbon-fixing phase that uses ATP and NADPH to convert CO_2 into glucose.

The light reactions are the primary focus here, as they provide the chemical energy necessary to drive the Calvin Cycle and ultimately enable the synthesis of sugars.

19.2 THE LIGHT REACTIONS: HARNESSING LIGHT ENERGY

The light reactions take place in the **thylakoid membranes** of chloroplasts in plants and algae, and in the cell membranes of photosynthetic bacteria. They rely on specialized pigments, protein complexes, and electron carriers to capture and convert solar energy into usable chemical energy. Two important products, **ATP** and **NADPH**, are generated during these reactions, both of which are essential for the subsequent Calvin Cycle.

19.2.1 Photosynthetic Pigments and Light Absorption

The light reactions begin when **photosynthetic pigments** absorb light. The main pigment involved in photosynthesis is **chlorophyll**, which is highly effective at capturing light in the blue and red portions of the spectrum. Accessory pigments, such as **carotenoids** and **phycobilins**, absorb light in different parts of the spectrum and help capture more solar energy, allowing organisms to maximize the light absorbed and drive photosynthesis efficiently.

When chlorophyll and other pigments absorb light, their electrons become "excited," meaning they reach a higher energy state. This energy is transferred to specialized chlorophyll molecules within the **photosystems** (Photosystem I and Photosystem II) embedded in the thylakoid membrane. These photosystems are multi-protein complexes that play a crucial role in converting light energy into chemical energy.

19.2.2 Photosystem II: The First Step in the Light Reactions

The light reactions start in **Photosystem II** (PSII), where light energy excites electrons in chlorophyll molecules. PSII absorbs light most effectively at a wavelength of 680 nm, which is why its reaction center is called **P680**. Once the P680 chlorophyll molecules absorb light, they release high-energy electrons, which are passed to a primary electron acceptor. This loss of electrons leaves the P680 reaction center oxidized and positively charged.

To replace these lost electrons, PSII splits water molecules in a process known as **photolysis.** Photolysis breaks down water (H_2O) into oxygen, protons (H^+ ions), and electrons. This reaction can be represented as follows:

$$2H_2O \rightarrow 4H^+ + 4e^- + O_2$$

The oxygen produced is released as a byproduct, which is essential for aerobic life on Earth. The electrons released from photolysis are then transferred to the oxidized P680, restoring its neutrality and allowing the light reactions to continue.

19.2.3 Electron Transport Chain: Establishing a Proton Gradient

After leaving PSII, the excited electrons travel along an **electron transport chain** (ETC), a series of protein complexes and electron carriers that facilitate the controlled release of energy from electrons. As electrons move through the ETC, they lose energy, which is used to pump protons from the stroma (the surrounding chloroplast fluid) into the thylakoid lumen. This creates a **proton gradient** across the thylakoid membrane, with a higher concentration of protons inside the lumen compared to the stroma. The creation of this proton gradient is essential because it provides the energy for ATP synthesis, a process known as **chemiosmosis.** The difference in proton concentration drives protons back into the stroma through an enzyme called **ATP synthase**, which uses the flow of protons to convert ADP and inorganic phosphate (Pi) into ATP. This ATP will then be used as an energy source in the Calvin Cycle.

19.2.4 Photosystem I: Producing NADPH

Following the electron transport chain, electrons reach **Photosystem I** (PSI), which absorbs light most effectively at a wavelength of 700 nm (hence, its reaction center is called **P700**). Light energy absorbed by PSI re-energizes the electrons, raising them to an even higher energy level. The excited electrons are then passed to a molecule called **ferredoxin**, an electron acceptor in PSI. The high-energy electrons in ferredoxin are transferred to NADP+ (nicotinamide adenine dinucleotide phosphate), a coenzyme that functions as an electron carrier. This reaction, catalyzed by the enzyme **NADP+ reductase**, results in the formation of NADPH:

$$NADP^{+} + 2e^{-} + H^{+} \rightarrow NADPH$$

NADPH serves as a reducing agent in the Calvin Cycle, providing the high-energy electrons necessary to convert carbon dioxide into glucose. With both ATP and NADPH produced, the light reactions come to a close, supplying the energy and reducing power needed for the carbon-fixing reactions.

19.3 IMPORTANCE OF THE LIGHT REACTIONS IN PHOTOSYNTHESIS

The light reactions are critical for photosynthesis because they generate ATP and NADPH, which drive the Calvin Cycle. Without these molecules, photoautotrophs would be unable to fix carbon dioxide into organic compounds, and the production of oxygen – a byproduct of photolysis – would cease.

These reactions play a crucial ecological role, as the oxygen released by photosynthetic organisms is essential for aerobic respiration in almost all forms of life. Additionally, the light reactions are central to the global carbon cycle, as they provide the energy needed for plants to absorb atmospheric carbon dioxide and convert it into carbohydrates. This process contributes to carbon sequestration, mitigating the greenhouse effect and climate change.

19.4 EVOLUTIONARY SIGNIFICANCE OF PHOTOTROPHY

Phototrophy likely evolved billions of years ago in ancient prokaryotes, such as cyanobacteria, which were the first organisms capable of oxygenic photosynthesis. The evolution of this process transformed the Earth's atmosphere, increasing oxygen levels and enabling the evolution of aerobic organisms. Modern plants and algae are descendants of these early phototrophs and continue to play a central role in maintaining the oxygen levels required for complex life.

The study of phototrophy has also expanded our understanding of life's adaptability and energy acquisition in extreme environments. Some bacteria use anoxygenic photosynthesis, a variant of photosynthesis that does not produce oxygen and uses other electron donors, such as hydrogen sulfide. This variation highlights the flexibility of phototrophy in adapting to diverse environmental conditions, including high salinity, low light, and nutrient-poor environments.

19.5 FUTURE DIRECTIONS IN PHOTOTROPHY RESEARCH

Advances in phototrophy research are opening up exciting possibilities in bioengineering and renewable energy. For instance, researchers are exploring ways to optimize photosynthetic pathways to increase crop yields, potentially addressing global food security issues. Additionally, bioengineers are investigating the development of **artificial photosynthesis** systems that mimic the light reactions to generate sustainable fuels and clean energy.

These technologies hold potential for reducing our reliance on fossil fuels by harnessing solar energy to produce biofuels, hydrogen, and other renewable sources. By better understanding the mechanisms of phototrophy and photosynthesis, scientists may unlock innovative solutions to some of the world's most pressing energy and environmental challenges.

19.6 CONCLUSION

Phototrophy, and specifically the light reactions of photosynthesis, is a cornerstone of life on Earth. Through a complex series of reactions, plants, algae, and some bacteria convert light energy into chemical energy, producing oxygen and forming the foundation of the food web. The light reactions capture solar energy and convert it into ATP and NADPH, driving the synthesis of organic molecules that sustain ecosystems and support the global carbon cycle.

The study of these processes highlights the intricate and interdependent nature of life, emphasizing the importance of preserving phototrophic organisms and ecosystems. As research continues, advancements in phototrophy and photosynthesis offer hope for sustainable energy solutions and a deeper understanding of life's adaptability.

CHAPTER 20

THE CALVIN-BENSON CYCLE AND THE PENTOSE PHOSPHATE PATHWAY

20.1 THE CALVIN-BENSON CYCLE

The Calvin cycle, light-independent reactions, bio synthetic phase, dark reactions, or photosynthetic carbon reduction (PCR) cycle of photosynthesis is a series of chemical reactions that convert carbon dioxide and hydrogen-carrier compounds into glucose.

The Calvin cycle is present in all photosynthetic eukaryotes and also many photosynthetic bacteria. In plants, these reactions occur in the stroma, the fluid-filled region of a chloroplast outside the thylakoid membranes. These reactions take the products (ATP and NADPH) of light-dependent reactions and perform further chemical processes on them.

The Calvin cycle uses the chemical energy of ATP and reducing power of NADPH from the light dependent reactions to produce sugars for the plant to use. These substrates are used in a series of reduction-oxidation (redox) reactions to produce sugars in a step-wise process; there is no direct reaction that converts several molecules of CO_2 to a sugar.

There are three phases to the light-independent reactions, collectively called the Calvin cycle: carboxylation, reduction reactions, and ribulose 1,5-bisphosphate (RuBP) regeneration.

Though it is also called the "dark reaction", the Calvin cycle does not actually occur in the dark or during night time.

This is because the process requires NADPH, which is short-lived and comes from light-dependent reactions. In the dark, plants instead release sucrose into the phloem from their starch reserves to provide energy for the plant.

The Calvin cycle thus happens when light is available independent of the kind of photosynthesis (C3 carbon fixation, C4 carbon fixation, and crassulacean acid metabolism (CAM)); CAM plants store malic acid in their vacuoles every night and release it by day to make this process work.

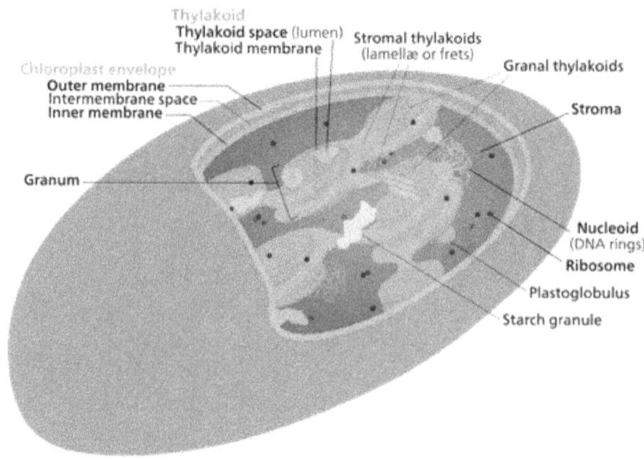

Fig 1: The Internal Structure of a Chloroplast

20.1.1 Coupling to other Metabolic Pathways

The reactions of the Calvin cycle are closely coupled to the thylakoid electron transport chain, as the energy required to reduce the carbon dioxide is provided by NADPH produced during the light dependent reactions. The process of photorespiration, also known as C2 cycle, is also coupled to the Calvin cycle, as it results from an alternative reaction of the RuBisCO enzyme, and its final byproduct is another glyceraldehyde-3-P molecule.

20.1.2 Calvin Cycle

The Calvin cycle, Calvin–Benson–Bassham (CBB) cycle, reductive pentose phosphate cycle (RPP cycle) or C3 cycle is a series of biochemical redox reactions that take place in the stroma of chloroplast in photosynthetic organisms. The cycle was discovered in 1950 by Melvin Calvin, James Bassham, and Andrew Benson at the University of California, Berkeley by using the radioactive isotope carbon-14.

Photosynthesis occurs in two stages in a cell. In the first stage, light-dependent reactions capture the energy of light and use it to make the energy-storage molecule ATP and the moderate-energy hydrogen carrier NADPH. The Calvin cycle uses these compounds to convert carbon dioxide and water into organic compounds that can be used by the organism (and by animals that feed on it). This set of reactions is also called *carbon fixation*.

The key enzyme of the cycle is called RuBisCO. In the following biochemical equations, the chemical species (phosphates and carboxylic acids) exist in equilibria among their various ionized states as governed by the pH. The enzymes in the Calvin cycle are functionally equivalent to most enzymes used in other metabolic pathways such as gluconeogenesis and the pentose phosphate pathway, but the enzymes in the Calvin cycle are found in the chloroplast stroma instead of the cell cytosol, separating the reactions.

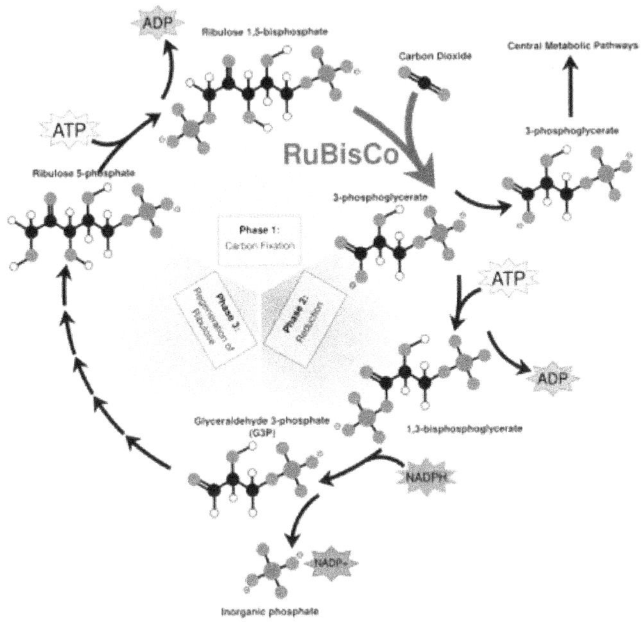

Fig 2: Overview of the Calvin Cycle and Carbon Fixation

They are activated in the light (which is why the name "dark reaction" is misleading), and also by products of the light-dependent reaction.

These regulatory functions prevent the Calvin cycle from being respired to carbon dioxide. Energy (in the form of ATP) would be wasted in carrying out these reactions when they have no net productivity.

The sum of reactions in the Calvin cycle is the following:

- $3\ CO_2 + 6\ NADPH + 9\ ATP + 5\ H_{20} \rightarrow$ glyceraldehyde-3-phosphate (G3P) + $6\ NADP^+ + 9\ ADP + 8\ P_i$ (P_i = inorganic phosphate)

Hexose (six-carbon) sugars are not products of the Calvin cycle. Although many texts list a product of photosynthesis as C_{6H12O6}, this is mainly for convenience to match the equation of aerobic respiration, where six-carbon sugars are oxidized in mitochondria.

The carbohydrate products of the Calvin cycle are three-carbon sugar phosphate molecules, or "triose phosphates", namely, glyceraldehyde-3-phosphate (G3P).

20.1.3 Steps

In the first stage of the Calvin cycle, a CO_2 molecule is incorporated into one of two three-carbon molecules (glyceraldehyde 3-phosphate or G3P), where it uses up two molecules of ATP and two molecules of NADPH, which had been produced in the light-dependent stage. The three steps involved are:

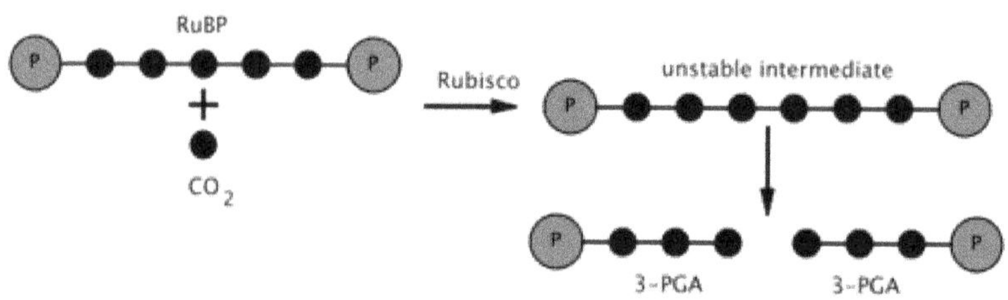

Fig: Calvin Cycle Step 1 (Black Circles Represent Carbon Atoms)

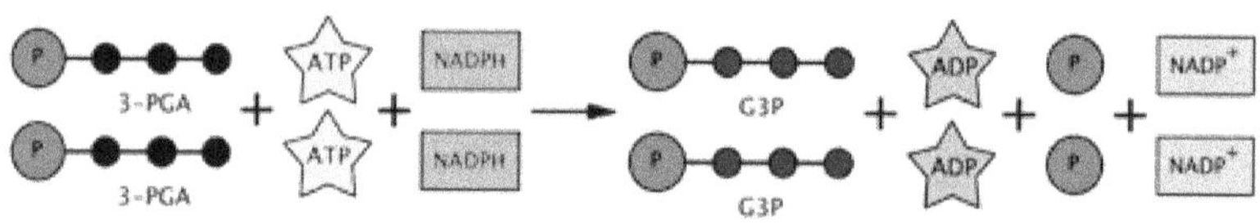

Fig 3: Calvin Cycle Steps 2 and 3 Combined

- The enzyme RuBisCO catalyses the carboxylation of ribulose-1,5-bisphosphate, RuBP, a 5-carbon compound, by carbon dioxide (a total of 6 carbons) in a two-step reaction. The product of the first step is enediol-enzyme complex that can capture CO_2 or O_2. Thus, enediol-enzyme complex is the real carboxylase/oxygenase. The CO_2 that is captured by enediol in second step produces an unstable six-carbon compound called 2-carboxy 3-keto 1,5-biphosphoribotol (CKABP) (or 3-keto-2-carboxyarabinitol 1,5-bisphosphate) that immediately splits into 2 molecules of 3-phosphoglycerate (also written as 3-phosphoglyceric acid, PGA, 3PGA, or 3-PGA), a 3-carbon compound.

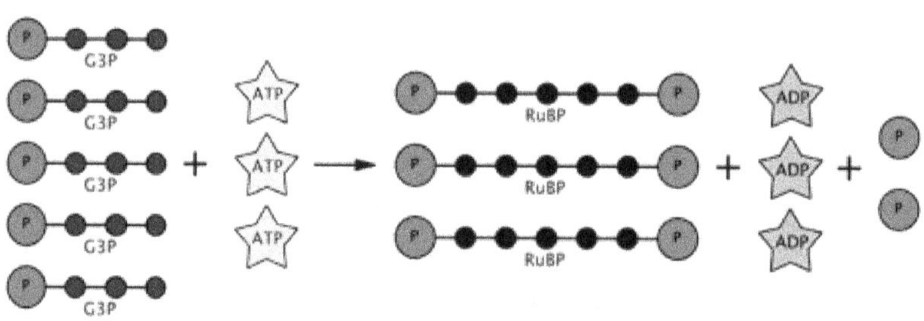

Fig 4: Regeneration Stage of the Calvin Cycle

- The enzyme phosphoglycerate kinase catalyses the phosphorylation of 3-PGA by ATP (which was produced in the light-dependent stage). 1,3-Bisphosphoglycerate (glycerate-1,3-bisphosphate) and ADP are the products. (However, note that two 3-PGAs are produced for every CO_2 that enters the cycle, so this step utilizes two ATP per CO_2 fixed.)
- The enzyme glyceraldehyde 3-phosphate dehydrogenase catalyses the reduction of 1,3BPGA by NADPH (which is another product of the light-dependent stage). Glyceraldehyde 3-phosphate (also called G3P, GP, TP,

PGAL, GAP) is produced, and the NADPH itself is oxidized and becomes $NADP^+$. Again, two NADPH are utilized per CO_2 fixed.

The next stage in the Calvin cycle is to regenerate RuBP. Five G3P molecules produce three RuBP molecules, using up three molecules of ATP. Since each CO_2 molecule produces two G3P molecules, three CO_2 molecules produce six G3P molecules, of which five are used to regenerate RuBP, leaving a net gain of one G3P molecule per three CO_2 molecules (as would be expected from the number of carbon atoms involved).

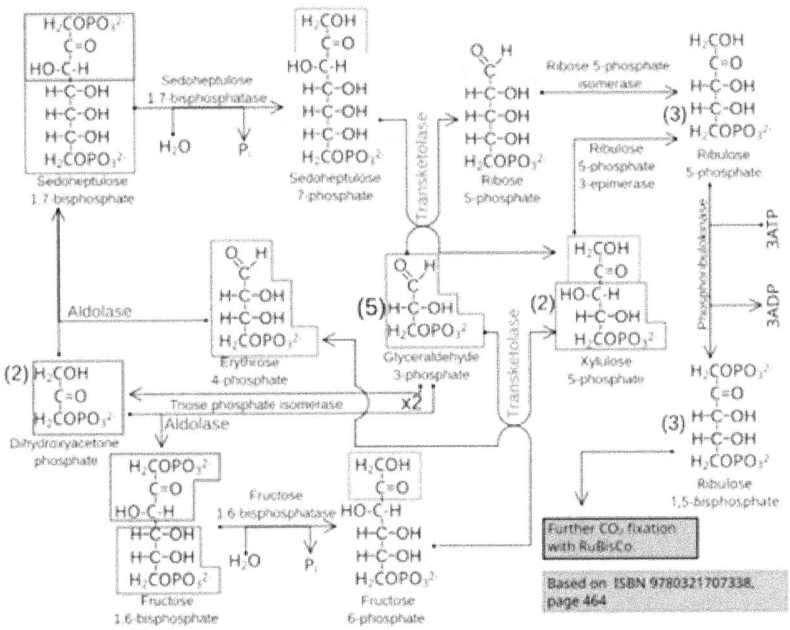

Fig 5: The Regeneration Stage of Calvin's Cycle. Substances and their Parts are Outlined in Colors: Green—Carbon Accepting Aldoses, Pink—Ketoses-Donors of Three-Carbon Groups, Yellow—Parts of Ketoses Remaining after Donation of Two-Carbon Keto-Groups Highlighted in Orange. Enzymes are also Highlighted: Aldolases in Purple and Transketolases in Red.

The regeneration stage can be broken down into a series of steps.

- Triose phosphate isomerase converts one of the G3P reversibly into dihydroxyacetone phosphate (DHAP), also a 3-carbon molecule.
- Aldolase and fructose-1,6-bisphosphatase convert a G3P and a DHAP into fructose 6-phosphate (6C). A phosphate ion is lost into solution.
- Then fixation of another CO_2 generates two more G3P.
- F6P has two carbons removed by transketolase, giving erythrose-4-phosphate (E4P). The two carbons on transketolase are added to a G3P, giving the ketose xylulose-5-phosphate (Xu5P).

- E4P and a DHAP (formed from one of the G3P from the secondCO_2 fixation) are converted into sedoheptulose-1,7-bisphosphate (7C) by aldolase enzyme.

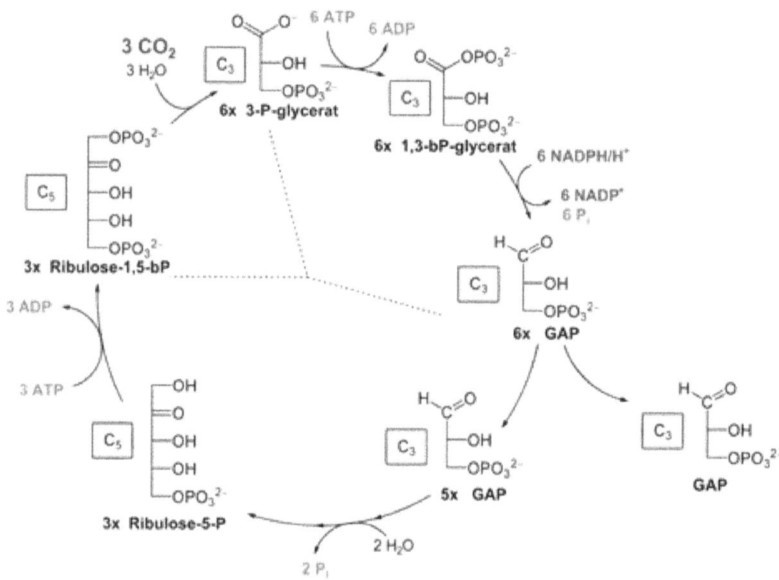

Fig 6: Simplified C3 Cycle with Structural Formulas

- Sedoheptulose-1,7-bisphosphatase (one of only three enzymes of the Calvin cycle that are unique to plants) cleaves sedoheptulose-1,7-bisphosphate into sedoheptulose-7-phosphate, releasing an inorganic phosphate ion into solution.
- Fixation of a third CO_2 generates two more G3P. The ketose S7P has two carbons removed by transketolase, giving ribose-5-phosphate (R5P), and the two carbons remaining on transketolase are transferred to one of the G3P, giving another Xu5P. This leaves one G3P as the product of fixation of 3 CO_2, with generation of three pentoses that can be converted to Ru5P.
- R5P is converted into ribulose-5-phosphate (Ru5P, RuP) by phosphopentose isomerase. Xu5P is converted into RuP by phosphopentose epimerase.
- Finally, phosphoribulokinase (another plant-unique enzyme of the pathway) phosphorylates RuP into RuBP, ribulose-1,5-bisphosphate, completing the Calvin *cycle*. This requires the input of one ATP.

Thus, of six G3P produced, five are used to make three RuBP (5C) molecules (totaling 15 carbons), with only one G3P available for subsequent conversion to hexose. This requires nine ATP molecules and six NADPH molecules per three CO_2 molecules.

The equation of the overall Calvin cycle is shown diagrammatically below.

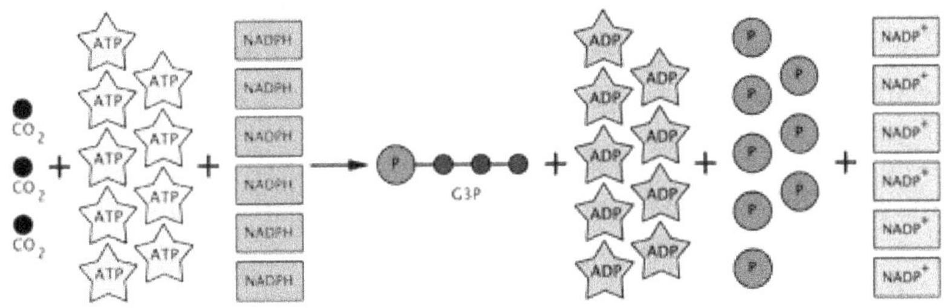

Fig 7: The Overall Equation of the Calvin Cycle (Black Circles Represent Carbon Atoms)

RuBisCO also reacts competitively with O_2 instead of CO_2 in photorespiration. The rate of photorespiration is higher at high temperatures. Photorespiration turns RuBP into 3-PGA and 2-phosphoglycolate, a 2-carbon molecule that can be converted via glycolate and glyoxalate to glycine. Via the glycine cleavage system and tetrahydrofolate, two glycines are converted into serine plus CO_2. Serine can be converted back to 3-phosphoglycerate. Thus, only 3 of 4 carbons from two phosphoglycolates can be converted back to 3-PGA. It can be seen that photorespiration has very negative consequences for the plant, because, rather than fixing CO_2, this process leads to loss of CO_2. C4 carbon fixation evolved to circumvent photorespiration, but can occur only in certain plants native to very warm or tropical climates—corn, for example. Furthermore, RuBisCOs catalyzing the light-independent reactions of photosynthesis generally exhibit an improved specificity for CO_2 relative to O_2, in order to minimize the oxygenation reaction. This improved specificity evolved after RuBisCO incorporated a new protein subunit.

20.1.4 Products

The immediate products of one turn of the Calvin cycle are 2 glyceraldehyde-3-phosphate (G3P) molecules, 3 ADP, and 2 $NADP^+$. (ADP and $NADP^+$ are not really "products". They are regenerated and later used again in the light-dependent reactions). Each G3P molecule is composed of 3 carbons. For the Calvin cycle to continue, RuBP (ribulose 1,5-bisphosphate) must be regenerated. So, 5 out of 6 carbons from the 2 G3P molecules are used for this purpose. Therefore, there is only 1 net carbon produced to play with for each turn. To create 1 surplus G3P requires 3 carbons, and therefore 3 turns of the Calvin cycle. To make one glucose molecule (which can be created from 2 G3P molecules) would require 6 turns of the Calvin cycle. Surplus G3P can also be used to form other carbohydrates such as starch, sucrose, and cellulose, depending on what the plant needs.

20.2 LIGHT-DEPENDENT REGULATION

These reactions do not occur in the dark or at night. There is a light-dependent regulation of the cycle enzymes, as the third step requires NADPH. There are two regulation systems at work when the cycle must be turned on or off: the thioredoxin/ferredoxin activation system, which activates some of the cycle enzymes; and the RuBisCo enzyme activation, active in the Calvin cycle, which involves its own activase.

The thioredoxin/ferredoxin system activates the enzymes glyceraldehyde-3-P dehydrogenase, glyceraldehyde-3-P phosphatase, fructose-1,6-bisphosphatase, sedoheptulose-1,7-bisphosphatase, and ribulose-5-phosphatase kinase, which are key points of the process. This happens when light is available, as the ferredoxin protein is reduced in the photosystem I complex of the thylakoid electron chain when electrons are circulating through it. Ferredoxin then binds to and reduces the thioredoxin protein, which activates the cycle enzymes by severing a cystine bond found in all these enzymes. This is a dynamic process as the same bond is formed again by other proteins that deactivate the enzymes. The implications of this process are that the enzymes remain mostly activated by day and are deactivated in the dark when there is no more reduced ferredoxin available.

The enzyme RuBisCo has its own, more complex activation process. It requires that a specific lysine amino acid be carbamylated to activate the enzyme. This lysine binds to RuBP and leads to a non-functional state if left uncarbamylated. A specific activase enzyme, called RuBisCo activase, helps this carbamylation process by removing one proton from the lysine and making the binding of the carbon dioxide molecule possible. Even then the RuBisCo enzyme is not yet functional, as it needs a magnesium ion bound to the lysine to function. This magnesium ion is released from the thylakoid lumen when the inner pH drops due to the active pumping of protons from the electron flow. RuBisCo activase itself is activated by increased concentrations of ATP in the stroma caused by its phosphorylation.

20.3 THE PENTOSE PHOSPHATE PATHWAY

The pentose phosphate pathway (also called the phosphogluconate pathway and the hexose monophosphate shunt or HMP shunt) is a metabolic pathway parallel to glycolysis. It generates NADPH and pentoses (five-carbon sugars) as well as ribose 5-phosphate, a precursor for the synthesis of nucleotides. While the pentose phosphate pathway does involve oxidation of glucose, its primary role is anabolic rather than catabolic. The pathway is especially important in red blood cells (erythrocytes). The reactions of the pathway were elucidated in the early 1950s by Bernard Horecker and co-workers.

There are two distinct phases in the pathway. The first is the oxidative phase, in which NADPH is generated, and the second is the non-oxidative synthesis of five-carbon sugars. For most organisms, the pentose phosphate pathway takes place in the cytosol; in plants, most steps take place in plastids.

Like glycolysis, the pentose phosphate pathway appears to have a very ancient evolutionary origin. The reactions of this pathway are mostly enzyme catalyzed in modern cells, however, they also occur non-enzymatically under conditions that replicate those of the Archean ocean, and are catalyzed by metal ions, particularly ferrous ions (Fe(II)). This suggests that the origins of the pathway could date back to the prebiotic world.

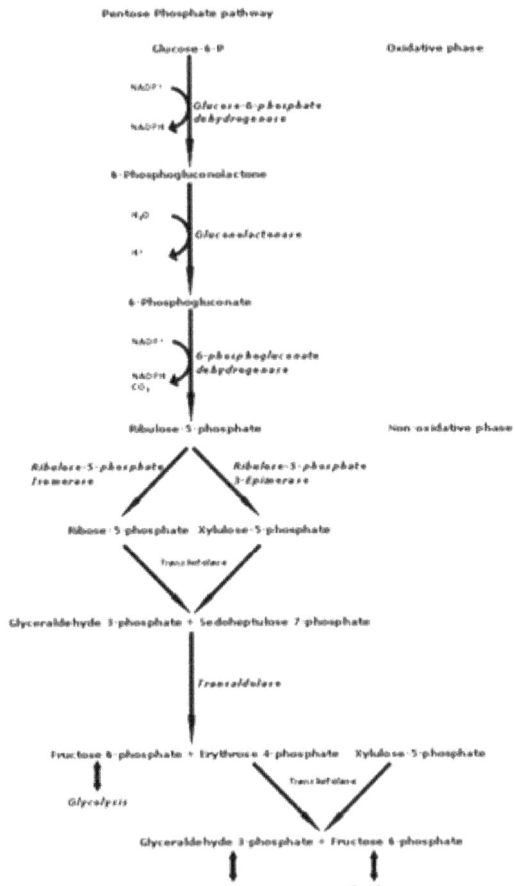

Fig 8: The Pentose Phosphate Pathway

20.3.1 Outcome

The primary results of the pathway are:

- The generation of reducing equivalents, in the form of NADPH, used in reductive biosynthesis reactions within cells (e.g. fatty acid synthesis).
- Production of ribose 5-phosphate (R5P), used in the synthesis of nucleotides and nucleic acids.
- Production of erythrose 4-phosphate (E4P) used in the synthesis of aromatic amino acids.

Aromatic amino acids, in turn, are precursors for many biosynthetic pathways, including the lignin in wood. Dietary pentose sugars derived from the digestion of nucleic acids may be metabolized through the pentose phosphate pathway, and the carbon skeletons of dietary carbohydrates may be converted into glycolytic/gluconeogenic intermediates. In mammals, the PPP occurs exclusively in the cytoplasm. In humans, it is found to be most active in the liver, mammary glands, and adrenal cortex. The PPP is one of the three main ways the body creates molecules with reducing power, accounting for approximately 60% of NADPH production in humans.

One of the uses of NADPH in the cell is to prevent oxidative stress. It reduces glutathione via glutathione reductase, which converts reactive H_2O_2 into H_2O by glutathione peroxidase. If absent, the H_2O_2 would be converted to hydroxyl free radicals by Fenton chemistry, which can attack the cell. Erythrocytes, for example, generate a large amount of NADPH through the pentose phosphate pathway to use in the reduction of glutathione. Hydrogen peroxide is also generated for phagocytes in a process often referred to as a respiratory burst.

20.3.2 Phases

Oxidative Phase

In this phase, two molecules of $NADP^+$ are reduced to NADPH, utilizing the energy from the conversion of glucose-6-phosphate into ribulose 5-phosphate.

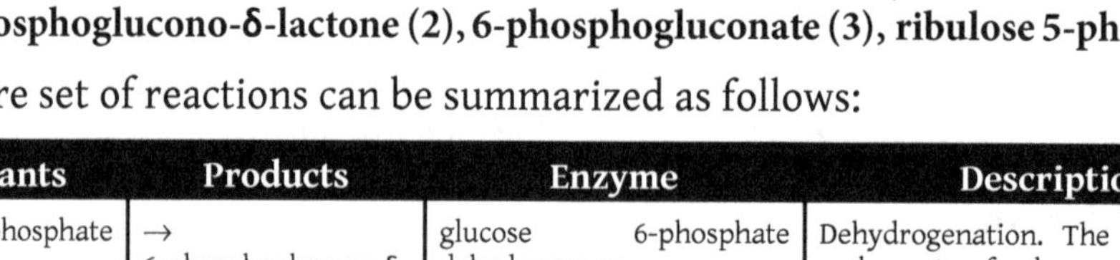

Fig 9: Oxidative phase of pentose phosphate pathway. Glucose-6-phosphate (1), 6-phosphoglucono-δ-lactone (2), 6-phosphogluconate (3), ribulose 5-phosphate (4)

The entire set of reactions can be summarized as follows:

Reactants	Products	Enzyme	Description
Glucose 6-phosphate + NADP+	→ 6-phosphoglucono-δ-lactone + NADPH	glucose 6-phosphate dehydrogenase	Dehydrogenation. The hydroxyl on carbon 1 of glucose 6-phosphate turns into a carbonyl, generating a lactone, and, in the process, NADPH is generated.
6-phosphoglucono-δ-lactone + H_2O	→ 6-phosphogluconate + H^+	6-phosphogluconolactonase	Hydrolysis
6-phosphogluconate + $NADP^+$	→ ribulose 5-phosphate + NADPH + CO_2	6-phosphogluconate dehydrogenase	Oxidative decarboxylation. $NADP^+$ is the electron acceptor, generating another molecule of NADPH, a CO_2, and ribulose 5-phosphate.

The overall reaction for this process is:

$$Glucose\ 6-phosphate\ +\ 2\ NADP^{+} + H_2O\ \rightarrow\ ribulose\ 5-phosphate\ +\ 2\ NADPH\ +\ 2\ H^{+} + CO_2$$

Non-Oxidative Phase

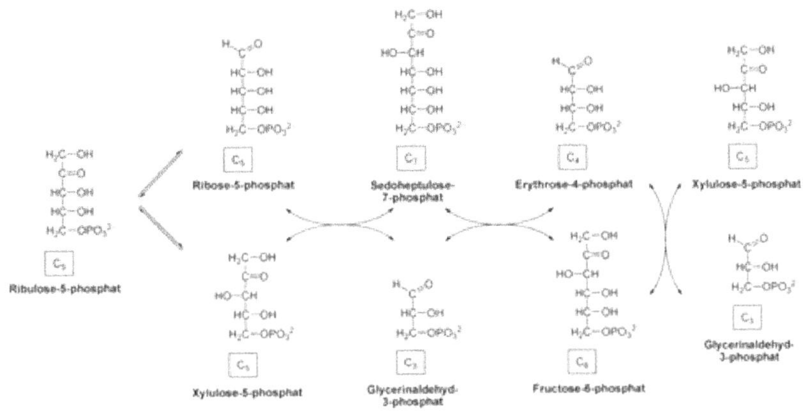

Fig 10: The Pentose Phosphate Pathway's Nonoxidative Phase

Reactants	Products	Enzymes
ribulose 5-phosphate	→ ribose 5-phosphate	ribose-5-phosphate isomerase
ribulose 5-phosphate	→ xylulose 5-phosphate	ribulose 5-phosphate 3-epimerase
xylulose 5-phosphate + ribose 5-phosphate	→ glyceraldehyde 3-phosphate + sedoheptulose 7-phosphate	transketolase
sedoheptulose 7-phosphate + glyceraldehyde 3-phosphate	→ erythrose 4-phosphate + fructose 6-phosphate	transaldolase
xylulose 5-phosphate + erythrose 4-phosphate	→ glyceraldehyde 3-phosphate + fructose 6-phosphate	transketolase

Net reaction: 3 ribulose-5-phosphate → 1 ribose-5-phosphate + 2 xylulose-5-phosphate → 2 fructose-6-phosphate + glyceraldehyde-3-phosphate

Regulation

Glucose-6-phosphate dehydrogenase is the rate-controlling enzyme of this pathway. It is allosterically stimulated by $NADP^{+}$ and strongly inhibited by NADPH. The ratio of NADPH:$NADP^{+}$ is the primary mode of regulation for the enzyme and is normally about 100:1 in liver cytosol.

This makes the cytosol a highly-reducing environment. An NADPH-utilizing pathway forms $NADP^{+}$, which stimulates Glucose-6-phosphate dehydrogenase to produce more NADPH. This step is also inhibited by acetyl CoA.

G6PD activity is also post-translationally regulated by cytoplasmic deacetylase SIRT2. SIRT2-mediated deacetylation and activation of G6PD stimulates oxidative branch of PPP to supply cytosolic NADPH to counteract oxidative damage or support *de novo* lipogenesis.

20.3.3 Erythrocytes

Several deficiencies in the level of activity (not function) of glucose-6-phosphate dehydrogenase have been observed to be associated with resistance to the malarial parasite *Plasmodium falciparum* among individuals of Mediterranean and African descent.

The basis for this resistance may be a weakening of the red cell membrane (the erythrocyte is the host cell for the parasite) such that it cannot sustain the parasitic life cycle long enough for productive growth.

CHAPTER 21
GLYCOGEN METABOLISM

21.1 AN INTRODUCTION

Glycogen is a multibranched polysaccharide of glucose that serves as a form of energy storage in animals, fungi, and bacteria. It is the main storage form of glucose in the human body.

Glycogen functions as one of three regularly used forms of energy reserves, creatine phosphate being for very short-term, glycogen being for short-term and the triglyceride stores in adipose tissue (i.e., body fat) being for long-term storage. Protein, broken down into amino acids, is seldom used as a main energy source except during starvation and glycolytic crisis.

In humans, glycogen is made and stored primarily in the cells of the liver and skeletal muscle. In the liver, glycogen can make up 5–6% of the organ's fresh weight: the liver of an adult, weighing 1.5 kg, can store roughly 100–120 grams of glycogen. In skeletal muscle, glycogen is found in a low concentration (1–2% of the muscle mass): the skeletal muscle of an adult weighing 70 kg stores roughly 400 grams of glycogen. Small amounts of glycogen are also found in other tissues and cells, including the kidneys, red blood cells, white blood cells, and glial cells in the brain. The uterus also stores glycogen during pregnancy to nourish the embryo.

The amount of glycogen stored in the body mostly depends on oxidative type 1 fibres, physical training, basal metabolic rate, and eating habits. Different levels of resting muscle glycogen are reached by changing the number of glycogen particles, rather than increasing the size of existing particles though most glycogen particles at rest are smaller than their theoretical maximum.

Approximately 4 grams of glucose are present in the blood of humans at all times; in fasting individuals, blood glucose is maintained constant at this level at the expense of glycogen stores, primarily from the liver (glycogen in skeletal muscle is mainly used as an immediate source of energy for that muscle rather than being used to maintain physiological blood glucose levels). Glycogen stores in skeletal muscle serve as a form of energy storage for the muscle itself; however, the breakdown of muscle glycogen impedes muscle glucose uptake from the blood, thereby increasing the amount of blood glucose available for use in other tissues. Liver glycogen stores serve as a store of glucose for use throughout the body, particularly the central nervous system. The human brain consumes approximately 60% of blood glucose in fasted, sedentary individuals.

Glycogen is an analogue of starch, a glucose polymer that functions as energy storage in plants. It has a structure similar to amylopectin (a component of starch), but is more extensively branched and compact than starch. Both are white powders in their dry state. Glycogen is found in the form of granules in the cytosol/cytoplasm in many cell types, and plays an important role in the glucose cycle. Glycogen forms an energy reserve that can be quickly mobilized to meet a sudden need for glucose, but one that is less compact than the energy reserves of triglycerides (lipids). As such it is also found as storage reserve in many parasitic protozoa.

21.2 HISTORY

Glycogen was discovered by Claude Bernard. His experiments showed that the liver contained a substance that could give rise to reducing sugar by the action of a "ferment" in the liver. By 1857, he described the isolation of a substance he called "*la matière glycogène*", or "sugar-forming substance". Soon after the discovery of glycogen in the liver, M.A. Sanson found that muscular tissue also contains glycogen. The empirical formula for glycogen of $(C_{6H10O5})_n$ was established by August Kekulé in 1858.

21.3 STRUCTURE

Glycogen is a branched biopolymer consisting of linear chains of glucose residues with an average chain length of approximately 8–12 glucose units and 2,000-60,000 residues per one molecule of glycogen.

Like amylopectin, glucose units are linked together linearly by $\alpha(1\rightarrow4)$ glycosidic bonds from one glucose to the next. Branches are linked to the chains from which they are branching off by $\alpha(1\rightarrow6)$ glycosidic bonds between the first glucose of the new branch and a glucose on the stem chain.

Each glycogen is essentially a ball of glucose trees, with around 12 layers, centered on a glycogenin protein, with three kinds of glucose chains: A, B, and C. There is only one C-chain, attached to the glycogenin. This C-chain is formed by the self-glucosylation of the glycogenin, forming a short primer chain. From the C-chain grows out B-chains, and from B-chains branch out B- and A-chains. The B-chains have on average 2 branch points, while the A-chains are terminal, thus unbranched. On average, each chain has length 12, tightly constrained to be between 11 and 15. All A-chains reach the spherical surface of the glycogen.

Glycogen in muscle, liver, and fat cells is stored in a hydrated form, composed of three or four parts of water per part of glycogen associated with 0.45 millimoles (18 mg) of potassium per gram of glycogen.

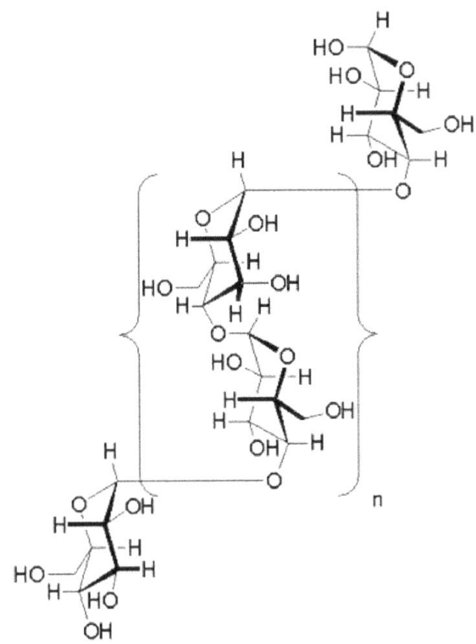

Fig 1: α(1→4)-Glycosidic Linkages in the Glycogen Oligomer

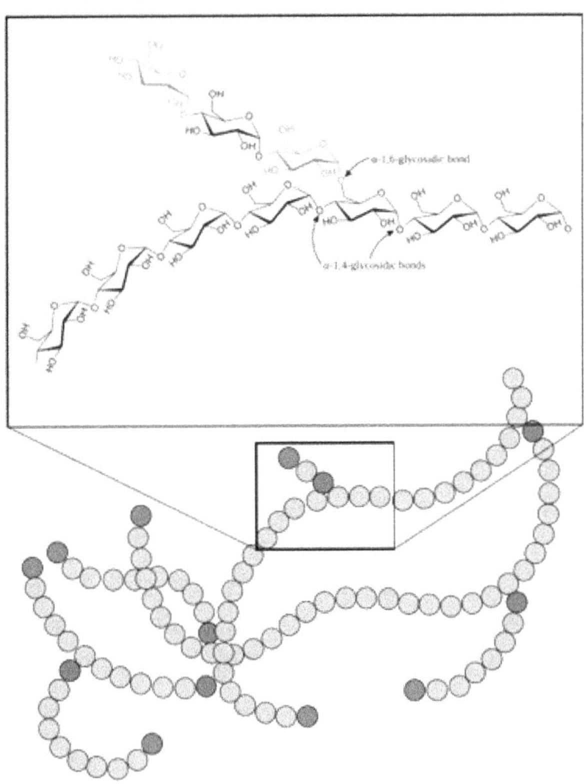

Fig 2: α(1→ 4)-Glycosidic and α(1→ 6)-Glycosidic Linkages in the Glycogen Oligomer

Glucose is an osmotic molecule, and can have profound effects on osmotic pressure in high concentrations possibly leading to cell damage or death if stored in the cell without being modified. Glycogen is a non-osmotic molecule, so it can be used as a solution to storing glucose in the cell without disrupting osmotic pressure.

21.4 FUNCTIONS

21.4.1 Liver

As a meal containing carbohydrates or protein is eaten and digested, blood glucose levels rise, and the pancreas secretes insulin. Blood glucose from the portal vein enters liver cells (hepatocytes). Insulin acts on the hepatocytes to stimulate the action of several enzymes, including glycogen synthase. Glucose molecules are added to the chains of glycogen as long as both insulin and glucose remain plentiful. In this postprandial or "fed" state, the liver takes in more glucose from the blood than it releases.

After a meal has been digested and glucose levels begin to fall, insulin secretion is reduced, and glycogen synthesis stops. When it is needed for energy, glycogen is broken down and converted again to glucose. Glycogen phosphorylase is the primary enzyme of glycogen breakdown. For the next 8–12 hours, glucose derived from liver glycogen is the primary source of blood glucose used by the rest of the body for fuel.

Glucagon, another hormone produced by the pancreas, in many respects serves as a countersignal to insulin. In response to insulin levels being below normal (when blood levels of glucose begin to fall below the normal range), glucagon is secreted in increasing amounts and stimulates both glycogenolysis (the breakdown of glycogen) and gluconeogenesis (the production of glucose from other sources).

21.4.2 Muscle

Muscle glycogen appears to function as a reserve of quickly available phosphorylated glucose, in the form of glucose-1-phosphate, for muscle cells. Glycogen contained within skeletal muscle cells are primarily in the form of β particles. Other cells that contain small amounts use it locally as well. As muscle cells lack glucose-6-phosphatase, which is required to pass glucose into the blood, the glycogen they store is available solely for internal use and is not shared with other cells. This is in contrast to liver cells, which, on demand, readily do break down their stored glycogen into glucose and send it through the blood stream as fuel for other organs.

Skeletal muscle needs ATP (provides energy) for muscle contraction and relaxation, in what is known as the sliding filament theory. Skeletal muscle relies predominantly on glycogenolysis for the first few minutes as it transitions from rest to activity, as well as throughout high-intensity aerobic activity and all anaerobic activity. During anaerobic activity, such as weightlifting and isometric exercise, the phosphagen system (ATP-PCr) and muscle glycogen are the only substrates used as they do not require oxygen nor blood flow.

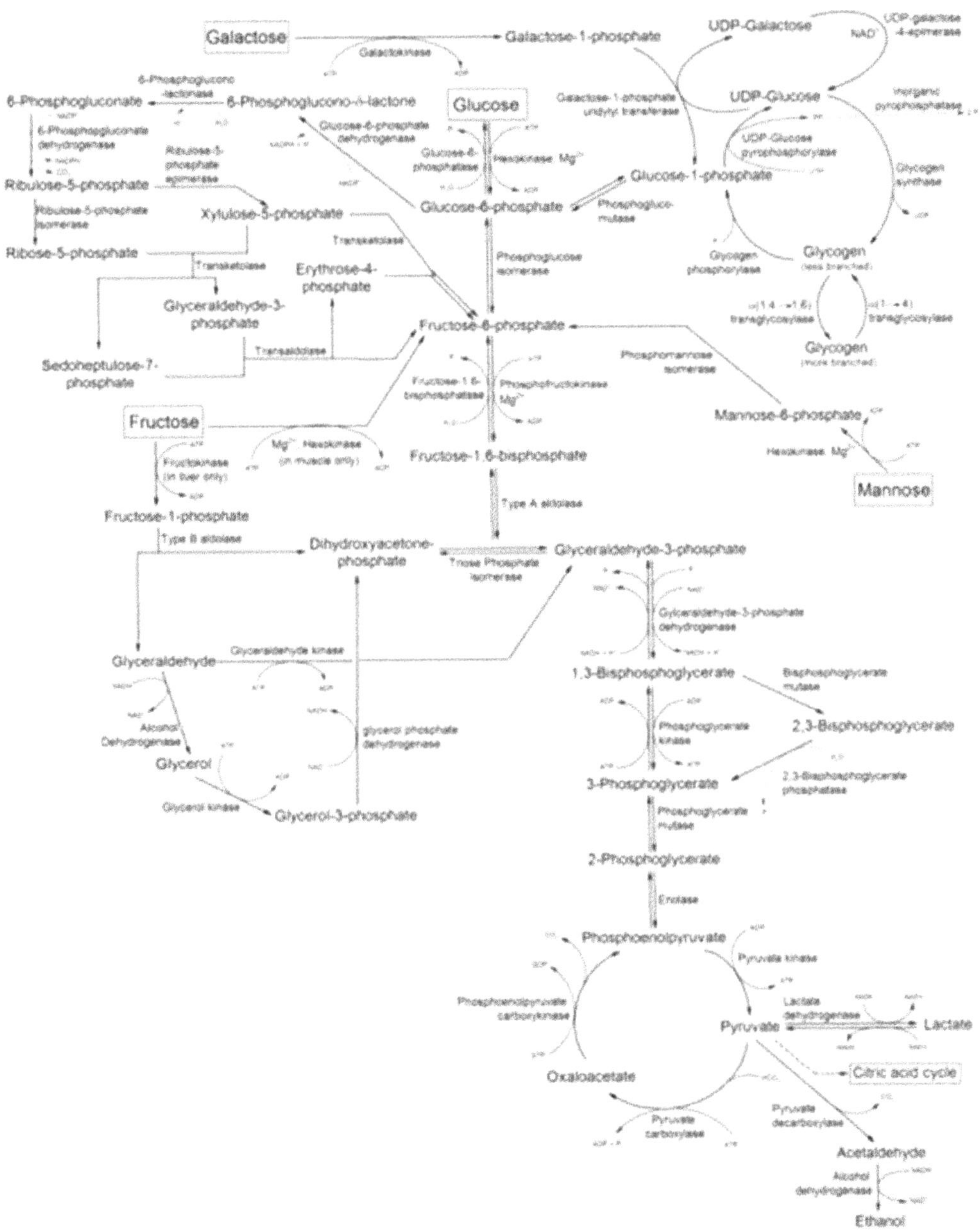

Fig 3: Metabolism of Common Monosaccharides

Different bioenergetic systems produce ATP at different speeds, with ATP produced from muscle glycogen being much faster than fatty acid oxidation. The level of exercise intensity determines how much of which substrate (fuel) is used for ATP synthesis also. Muscle glycogen can supply a much higher rate of substrate for ATP synthesis than blood glucose. During maximum intensity exercise, muscle glycogen can supply 40 mmol glucose/kg wet weight/minute, whereas blood glucose can supply 4 - 5 mmol. Due to its high supply rate and quick ATP synthesis, during high-intensity aerobic activity (such as brisk walking, jogging, or running), the higher the exercise intensity, the more the muscle cell produces ATP from muscle glycogen. This reliance on muscle glycogen is not only to provide the muscle with enough ATP during high-intensity exercise, but also to maintain blood glucose homeostasis (that is, to not become hypoglycaemic by the muscles needing to extract far more glucose from the blood than the liver can provide).

A deficit of muscle glycogen leads to muscle fatigue known as “hitting the wall” or “the bonk”.

21.5 STRUCTURE TYPE

In 1999, Meléndez et al claimed that the structure of glycogen is optimal under a particular metabolic constraint model, where the structure was suggested to be “fractal” in nature. However, research by Besford et al used small angle X-ray scattering experiments accompanied by branching theory models to show that glycogen is a randomly hyperbranched polymer nanoparticle. Glycogen is not fractal in nature. This has been subsequently verified by others who have performed Monte Carlo simulations of glycogen particle growth, and shown that the molecular density reaches a maximum near the centre of the nanoparticle structure, not at the periphery (contradicting a fractal structure that would have greater density at the periphery).

21.6 METABOLISM

21.6.1 Synthesis

Glycogen synthesis is, unlike its breakdown, endergonic—it requires the input of energy. Energy for glycogen synthesis comes from uridine triphosphate (UTP), which reacts with glucose-1-phosphate, forming UDP-glucose, in a reaction catalysed by UTP—glucose-1-phosphate uridylyltransferase. Glycogen is synthesized from monomers of UDP-glucose initially by the protein glycogenin, which has two tyrosine anchors for the reducing end of glycogen, since glycogenin is a homodimer. After about eight glucose molecules have been added to a tyrosine residue, the enzyme glycogen synthase progressively lengthens the glycogen chain using UDP-glucose, adding $\alpha(1\rightarrow4)$-bonded glucose to the nonreducing end of the glycogen chain.

The glycogen branching enzyme catalyzes the transfer of a terminal fragment of six or seven glucose residues from a nonreducing end to the C-6 hydroxyl group of a glucose residue deeper into the interior of the glycogen molecule. The branching enzyme can act upon only a branch having at least 11 residues, and the enzyme may transfer to the same glucose chain or adjacent glucose chains.

21.6.2 Breakdown

Glycogen is cleaved from the nonreducing ends of the chain by the enzyme glycogen phosphorylase to produce monomers of glucose-1-phosphate:

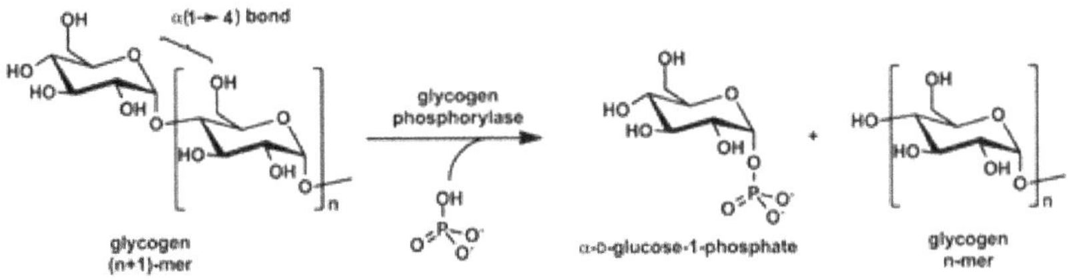

In vivo, phosphorolysis proceeds in the direction of glycogen breakdown because the ratio of phosphate and glucose-1-phosphate is usually greater than 100. Glucose-1-phosphate is then converted to glucose 6 phosphate (G6P) by phosphoglucomutase. A special debranching enzyme is needed to remove the $\alpha(1\rightarrow6)$ branches in branched glycogen and reshape the chain into a linear polymer. The G6P monomers produced have three possible fates:

- G6P can continue on the glycolysis pathway and be used as fuel.
- G6P can enter the pentose phosphate pathway via the enzyme glucose-6-phosphate dehydrogenase to produce NADPH and 5 carbon sugars.
- In the liver and kidney, G6P can be dephosphorylated back to glucose by the enzyme glucose 6-phosphatase. This is the final step in the gluconeogenesis pathway.

21.7 CLINICAL RELEVANCE

21.7.1 Disorders of Glycogen Metabolism

The most common disease in which glycogen metabolism becomes abnormal is diabetes, in which, because of abnormal amounts of insulin, liver glycogen can be abnormally accumulated or depleted. Restoration of normal glucose metabolism usually normalizes glycogen metabolism, as well.

In hypoglycemia caused by excessive insulin, liver glycogen levels are high, but the high insulin levels prevent the glycogenolysis necessary to maintain normal blood sugar levels. Glucagon is a common treatment for this type of hypoglycemia.

Various inborn errors of carbohydrate metabolism are caused by deficiencies of enzymes or transport proteins necessary for glycogen synthesis or breakdown. These are collectively referred to as glycogen storage diseases.

21.7.2 Glycogen Depletion and Endurance Exercise

Long-distance athletes, such as marathon runners, cross-country skiers, and cyclists, often experience glycogen depletion, where almost all of the athlete's glycogen stores are depleted after long periods of exertion without sufficient carbohydrate consumption. This phenomenon is referred to as "hitting the wall" in running and "bonking" in cycling. Glycogen depletion can be forestalled in three possible ways:

- **First**, during exercise, carbohydrates with the highest possible rate of conversion to blood glucose (high glycemic index) are ingested continuously. The best possible outcome of this strategy replaces about 35% of glucose consumed at heart rates above about 80% of maximum.
- **Second**, through endurance training adaptations and specialized regimens (e.g. fasting, low-intensity endurance training), the body can condition type I muscle fibers to improve both fuel use efficiency and workload capacity to increase the percentage of fatty acids used as fuel, sparing carbohydrate use from all sources.
- **Third**, by consuming large quantities of carbohydrates after depleting glycogen stores as a result of exercise or diet, the body can increase storage capacity of intramuscular glycogen stores. This process is known as carbohydrate loading. In general, glycemic index of carbohydrate source does not matter since muscular insulin sensitivity is increased as a result of temporary glycogen depletion.

When athletes ingest both carbohydrate and caffeine following exhaustive exercise, their glycogen stores tend to be replenished more rapidly; however, the minimum dose of caffeine at which there is a clinically significant effect on glycogen repletion has not been established.

CHAPTER 22

FATTY ACID AND TRIACYLGLYCEROL METABOLISM

22.1 FATTY ACID

In chemistry, particularly in biochemistry, a fatty acid is a carboxylic acid with an aliphatic chain, which is either saturated or unsaturated. Most naturally occurring fatty acids have an unbranched chain of an even number of carbon atoms, from 4 to 28. Fatty acids are a major component of the lipids (up to 70% by weight) in some species such as microalgae but in some other organisms are not found in their standalone form, but instead exist as three main classes of esters: triglycerides, phospholipids, and cholesteryl esters. In any of these forms, fatty acids are both important dietary sources of fuel for animals and important structural components for cells.

22.1.1 History

The concept of fatty acid (*acide gras*) was introduced in 1813 by Michel Eugène Chevreul, though he initially used some variant terms: *graisse acide* and *acide huileux* ("acid fat" and "oily acid").

22.1.2 Types of Fatty Acids

Length of Fatty Acids

- **Short-chain fatty acids (SCFAs)** are fatty acids with aliphatic tails of five or fewer carbons (e.g. butyric acid).
- **Medium-chain fatty acids (MCFAs)** are fatty acids with aliphatic tails of 6 to 12 carbons, which can form medium-chain triglycerides.
- **Long-chain fatty acids (LCFAs)** are fatty acids with aliphatic tails of 13 to 21 carbons.
- **Very long chain fatty acids (VLCFAs)** are fatty acids with aliphatic tails of 22 or more carbons.

22.1.3 Saturated Fatty Acids

Saturated fatty acids have no C=C double bonds. They have the formula $CH_3(CH_2)_nCOOH$, for different *n*. An important saturated fatty acid is stearic acid ($n = 16$), which when neutralized with sodium hydroxide is the most common form of soap.

Table 1: Examples of Saturated Fatty Acids

Common Name	Chemical Structure	C:D
Caprylic acid	$CH_3(CH_2)_6COOH$	8:0
Capric acid	$CH_3(CH_2)_8COOH$	10:0
Lauric acid	$CH_3(CH_2)_{10}COOH$	12:0
Myristic acid	$CH_3(CH_2)_{12}COOH$	14:0
Palmitic acid	$CH_3(CH_2)_{14}COOH$	16:0
Stearic acid	$CH_3(CH_2)_{16}COOH$	18:0
Arachidic acid	$CH_3(CH_2)_{18}COOH$	20:0
Behenic acid	$CH_3(CH_2)_{20}COOH$	22:0
Lignoceric acid	$CH_3(CH_2)_{22}COOH$	24:0
Cerotic acid	$CH_3(CH_2)_{24}COOH$	26:0

22.1.4 Unsaturated Fatty Acids

Unsaturated fatty acids have one or more C=C double bonds. The C=C double bonds can give either *cis* or *trans* isomers.

- **Cis:** A *cis* configuration means that the two hydrogen atoms adjacent to the double bond stick out on the same side of the chain. The rigidity of the double bond freezes its conformation and, in the case of the *cis* isomer, causes the chain to bend and restricts the conformational freedom of the fatty acid. The more double bonds the chain has in the *cis* configuration, the less flexibility it has. When a chain has many *cis* bonds, it becomes quite curved in its most accessible conformations. For example, oleic acid, with one double bond, has a "kink" in it, whereas linoleic acid, with two double bonds, has a more pronounced bend. α-Linolenic acid, with three double bonds, favors a hooked shape. The effect of this is that, in restricted environments, such as when fatty acids are part of a phospholipid in a lipid bilayer or triglycerides in lipid droplets, cis bonds limit the ability of fatty acids to be closely packed, and therefore can affect the melting temperature of the membrane or of the fat. Cis unsaturated fatty acids, however, increase cellular membrane fluidity, whereas trans unsaturated fatty acids do not.
- **Trans:** A *trans* configuration, by contrast, means that the adjacent two hydrogen atoms lie on *opposite* sides of the chain. As a result, they do not cause the chain to bend much, and their shape is similar to straight saturated fatty acids.

In most naturally occurring unsaturated fatty acids, each double bond has three (n-3), six (n-6), or nine (n-9) carbon atoms after it, and all double bonds have a cis configuration. Most fatty acids in the *trans* configuration (trans fats) are not found in nature and are the result of human processing (e.g., hydrogenation). Some trans fatty acids also occur naturally in the milk and meat of ruminants (such as cattle and sheep). They are produced, by fermentation, in the rumen of these animals. They are also found in dairy products from milk of ruminants, and may be also found in breast milk of women who obtained them from their diet.

22.1.5 Even- vs. Odd-Chained Fatty Acids

Most fatty acids are even-chained, e.g. stearic (C18) and oleic (C18), meaning they are composed of an even number of carbon atoms. Some fatty acids have odd numbers of carbon atoms; they are referred to as odd-chained fatty acids (OCFA). The most common OCFA are the saturated C15 and C17 derivatives, pentadecanoic acid and heptadecanoic acid respectively, which are found in dairy products. On a molecular level, OCFAs are biosynthesized and metabolized slightly differently from the even-chained relatives.

22.1.6 Branching

Most common fatty acids are straight-chain compounds, with no additional carbon atoms bonded as side groups to the main hydrocarbon chain. Branched-chain fatty acids contain one or more methyl groups bonded to the hydrocarbon chain.

22.2 NOMENCLATURE

22.2.1 Carbon Atom Numbering

Most naturally occurring fatty acids have an unbranched chain of carbon atoms, with a carboxyl group (–COOH) at one end, and a methyl group (–CH3) at the other end.

The position of each carbon atom in the backbone of a fatty acid is usually indicated by counting from 1 at the -COOH end. Carbon number *x* is often abbreviated C-*x* (or sometimes C*x*), with *x* = 1, 2, 3, etc. This is the numbering scheme recommended by the IUPAC.

Another convention uses letters of the Greek alphabet in sequence, starting with the first carbon *after* the carboxyl group. Thus carbon α (alpha) is C-2, carbon β (beta) is C-3, and so forth.

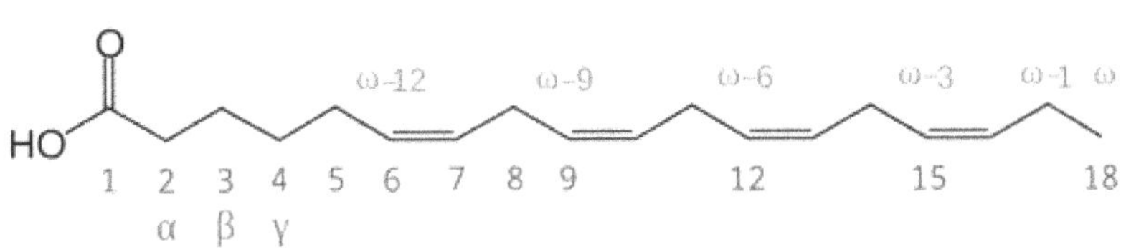

Fig 1: Numbering of Carbon Atoms

The systematic (IUPAC) C-*x* numbers are in blue. The omega-minus "ω−*x*" labels are in red. The Greek letter labels are in green. Note that unsaturated fatty acids with a *cis* configuration are actually "kinked" rather than straight as shown here.

Although fatty acids can be of diverse lengths, in this second convention the last carbon in the chain is always labelled as ω (omega), which is the last letter in the Greek alphabet. A third numbering convention counts the carbons from that end, using the labels "ω", "ω−1", "ω−2".

Alternatively, the label "ω−*x*" is written "n−*x*", where the "n" is meant to represent the number of carbons in the chain.

In either numbering scheme, the position of a double bond in a fatty acid chain is always specified by giving the label of the carbon closest to the carboxyl end. Thus, in an 18 carbon fatty acid, a double bond between C-12 (or ω−6) and C-13 (or ω−5) is said to be "at" position C-12 or ω−6.

The IUPAC naming of the acid, such as "octadec-12-enoic acid" (or the more pronounceable variant "12-octadecanoic acid") is always based on the "C" numbering.

The notation $\Delta^{x,y,...}$ is traditionally used to specify a fatty acid with double bonds at positions *x*,*y*,.... (The capital Greek letter "Δ" (delta) corresponds to Roman "D", for Double bond). Thus, for example, the 20-carbon arachidonic acid is $\Delta^{5,8,11,14}$, meaning that it has double bonds between carbons 5 and 6, 8 and 9, 11 and 12, and 14 and 15.

In the context of human diet and fat metabolism, unsaturated fatty acids are often classified by the position of the double bond closest between to the ω carbon (only), even in the case of multiple double bonds such as the essential fatty acids. Thus linoleic acid (18 carbons, $\Delta^{9,12}$), γ-linolenic acid (18-carbon, $\Delta^{6,9,12}$), and arachidonic acid (20-carbon, $\Delta^{5,8,11,14}$) are all classified as "ω−6" fatty acids; meaning that their formula ends with $-CH{=}CH-CH_2-CH_2-CH_2-CH_2-CH_3$.

Fatty acids with an odd number of carbon atoms are called odd-chain fatty acids, whereas the rest are even-chain fatty acids. The difference is relevant to gluconeogenesis.

22.2.2 Free Fatty Acids

When circulating in the plasma (plasma fatty acids), not in their ester, fatty acids are known as non-esterified fatty acids (NEFAs) or free fatty acids (FFAs). FFAs are always bound to a transport protein, such as albumin.

FFAs also form from triglyceride food oils and fats by hydrolysis, contributing to the characteristic rancid odor. An analogous process happens in biodiesel with risk of part corrosion.

22.2.3 Naming of Fatty Acids

The following table describes the most common systems of naming fatty acids.

Nomenclature	Examples	Explanation
Trivial	Palmitoleic acid	Trivial names (or common names) are non-systematic historical names, which are the most frequent naming system used in literature. Most common fatty acids have trivial names in addition to their *systematic names*. These names frequently do not follow any pattern, but they are concise and often unambiguous.
Systematic	cis-9-octadec-9-enoic acid (9Z)-octadec-9-enoic acid	Systematic names (or IUPAC names) derive from the standard *IUPAC Rules for the Nomenclature of Organic Chemistry*, published in 1979, along with a recommendation published specifically for lipids in 1977. Carbon atom numbering begins from the carboxylic end of the molecule backbone. Double bonds are labelled with *cis-/trans-* notation or *E-/Z-* notation, where appropriate. This notation is generally more verbose than common nomenclature, but has the advantage of being more technically clear and descriptive.
Δ^x	*cis*-Δ^9, *cis*-Δ^{12} octadecadienoic acid	In Δ^x (or delta-*x*) nomenclature, each double bond is indicated by Δ^x, where the double bond begins at the xth carbon–carbon bond, counting from carboxylic end of the molecule backbone. Each double bond is preceded by a *cis-* or *trans-* prefix, indicating the configuration of the molecule around the bond. For example, linoleic acid is designated "*cis*-Δ^9, *cis*-Δ^{12} octadecadienoic acid". This nomenclature has the advantage of being less verbose than systematic nomenclature, but is no more technically clear or descriptive.
n – *x* (or ω-*x*)	*n*-3 (or ω3-)	*n*-*x* (*n* minus *x*; also **ω**-*x* or omega-*x*) nomenclature both provides names for individual compounds and classifies them by their likely biosynthetic properties in animals. A double bond is located on the xth carbon–carbon bond, counting from the methyl end of the molecule backbone. For example, α-linolenic acid is classified as a *n*-3 or omega-3 fatty acid, and so it is likely to share a biosynthetic pathway with other compounds of this type. The ω-*x*, omega-*x*, or "omega" notation is common in popular nutritional literature, but IUPAC has deprecated it in favor of *n*-*x* notation in technical documents. The most commonly researched fatty acid biosynthetic pathways are *n*-3 and *n*-6.

Nomenclature	Examples	Explanation
Lipid numbers	18:3 18:3n3 18:3, *cis*,*cis*,*cis*-Δ^9,Δ^{12},Δ^{15} 18:3(9,12,15)	Lipid numbers take the form *C*:*D*, where *C* is the number of carbon atoms in the fatty acid and *D* is the number of double bonds in the fatty acid. If D is more than one, the double bonds are assumed to be interrupted by CH_2 units, *i.e.*, at intervals of 3 carbon atoms along the chain. For instance, α-linolenic acid is an 18:3 fatty acid and its three double bonds are located at positions Δ^9, Δ^{12}, and Δ^{15}. This notation can be ambiguous, as some different fatty acids can have the same *C*:*D* numbers. Consequently, when ambiguity exists this notation is usually paired with either a Δ^x or *n*−*x* term. For instance, although α-linolenic acid and γ-linolenic acid are both 18:3, they may be unambiguously described as 18:3n3 and 18:3n6 fatty acids, respectively. For the same purpose, IUPAC recommends using a list of double bond positions in parentheses, appended to the C:D notation. For instance, IUPAC recommended notations for α- and γ-linolenic acid are 18:3(9,12,15) and 18:3(6,9,12), respectively.

22.2.4 Production

Industrial

Fatty acids are usually produced industrially by the hydrolysis of triglycerides, with the removal of glycerol. Phospholipids represent another source. Some fatty acids are produced synthetically by hydrocarboxylation of alkenes.

By Animals

In animals, fatty acids are formed from carbohydrates predominantly in the liver, adipose tissue, and the mammary glands during lactation.

Carbohydrates are converted into pyruvate by glycolysis as the first important step in the conversion of carbohydrates into fatty acids. Pyruvate is then decarboxylated to form acetyl-CoA in the mitochondrion. However, this acetyl CoA needs to be transported into cytosol where the synthesis of fatty acids occurs. This cannot occur directly.

To obtain cytosolic acetyl-CoA, citrate (produced by the condensation of acetyl-CoA with oxaloacetate) is removed from the citric acid cycle and carried across the inner mitochondrial membrane into the cytosol. There it is cleaved by ATP citrate lyase into acetyl-CoA and oxaloacetate. The oxaloacetate is returned to the mitochondrion as malate. The cytosolic acetyl-CoA is carboxylated by acetyl-CoA carboxylase into malonyl-CoA, the first committed step in the synthesis of fatty acids.

Malonyl-CoA is then involved in a repeating series of reactions that lengthens the growing fatty acid chain by two carbons at a time. Almost all natural fatty acids, therefore, have even numbers of carbon atoms. When synthesis is complete the free fatty acids are nearly always combined with glycerol (three fatty acids to one glycerol molecule) to form triglycerides, the main storage form of fatty acids, and thus of energy in animals. However, fatty acids are also important components of the phospholipids that form the phospholipid bilayers out of which all the membranes of the cell are constructed (the cell wall, and the membranes that enclose all the organelles within the cells, such as the nucleus, the mitochondria, endoplasmic reticulum, and the Golgi apparatus).

The "uncombined fatty acids" or "free fatty acids" found in the circulation of animals come from the breakdown (or lipolysis) of stored triglycerides. Because they are insoluble in water, these fatty acids are transported bound to plasma albumin. The levels of "free fatty acids" in the blood are limited by the availability of albumin binding sites. They can be taken up from the blood by all cells that have mitochondria (with the exception of the cells of the central nervous system). Fatty acids can only be broken down in mitochondria, by means of beta-oxidation followed by further combustion in the citric acid cycle to CO_2 and water. Cells in the central nervous system, although they possess mitochondria, cannot take free fatty acids up from the blood, as the blood–brain barrier is impervious to most free fatty acids, excluding short-chain fatty acids and medium-chain fatty acids. These cells have to manufacture their own fatty acids from carbohydrates, as described above, in order to produce and maintain the phospholipids of their cell membranes, and those of their organelles.

Variation between Animal Species

Studies on the cell membranes of mammals and reptiles discovered that mammalian cell membranes are composed of a higher proportion of polyunsaturated fatty acids (DHA, omega-3 fatty acid) than reptiles. Studies on bird fatty acid composition have noted similar proportions to mammals but with 1/3rd less omega-3 fatty acids as compared to omega-6 for a given body size.

This fatty acid composition results in a more fluid cell membrane but also one that is permeable to various ions (H^+ & Na^+), resulting in cell membranes that are more costly to maintain. This maintenance cost has been argued to be one of the key causes for the high metabolic rates and concomitant warm-bloodedness of mammals and birds. However polyunsaturation of cell membranes may also occur in response to chronic cold temperatures as well. In fish increasingly cold environments lead to increasingly high cell membrane content of both monounsaturated and polyunsaturated fatty acids, to maintain greater membrane fluidity (and functionality) at the lower temperatures.

22.2.5 Fatty Acids in Dietary Fats

The following table gives the fatty acid, vitamin E and cholesterol composition of some common dietary fats.

	Saturated	Monounsaturated	Polyunsaturated	Cholesterol	Vitamin E
	g/100g	g/100g	g/100g	mg/100g	mg/100g
Animal fats					
Duck fat	33.2	49.3	12.9	100	2.70
Lard	40.8	43.8	9.6	93	0.60
Tallow	49.8	41.8	4.0	109	2.70
Butter	54.0	19.8	2.6	230	2.00
Vegetable fats					
Coconut oil	85.2	6.6	1.7	0	.66
Cocoa butter	60.0	32.9	3.0	0	1.8
Palm kernel oil	81.5	11.4	1.6	0	3.80
Palm oil	45.3	41.6	8.3	0	33.12
Cottonseed oil	25.5	21.3	48.1	0	42.77
Wheat germ oil	18.8	15.9	60.7	0	136.65
Soybean oil	14.5	23.2	56.5	0	16.29
Olive oil	14.0	69.7	11.2	0	5.10
Corn oil	12.7	24.7	57.8	0	17.24
Sunflower oil	11.9	20.2	63.0	0	49.00
Safflower oil	10.2	12.6	72.1	0	40.68
Hemp oil	10	15	75	0	12.34
Canola/ Rapeseed oil	5.3	64.3	24.8	0	22.21

22.2.6 Reactions of Fatty Acids

Fatty acids exhibit reactions like other carboxylic acids, i.e. they undergo esterification and acid-base reactions.

Acidity

Fatty acids do not show a great variation in their acidities, as indicated by their respective pK_a. Nonanoic acid, for example, has a pK_a of 4.96, being only slightly weaker than acetic acid (4.76). As the chain length increases, the solubility of the fatty acids in water decreases, so that the longer-chain fatty acids have minimal effect on the pH of an aqueous solution. Near neutral pH, fatty acids exist at their conjugate bases, i.e. oleate, etc.

Solutions of fatty acids in ethanol can be titrated with sodium hydroxide solution using phenolphthalein as an indicator.

This analysis is used to determine the free fatty acid content of fats; i.e., the proportion of the triglycerides that have been hydrolyzed. Neutralization of fatty acids, one form of saponification (soap-making), is a widely practiced route to metallic soaps.

Hydrogenation and Hardening

Hydrogenation of unsaturated fatty acids is widely practiced. Typical conditions involve 2.0–3.0 MPa of H_2 pressure, 150 °C, and nickel supported on silica as a catalyst. This treatment affords saturated fatty acids. The extent of hydrogenation is indicated by the iodine number.

Hydrogenated fatty acids are less prone toward rancidification. Since the saturated fatty acids are higher melting than the unsaturated precursors, the process is called hardening. Related technology is used to convert vegetable oils into margarine. The hydrogenation of triglycerides (vs. fatty acids) is advantageous because the carboxylic acids degrade the nickel catalysts, affording nickel soaps. During partial hydrogenation, unsaturated fatty acids can be isomerized from *cis* to *trans* configuration.

More forcing hydrogenation, i.e. using higher pressures of H_2 and higher temperatures, converts fatty acids into fatty alcohols. Fatty alcohols are, however, more easily produced from fatty acid esters. In the Varrentrapp reaction certain unsaturated fatty acids are cleaved in molten alkali, a reaction which was, at one point of time, relevant to structure elucidation.

Auto-Oxidation and Rancidity

Unsaturated fatty acids and their esters undergo auto-oxidation, which involves replacement of a C-H bond with C-O bond. The process requires oxygen (air) and is accelerated by the presence of traces of metals, which serve as catalysts. Doubly unsaturated fatty acids are particularly prone to this reaction. Vegetable oils resist this process to a small degree because they contain antioxidants, such as tocopherol. Fats and oils often are treated with chelating agents such as citric acid to remove the metal catalysts.

Ozonolysis

Unsaturated fatty acids are susceptible to degradation by ozone. This reaction is practiced in the production of azelaic acid ($(CH_2)_7(CO_2H)_2$) from oleic acid.

22.2.7 Circulation

Digestion and Intake

Short- and medium-chain fatty acids are absorbed directly into the blood via intestine capillaries and travel through the portal vein just as other absorbed nutrients do. However, long-chain fatty acids are not directly released into the intestinal capillaries. Instead they are absorbed into the fatty walls of the intestine villi and reassemble again into triglycerides. The triglycerides are coated with cholesterol and protein (protein coat) into a compound called a chylomicron.

From within the cell, the chylomicron is released into a lymphatic capillary called a lacteal, which merges into larger lymphatic vessels. It is transported via the lymphatic system and the thoracic duct up to a location near the heart (where the arteries and veins are larger). The thoracic duct empties the chylomicrons into the bloodstream via the left subclavian vein. At this point the chylomicrons can transport the triglycerides to tissues where they are stored or metabolized for energy.

Metabolism

Fatty acids are broken down to CO_2 and water by the intra-cellular mitochondria through beta oxidation and the citric acid cycle. In the final step (oxidative phosphorylation), reactions with oxygen release a lot of energy, captured in the form of large quantities of ATP. Many cell types can use either glucose or fatty acids for this purpose, but fatty acids release more energy per gram. Fatty acids (provided either by ingestion or by drawing on triglycerides stored in fatty tissues) are distributed to cells to serve as a fuel for muscular contraction and general metabolism.

Essential Fatty Acids

Fatty acids that are required for good health but cannot be made in sufficient quantity from other substrates, and therefore must be obtained from food, are called essential fatty acids. There are two series of essential fatty acids: one has a double bond three carbon atoms away from the methyl end; the other has a double bond six carbon atoms away from the methyl end. Humans lack the ability to introduce double bonds in fatty acids beyond carbons 9 and 10, as counted from the carboxylic acid side. Two essential fatty acids are linoleic acid (LA) and alpha-linolenic acid (ALA). These fatty acids are widely distributed in plant oils. The human body has a limited ability to convert ALA into the longer-chain omega-3 fatty acids – eicosapentaenoic acid (EPA) and docosahexaenoic acid (DHA), which can also be obtained from fish. Omega-3 and omega-6 fatty acids are biosynthetic precursors to endocannabinoids with antinociceptive, anxiolytic, and neurogenic properties.

Distribution

Blood fatty acids adopt distinct forms in different stages in the blood circulation. They are taken in through the intestine in chylomicrons, but also exist in very low density lipoproteins (VLDL) and low density lipoproteins (LDL) after processing in the liver. In addition, when released from adipocytes, fatty acids exist in the blood as free fatty acids. It is proposed that the blend of fatty acids exuded by mammalian skin, together with lactic acid and pyruvic acid, is distinctive and enables animals with a keen sense of smell to differentiate individuals.

22.2.8 Skin

The stratum corneum – the outermost layer of the epidermis – is composed of terminally differentiated and enucleated corneocytes within a lipid matrix. Together with cholesterol and ceramides, free fatty acids form a water-impermeable barrier that prevents evaporative water loss. Generally, the epidermal lipid matrix is composed of an equimolar mixture of ceramides (about 50% by weight), cholesterol (25%), and free fatty acids (15%). Saturated fatty acids 16 and 18 carbons in length are the dominant types in the epidermis, while unsaturated fatty acids and saturated fatty acids of various other lengths are also present. The relative abundance of the different fatty acids in the epidermis is dependent on the body site the skin is covering. There are also characteristic epidermal fatty acid alterations that occur in psoriasis, atopic dermatitis, and other inflammatory conditions.

22.2.9 Analysis

The chemical analysis of fatty acids in lipids typically begins with an interesterification step that breaks down their original esters (triglycerides, waxes, phospholipids etc.) and converts them to methyl esters, which are then separated by gas chromatography or analyzed by gas chromatography and mid-infrared spectroscopy. Separation of unsaturated isomers is possible by silver ion complemented thin-layer chromatography. Other separation techniques include high-performance liquid chromatography (with short columns packed with silica gel with bonded phenylsulfonic acid groups whose hydrogen atoms have been exchanged for silver ions). The role of silver lies in its ability to form complexes with unsaturated compounds.

22.2.10 Industrial Uses

Fatty acids are mainly used in the production of soap, both for cosmetic purposes and, in the case of metallic soaps, as lubricants. Fatty acids are also converted, via their methyl esters, to fatty alcohols and fatty amines, which are precursors to surfactants, detergents, and lubricants. Other applications include their use as emulsifiers, texturizing agents, wetting agents, anti-foam agents, or stabilizing agents.

Esters of fatty acids with simpler alcohols (such as methyl-, ethyl-, n-propyl-, isopropyl- and butyl esters) are used as emollients in cosmetics and other personal care products and as synthetic lubricants. Esters of fatty acids with more complex alcohols, such as sorbitol, ethylene glycol, diethylene glycol, and polyethylene glycol are consumed in food, or used for personal care and water treatment, or used as synthetic lubricants or fluids for metal working.

22.3 TRIACYLGLYCEROL METABOLISM

Triacylglycerols, also known as triglycerides, are the primary storage form of fats in animals and play a crucial role in energy metabolism. Comprised of three fatty acid molecules attached to a glycerol backbone, triacylglycerols are highly energy-dense molecules, making them ideal for long-term energy storage. Their metabolism is critical for energy balance, as they provide fuel during fasting and extended periods of physical activity.

Triacylglycerol metabolism is a dynamic and complex process essential for energy storage and release in animals. Through tightly regulated synthesis, storage, and breakdown mechanisms, triacylglycerols provide a critical energy reserve, allowing the body to maintain energy balance and adapt to various metabolic states. Understanding these processes can provide insights into health conditions related to lipid metabolism and inspire therapeutic strategies for managing metabolic disorders. As researchers continue to explore triacylglycerol metabolism, the knowledge gained has the potential to enhance our understanding of energy balance and improve health outcomes.

22.3.1 Structure and Properties of Triacylglycerols

Triacylglycerols are composed of three fatty acids esterified to a glycerol molecule. Each fatty acid is connected to one of the three hydroxyl groups on the glycerol backbone, creating a triacylglycerol molecule. These molecules are hydrophobic, meaning they do not mix well with water.

This property allows them to be packed densely in specialized fat-storing cells, or adipocytes, without adding extra weight from water retention. The high energy density of triacylglycerols—providing about 9 kcal of energy per gram—makes them efficient for energy storage and a critical energy source for cells during metabolic stress.

In the human body, triacylglycerols are stored mainly in adipose tissue, although smaller amounts are found in the liver and muscle cells. The liver plays an essential role in triacylglycerol metabolism, synthesizing these molecules and distributing them to other tissues in the form of lipoproteins. When the body needs energy, enzymes break down stored triacylglycerols, releasing fatty acids into the bloodstream where they can be taken up by cells and metabolized for energy.

22.3.2 Synthesis of Triacylglycerols (Triacylglycerol Anabolism)

Triacylglycerol synthesis, or **lipogenesis**, occurs primarily in the liver and adipose tissue. The process begins with the synthesis of **fatty acids** through a process called **fatty acid synthesis**, in which acetyl-CoA is converted into long-chain fatty acids. This pathway is highly regulated by insulin, a hormone that promotes fat storage and synthesis when blood glucose levels are high.

Step 1: Glycerol Phosphate Formation

The synthesis of triacylglycerols starts with the production of **glycerol-3-phosphate**, which serves as the glycerol backbone for fatty acid attachment. Glycerol-3-phosphate is derived from two main sources:

- **Dihydroxyacetone phosphate (DHAP)**, a glycolysis intermediate, which is converted into glycerol-3-phosphate by the enzyme glycerol-3-phosphate dehydrogenase.
- **Free glycerol**, which is phosphorylated by the enzyme glycerol kinase in the liver.

Step 2: Esterification of Fatty Acids to Glycerol-3-Phosphate

The next step in triacylglycerol synthesis involves the sequential addition of fatty acids to glycerol-3-phosphate. Fatty acids are first activated to **fatty acyl-CoA** forms by the enzyme acyl-CoA synthetase. The activated fatty acids are then attached to the glycerol-3-phosphate backbone in three steps, each catalyzed by an acyltransferase enzyme, resulting in the formation of **triacylglycerol**.

Once synthesized, triacylglycerols are either stored in adipocytes or packaged into **very low-density lipoproteins** (VLDLs) in the liver and transported through the bloodstream to other tissues. In adipocytes, triacylglycerols are stored in large lipid droplets, where they remain until needed for energy.

22.3.3 Storage of Triacylglycerols in Adipose Tissue

The adipose tissue serves as the primary storage site for triacylglycerols. **Adipocytes**, or fat cells, are specialized cells that store fat in the form of large lipid droplets. When food intake provides excess energy, the liver synthesizes triacylglycerols and transports them to adipose tissue for storage. Insulin, a hormone released after meals, promotes the uptake of glucose and free fatty acids into adipocytes, enhancing triacylglycerol synthesis.

The accumulation of triacylglycerols in adipose tissue allows the body to have a readily available energy reserve for times of fasting or increased energy demands, such as physical activity. Unlike glycogen, which provides short-term energy storage, triacylglycerols provide a much larger energy reserve, capable of sustaining energy needs for days or even weeks during prolonged fasting.

22.3.4 Triacylglycerol Catabolism: Lipolysis and Energy Release

When the body needs energy, such as during fasting, exercise, or periods of low carbohydrate intake, stored triacylglycerols undergo **lipolysis**, the process of breaking down triacylglycerols into free fatty acids and glycerol. Lipolysis primarily occurs in adipose tissue, where triacylglycerols stored in lipid droplets are hydrolyzed by a series of enzymes, releasing fatty acids and glycerol into the bloodstream.

Step 1: Hormonal Regulation of Lipolysis

Lipolysis is tightly regulated by hormones that respond to energy needs. **Epinephrine** and **glucagon** are the main hormones that stimulate lipolysis by activating **hormone-sensitive lipase (HSL)**, an enzyme that catalyzes the initial breakdown of triacylglycerols. When energy levels are low, these hormones trigger the release of fatty acids into the blood to supply energy to various tissues.

Conversely, insulin inhibits lipolysis by deactivating hormone-sensitive lipase, thus promoting fat storage when energy is abundant. This hormonal regulation ensures that triacylglycerol breakdown is activated only when necessary, maintaining an efficient energy balance.

Step 2: Breakdown of Triacylglycerols

In the lipolysis process, triacylglycerols are broken down in three steps:

- **Hormone-sensitive lipase** removes the first fatty acid from the triacylglycerol, yielding a diacylglycerol.
- **Diacylglycerol lipase** removes a second fatty acid, producing a monoacylglycerol.
- **Monoacylglycerol lipase** completes the process by removing the final fatty acid, releasing glycerol.

The released fatty acids enter the bloodstream, where they bind to **albumin**, a protein that transports them to tissues such as muscle and liver. Glycerol, on the other hand, is transported to the liver, where it can enter glycolysis or gluconeogenesis to produce energy or glucose.

22.3.5 Oxidation of Fatty Acids in Mitochondria

Once fatty acids are delivered to cells in need of energy, they undergo **beta-oxidation** in the mitochondria. Beta-oxidation is a series of reactions that break down fatty acids into **acetyl-CoA** molecules, which then enter the citric acid cycle to produce ATP, the primary energy currency of cells.

Beta-Oxidation Pathway

The beta-oxidation pathway occurs in four steps, each removing two carbons from the fatty acid chain:

- **Oxidation:** The fatty acid is oxidized, forming a double bond and producing $FADH_2$.
- **Hydration:** Water is added to the double bond, converting it into a hydroxyl group.
- **Oxidation:** The hydroxyl group is oxidized to a keto group, yielding NADH.
- **Cleavage:** Acetyl-CoA is cleaved from the fatty acid, shortening it by two carbons.

Each cycle of beta-oxidation produces one molecule of $FADH_2$, one molecule of NADH, and one molecule of acetyl-CoA, which enter the electron transport chain and the citric acid cycle, respectively, to produce ATP.

22.3.6 Triacylglycerol Metabolism in Fasting and Starvation

During extended fasting or starvation, the body increasingly relies on triacylglycerol stores for energy.

Fatty acids released from adipose tissue serve as the primary energy source for most tissues, including muscle, liver, and kidneys. However, the brain, which primarily uses glucose, eventually adapts by using **ketone bodies**, which are produced from fatty acids in the liver.

Ketogenesis, the process of producing ketone bodies, occurs in the liver when acetyl-CoA levels from beta-oxidation exceed the capacity of the citric acid cycle. Ketone bodies, such as acetoacetate and beta-hydroxybutyrate, can be transported through the bloodstream to the brain and other tissues, where they are converted back to acetyl-CoA and used for energy production.

22.3.7 Physiological Significance of Triacylglycerol Metabolism

The metabolism of triacylglycerols is crucial for maintaining energy homeostasis and supporting cellular functions, particularly during fasting and prolonged physical activity.

Fat storage allows organisms to endure long periods without food by providing a concentrated energy source. However, imbalances in triacylglycerol metabolism, such as excessive storage or impaired breakdown, can lead to metabolic disorders.

Obesity and Insulin Resistance

Excessive storage of triacylglycerols in adipose tissue is a hallmark of obesity, a condition often associated with insulin resistance and type 2 diabetes.

In obesity, the overaccumulation of triacylglycerols disrupts metabolic regulation, leading to an increased risk of cardiovascular disease, liver disease, and other health issues.

Lipid Metabolism Disorders

Disorders in triacylglycerol metabolism, such as **lipodystrophy** or **hyperlipidemia**, affect lipid storage and can lead to abnormal fat distribution or elevated levels of lipids in the blood. These conditions are associated with metabolic syndrome and cardiovascular complications.

CHAPTER 23

PROTEIN TURNOVER AND AMINO ACID CATABOLISM

23.1 PROTEIN TURNOVER

In cell biology, protein turnover refers to the replacement of older proteins as they are broken down within the cell. Different types of proteins have very different turnover rates. A balance between protein synthesis and protein degradation is required for good health and normal protein metabolism. More synthesis than breakdown indicates an anabolic state that builds lean tissues, more breakdown than synthesis indicates a catabolic state that burns lean tissues. According to D.S. Dunlop, protein turnover occurs in brain cells the same as any other eukaryotic cells, but that "knowledge of those aspects of control and regulation specific or peculiar to brain is an essential element for understanding brain function." Protein turnover is believed to decrease with age in all senescent organisms including humans. This results in an increase in the amount of damaged protein within the body.

Table 1: Example Protein Half-Lives

Name	Half-Life
Collagen	117 years
Eye lens crystallin	>70 years
RFC1	9 hours
RPS8	3 hours
Ornithine decarboxylase	11 minutes

23.1.1 Protein Turnover in the Exercise Science

Four weeks of aerobic exercise has been shown to increase skeletal muscle protein turnover in previously unfit individuals. A diet high in protein increases whole body turnover in endurance athletes. Some bodybuilding supplements claim to reduce the protein breakdown by reducing or blocking the number of catabolic hormones within the body. This is believed to increase anabolism. However, if protein breakdown falls too low then the body would not be able to remove muscle cells that have been damaged during workouts which would in turn prevent the growth of new muscle cells.

23.2 AMINO ACID CATABOLISM

Amino acid catabolism is the metabolic breakdown of amino acids to generate energy, produce biochemical intermediates, and synthesize key molecules for cellular functions. Unlike fats and carbohydrates, which are primarily used as energy sources, amino acids are chiefly involved in building proteins. However, under certain physiological conditions, particularly during fasting, prolonged exercise, or dietary protein intake exceeding the body's requirements, amino acids are catabolized to fulfill energy and metabolic needs. This process involves deamination, the removal of amino groups, and conversion into various intermediates that enter the citric acid cycle or other metabolic pathways.

Amino acid catabolism is a complex yet highly regulated process essential for energy production, nitrogen disposal, and metabolic flexibility. Through transamination, deamination, and the urea cycle, amino acids are effectively broken down, linking protein metabolism with carbohydrate and lipid metabolism to maintain energy balance and metabolic health. Understanding the nuances of amino acid catabolism not only provides insight into cellular metabolism but also highlights potential areas for therapeutic intervention in metabolic disorders.

23.2.1 Overview of Amino Acid Catabolism

Amino acids are obtained from dietary proteins and the breakdown of cellular proteins. Unlike carbohydrates and fats, amino acids are not stored in the body in any significant quantities, so their catabolism is primarily demand-driven, depending on the body's metabolic needs. The breakdown of amino acids serves multiple purposes:

- **Energy production** during periods of fasting or caloric deficit.
- **Generation of metabolic intermediates** used in the synthesis of glucose, fatty acids, or ketone bodies.
- **Excretion of nitrogen** through the urea cycle to prevent nitrogen accumulation, which can be toxic.

Amino acid catabolism generally begins with the removal of the amino group, producing ammonia, which is then converted into urea for safe excretion. The remaining carbon skeletons are further processed to produce energy or synthesize other biomolecules.

23.2.2 Step 1: Transamination and Deamination

Transamination

Transamination is the first step in amino acid catabolism and involves the transfer of an amino group from an amino acid to an alpha-keto acid. This reaction is catalyzed by enzymes called **aminotransferases** (or transaminases), which require pyridoxal phosphate (a derivative of vitamin B6) as a cofactor. The transamination reaction produces glutamate and an alpha-keto acid that can enter various metabolic pathways.

For example, **alanine transaminase (ALT)** and **aspartate transaminase (AST)** are aminotransferases that catalyze reactions where alanine and aspartate transfer their amino groups to alpha-ketoglutarate, forming glutamate and pyruvate or oxaloacetate, respectively. These reactions are reversible and play a key role in nitrogen metabolism and amino acid biosynthesis.

Oxidative Deamination

Once glutamate is formed, the amino group must be removed to facilitate further metabolism. This occurs through **oxidative deamination**, where glutamate is converted back to alpha-ketoglutarate, releasing ammonia. This reaction is catalyzed by **glutamate dehydrogenase** in the mitochondria, which uses either NAD+ or NADP+ as a cofactor. The ammonia produced is toxic, especially to the brain, and must be rapidly converted into a less harmful form for excretion.

23.2.3 Step 2: The Urea Cycle

The ammonia generated from deamination is converted into urea via the **urea cycle**, a process that occurs primarily in the liver. The urea cycle is essential for maintaining nitrogen balance and preventing the buildup of toxic ammonia. The urea formed is transported to the kidneys, where it is excreted in the urine.

Key Steps of the Urea Cycle

- **Formation of Carbamoyl Phosphate:** Ammonia and bicarbonate are combined to form carbamoyl phosphate, catalyzed by **carbamoyl phosphate synthetase I** in the mitochondria. This reaction requires ATP, reflecting the energy cost associated with nitrogen excretion.
- **Ornithine and Citrulline Formation:** Carbamoyl phosphate combines with ornithine to form citrulline, catalyzed by **ornithine transcarbamylase**. Citrulline then moves from the mitochondria to the cytoplasm.
- **Synthesis of Argininosuccinate:** In the cytoplasm, citrulline combines with aspartate to form argininosuccinate, catalyzed by **argininosuccinate synthetase**. Aspartate provides an additional amino group for urea formation.
- **Cleavage of Argininosuccinate:** Argininosuccinate is then cleaved by **argininosuccinase** to produce arginine and fumarate. Fumarate enters the citric acid cycle, linking amino acid catabolism with energy production.

- **Formation of Urea:** Arginine is finally converted to urea and ornithine by **arginase.** Ornithine re-enters the mitochondria, allowing the urea cycle to continue.

The urea cycle is highly regulated, with levels of key enzymes adjusted according to protein intake and ammonia levels.

By converting ammonia into urea, the body safely excretes excess nitrogen while linking amino acid catabolism with the central metabolic pathways.

23.2.4 Step 3: Fate of the Carbon Skeletons

Once the amino group is removed, the carbon skeletons of amino acids, now in the form of alpha-keto acids, can follow different metabolic fates depending on the type of amino acid and the body's metabolic needs.

These carbon skeletons are classified as **glucogenic** or **ketogenic** based on their metabolic destinations:

- **Glucogenic Amino Acids:** These amino acids are converted into intermediates of the citric acid cycle, such as oxaloacetate or pyruvate, which can be used for glucose synthesis via gluconeogenesis. Examples of glucogenic amino acids include alanine, aspartate, and glutamine.
- **Ketogenic Amino Acids:** These amino acids are converted into acetyl-CoA or acetoacetate, precursors for ketone body synthesis. Examples include leucine and lysine, the only purely ketogenic amino acids.

Some amino acids, like isoleucine and phenylalanine, are both glucogenic and ketogenic, able to produce glucose or ketone bodies depending on the body's energy requirements.

23.2.5 Catabolism of Specific Amino Acids

Each amino acid has a unique catabolic pathway that reflects its structure and chemical properties. Below are examples of catabolic pathways for several amino acids:

Alanine and Glutamine

Alanine and glutamine are among the most abundant amino acids in circulation. Alanine can be converted into pyruvate through transamination and enter gluconeogenesis. Glutamine, which transports ammonia between tissues, is converted to glutamate and then to alpha-ketoglutarate, entering the citric acid cycle for energy production.

Phenylalanine and Tyrosine

Phenylalanine is catabolized to tyrosine, which then undergoes further breakdown to form acetoacetate and fumarate. This pathway is important as phenylalanine and tyrosine are precursors for dopamine, norepinephrine, and epinephrine, critical neurotransmitters in the brain.

Branched-Chain Amino Acids (BCAAs): Leucine, Isoleucine, and Valine

The branched-chain amino acids (BCAAs)—leucine, isoleucine, and valine—are catabolized mainly in muscle tissue rather than the liver. They provide an important energy source during muscle activity and fasting. Leucine is purely ketogenic, while valine is glucogenic, and isoleucine has both properties. These BCAAs are first transaminated to form keto acids, which undergo further reactions to produce energy.

23.2.6 Regulation of Amino Acid Catabolism

The regulation of amino acid catabolism is closely tied to the body's nutritional and physiological status. Hormones like **insulin** and **glucagon** play a significant role in determining the rate of amino acid breakdown. When dietary protein intake is high or during fasting, glucagon stimulates amino acid catabolism to produce glucose and maintain blood glucose levels. Conversely, during periods of high protein synthesis, such as growth, insulin favors protein synthesis over breakdown.

Allosteric regulation of key enzymes, like carbamoyl phosphate synthetase I, also helps control amino acid catabolism. The presence of amino acids like arginine stimulates the urea cycle, promoting nitrogen removal when amino acid breakdown is high.

23.2.7 Physiological Significance of Amino Acid Catabolism

Amino acid catabolism is crucial for maintaining nitrogen balance, producing energy, and generating intermediates for biosynthetic processes.

In states of high energy demand, such as fasting or prolonged exercise, amino acids provide an alternative fuel source and substrates for gluconeogenesis. Additionally, amino acid breakdown contributes to maintaining acid-base balance by producing bicarbonate and ammonia, which can buffer excess acids in the blood.

23.2.8 Pathological Aspects of Amino Acid Catabolism

Disruptions in amino acid catabolism can lead to metabolic disorders. For example, defects in the enzyme phenylalanine hydroxylase result in **phenylketonuria (PKU)**, an inherited disorder where phenylalanine accumulates in the blood, causing neurological damage if untreated.

Similarly, deficiencies in enzymes of the urea cycle lead to **hyperammonemia**, where ammonia accumulates, affecting brain function and potentially leading to fatal outcomes.

CHAPTER 24
INTEGRATION OF ENERGY METABOLISM

The integration of energy metabolism is essential for maintaining the body's energy balance, allowing it to respond effectively to various physiological states such as feeding, fasting, exercise, and rest. This complex coordination among carbohydrates, lipids, and proteins occurs through regulated pathways and hormones, ensuring that the energy needs of tissues are met while maintaining overall metabolic homeostasis. The metabolic integration enables organisms to adapt to changing energy demands and nutrient availability, and it is achieved through a well-orchestrated network involving key organs like the liver, muscle, adipose tissue, and the brain.

24.1 OVERVIEW OF ENERGY METABOLISM PATHWAYS

Energy metabolism relies on three primary nutrient sources: carbohydrates, fats, and proteins. These macromolecules undergo distinct but interconnected catabolic and anabolic pathways to meet the body's energy requirements.

- **Carbohydrates**: The breakdown of carbohydrates produces glucose, which enters glycolysis to generate ATP. When energy is abundant, glucose is stored as glycogen in the liver and muscles. In times of scarcity, gluconeogenesis synthesizes glucose to maintain blood sugar levels.
- **Lipids:** Fatty acids derived from dietary fats or mobilized from adipose tissue undergo beta-oxidation to yield acetyl-CoA, which enters the citric acid cycle. Excess acetyl-CoA can be converted into ketone bodies in the liver during prolonged fasting, providing an alternative fuel source.
- **Proteins:** Proteins are generally used for cellular structure and function rather than as a primary energy source. However, amino acids can be catabolized to provide intermediates for the citric acid cycle or gluconeogenesis, especially under conditions of low carbohydrate availability.

These pathways are integrated to ensure that energy supply matches the demand, and that metabolites can be redirected as needed to maintain blood glucose levels and support cellular functions.

24.2 KEY ORGANS IN ENERGY METABOLISM

Different organs specialize in various aspects of energy metabolism, working together to ensure systemic metabolic balance.

24.2.1 The Liver: Central Hub of Metabolic Integration

The liver plays a central role in metabolic regulation, acting as the primary site for gluconeogenesis, glycogen storage, and lipid metabolism. During the fed state, the liver absorbs glucose from the bloodstream, converting it into glycogen for storage and synthesizing fatty acids when glycogen stores are full. During fasting, the liver becomes a key site for gluconeogenesis and ketogenesis, producing glucose and ketone bodies to supply other tissues with energy.

24.2.2 Muscle Tissue: Energy Storage and Utilization

Muscle tissue primarily uses glucose and fatty acids for energy. Glycogen stored in muscle provides a rapid energy source during physical activity, and muscle can oxidize fatty acids from circulating lipids during prolonged exercise. While muscles cannot directly participate in gluconeogenesis, they release amino acids (particularly alanine and glutamine) into the bloodstream during fasting, which are transported to the liver for glucose synthesis.

24.2.3 Adipose Tissue: Storage and Mobilization of Fats

Adipose tissue stores energy in the form of triglycerides. During the fed state, insulin promotes lipogenesis, the synthesis of triglycerides from fatty acids and glycerol.

During fasting or exercise, the release of hormones like glucagon and epinephrine stimulates lipolysis, breaking down triglycerides into fatty acids and glycerol, which are then released into the bloodstream to be used by the liver and muscles for energy.

24.2.4 The Brain: Glucose as Primary Fuel

The brain has a high metabolic demand and relies almost exclusively on glucose for energy under normal conditions, as it cannot store glycogen or use fatty acids. During prolonged fasting, however, the brain adapts to utilize ketone bodies produced by the liver as an alternative energy source, reducing its dependence on glucose and sparing muscle tissue from excessive breakdown for gluconeogenesis.

24.3 HORMONAL REGULATION OF ENERGY METABOLISM

Hormones play a vital role in integrating energy metabolism by signaling organs to either store or mobilize energy substrates. The balance of anabolic and catabolic hormones, primarily insulin, glucagon, epinephrine, and cortisol, regulates the body's metabolic state.

24.3.1 Insulin: The Anabolic Hormone

Insulin is released by the pancreas in response to high blood glucose levels following a meal. It promotes glucose uptake in the liver, muscle, and adipose tissue, facilitating glycogen synthesis, lipogenesis, and protein synthesis. By stimulating pathways that store energy and inhibiting catabolic processes, insulin promotes energy storage and anabolic growth.

24.3.2 Glucagon: The Catabolic Hormone

Glucagon is released in response to low blood glucose levels during fasting. It stimulates glycogenolysis and gluconeogenesis in the liver, increasing blood glucose availability for tissues, particularly the brain. Glucagon also promotes lipolysis in adipose tissue, releasing fatty acids as an alternative fuel source.

24.3.3 Epinephrine and Cortisol: Stress Hormones

Epinephrine and cortisol are released during periods of physical or emotional stress. Epinephrine stimulates glycogenolysis in the liver and muscle, quickly raising blood glucose levels to meet immediate energy demands. Cortisol, a glucocorticoid, enhances gluconeogenesis and lipolysis while also mobilizing amino acids from muscle proteins. Together, these hormones ensure that the body can maintain energy supply during stress.

24.4 METABOLIC STATES: FED, FASTING, AND STARVATION

The body's metabolic pathways shift to meet energy demands based on dietary intake and physiological needs, with each state engaging different metabolic strategies.

24.4.1 Fed State: Anabolism and Storage

In the fed state, following food intake, blood glucose levels rise, triggering insulin release. Insulin facilitates glucose uptake by tissues and promotes glycogen synthesis in the liver and muscles. Additionally, excess glucose is converted to triglycerides for long-term storage in adipose tissue. Amino acids are directed towards protein synthesis, and lipogenesis occurs in both the liver and adipose tissue when glycogen stores are sufficient.

24.4.2 Fasting State: Mobilization and Maintenance

During fasting, blood glucose levels drop, leading to decreased insulin and increased glucagon levels. The liver responds by breaking down glycogen to release glucose and initiating gluconeogenesis using lactate, glycerol, and amino acids. Adipose tissue releases fatty acids via lipolysis, which tissues like the muscles and liver can use for energy. The liver also starts producing ketone bodies from fatty acids, supplying an alternative energy source to the brain and muscles as glycogen stores diminish.

24.4.3 Starvation: Adaptation to Extended Caloric Deficit

In prolonged fasting or starvation, the body relies heavily on fat stores for energy. Ketone body production increases significantly to support brain metabolism, reducing the need for gluconeogenesis from amino acids and conserving muscle tissue. The body prioritizes maintaining blood glucose for essential functions, sparing protein degradation as much as possible. This adaptive shift allows the body to survive on limited energy intake by maximizing fat oxidation and ketone utilization.

24.5 INTEGRATION OF CARBOHYDRATE, FAT, AND PROTEIN METABOLISM

The integration of carbohydrate, fat, and protein metabolism enables the body to respond dynamically to fluctuating energy demands. During the fed state, insulin promotes the storage of glucose as glycogen and excess glucose as fat. In fasting states, glycogen breakdown, gluconeogenesis, and lipolysis provide the needed energy. Protein catabolism contributes to gluconeogenesis, providing glucose precursors but is minimized to preserve muscle tissue.

24.5.1 Interconnected Pathways

Metabolic pathways are interconnected, with intermediates from one pathway fueling others:

- **Glycolysis and Citric Acid Cycle:** Pyruvate, the end product of glycolysis, can be converted to acetyl-CoA, which enters the citric acid cycle for ATP production. Alternatively, acetyl-CoA can be used for fatty acid synthesis.
- **Beta-Oxidation and Ketogenesis:** Fatty acids broken down through beta-oxidation yield acetyl-CoA, which fuels the citric acid cycle or is converted into ketone bodies in the liver during prolonged fasting.
- **Amino Acids and Gluconeogenesis:** Certain amino acids, like alanine and glutamine, are converted to glucose precursors in the liver, supporting blood glucose levels when carbohydrate availability is low.

24.6 PATHOPHYSIOLOGICALASPECTSOFENERGYMETABOLISM IMBALANCE

Disruptions in the integration of energy metabolism can lead to metabolic disorders, including diabetes, obesity, and metabolic syndrome. In type 2 diabetes, insulin resistance impairs glucose uptake, causing hyperglycemia and altering fat and protein metabolism. Obesity results from chronic energy imbalance, where excess calorie intake is stored as fat, often leading to insulin resistance and cardiovascular complications.

24.7 CONCLUSION

The integration of energy metabolism is a finely tuned process essential for maintaining homeostasis and adapting to fluctuating energy demands. Through coordinated pathways for carbohydrates, lipids, and proteins, and regulated by hormones like insulin and glucagon, the body can transition smoothly between different metabolic states.

This coordination ensures an energy supply for vital organs while preventing excess nutrient accumulation or depletion. Understanding these integrated processes provides valuable insight into metabolic health and highlights potential therapeutic strategies for metabolic disorders.

CHAPTER 25
BIOSYNTHESIS OF AMINO ACIDS

The biosynthesis of amino acids is a fundamental aspect of cellular metabolism, allowing organisms to produce the building blocks necessary for protein synthesis and numerous metabolic functions.

Amino acids are divided into essential and non-essential categories: non-essential amino acids can be synthesized by the body, while essential amino acids must be obtained from the diet.

Amino acid biosynthesis involves various metabolic pathways that produce different amino acids from common metabolic intermediates, requiring energy input and nitrogen incorporation.

25.1 OVERVIEW OF AMINO ACID BIOSYNTHESIS

The synthesis of amino acids is closely integrated with other metabolic processes, particularly the citric acid cycle, glycolysis, and the pentose phosphate pathway. These pathways provide precursors such as oxaloacetate, pyruvate, and alpha-ketoglutarate, which serve as starting points for amino acid synthesis. The biosynthetic pathways are organized into families based on shared metabolic intermediates and mechanisms of nitrogen incorporation.

- **Transamination:** The addition of an amino group to a keto acid is a central step in the synthesis of amino acids. This process is facilitated by enzymes called transaminases, which transfer amino groups from amino donors to alpha-keto acids.
- **Nitrogen Assimilation:** Incorporating nitrogen is essential, as amino acids contain nitrogen in their amino groups. Glutamate and glutamine serve as key nitrogen donors in the synthesis of other amino acids, particularly for incorporating nitrogen derived from ammonia.

25.2 FAMILIES OF AMINO ACIDS BASED ON PRECURSORS

Amino acid biosynthesis is often categorized by the metabolic precursors from which the amino acids are derived.

The major precursor families include alpha-ketoglutarate, oxaloacetate, pyruvate, 3-phosphoglycerate, phosphoenolpyruvate, and ribose-5-phosphate.

25.2.1 Alpha-Ketoglutarate Family

Alpha-ketoglutarate, a citric acid cycle intermediate, serves as the precursor for glutamate, glutamine, proline, and arginine.

- **Glutamate and Glutamine:** Glutamate is formed through the reductive amination of alpha-ketoglutarate catalyzed by glutamate dehydrogenase. Glutamine synthetase further converts glutamate to glutamine by adding an additional amino group, using ATP for energy. Glutamine and glutamate are primary nitrogen donors for synthesizing other amino acids.
- **Proline:** Proline is synthesized from glutamate through a series of reactions involving phosphorylation and cyclization, with NADH or NADPH providing the reducing power.
- **Arginine:** Arginine synthesis is more complex, proceeding through the urea cycle in mammals. Glutamate is converted to ornithine, which enters the urea cycle, where arginine is synthesized as an intermediate before being released.

25.2.2 Oxaloacetate Family

Oxaloacetate, another citric acid cycle intermediate, serves as the precursor for aspartate, asparagine, lysine, methionine, and threonine.

- **Aspartate and Asparagine:** Aspartate is synthesized through transamination of oxaloacetate, typically with glutamate as the nitrogen donor. Asparagine synthetase converts aspartate to asparagine by adding an amino group from glutamine in an ATP-dependent reaction.
- **Lysine, Methionine, and Threonine:** These essential amino acids are synthesized through complex pathways in microorganisms and plants, with aspartate as a common precursor. In animals, lysine, methionine, and threonine must be obtained from the diet, as they lack the necessary enzymes to produce them.

25.2.3 Pyruvate Family

Pyruvate, the end product of glycolysis, is the precursor for alanine, valine, leucine, and isoleucine.

- **Alanine:** Alanine is synthesized by a simple transamination reaction where pyruvate accepts an amino group from glutamate, forming alanine and alpha-ketoglutarate.
- **Valine, Leucine, and Isoleucine:** These branched-chain amino acids share initial steps in their biosynthetic pathways, involving the

conversion of pyruvate and other intermediates. They are essential amino acids in animals and are synthesized through complex pathways in plants and bacteria.

25.2.4 3-Phosphoglycerate Family

3-phosphoglycerate, an intermediate in glycolysis, is the precursor for serine, glycine, and cysteine.

- **Serine:** Serine is synthesized from 3-phosphoglycerate in three steps: oxidation, transamination, and dephosphorylation. NAD+ is reduced in the oxidation step, and the resulting 3-phosphohydroxypyruvate accepts an amino group from glutamate to form phosphoserine, which is dephosphorylated to serine.
- **Glycine:** Glycine is synthesized from serine in a single-step reaction catalyzed by serine hydroxymethyltransferase, which transfers a carbon group from serine to tetrahydrofolate.
- **Cysteine:** Cysteine synthesis begins with serine, which undergoes modification to incorporate sulfur from methionine through the transsulfuration pathway. Methionine is converted to homocysteine, which reacts with serine to form cysteine.

25.2.5 Phosphoenolpyruvate and Erythrose-4-Phosphate Family

Aromatic amino acids—phenylalanine, tyrosine, and tryptophan—are synthesized from phosphoenolpyruvate (PEP) and erythrose-4-phosphate, intermediates in glycolysis and the pentose phosphate pathway, respectively.

- **Phenylalanine and Tyrosine:** Phenylalanine and tyrosine share the same initial pathway, leading to the formation of chorismate. Phenylalanine is synthesized directly from chorismate, while tyrosine is derived from phenylalanine by hydroxylation, catalyzed by phenylalanine hydroxylase.
- **Tryptophan:** Tryptophan synthesis involves multiple steps from chorismate, with indole as an intermediate. This pathway is complex, requiring significant energy input, and thus tryptophan is an essential amino acid in animals.

25.2.6 Ribose-5-Phosphate Family

Histidine synthesis is unique among amino acids, requiring ribose-5-phosphate from the pentose phosphate pathway as its precursor.

- **Histidine:** The histidine biosynthesis pathway involves the condensation of phosphoribosyl pyrophosphate (PRPP) with ATP, leading to a series of reactions that form histidine. Due to its unique precursor and pathway complexity, histidine is often synthesized only in plants and microorganisms, making it essential in animal diets.

25.3 REGULATION OF AMINO ACID BIOSYNTHESIS

Amino acid biosynthesis is tightly regulated to ensure a balanced supply of amino acids without overproduction. This regulation is achieved through feedback inhibition, enzyme regulation, and gene expression control.

25.3.1 Feedback Inhibition

Feedback inhibition is a common mechanism in amino acid biosynthesis, where the end product of a pathway inhibits the first enzyme in the pathway, preventing excess accumulation. For example, in the aspartate pathway, threonine, lysine, and methionine each inhibit enzymes early in the pathway, adjusting their production according to demand.

25.3.2 Allosteric Regulation

Allosteric regulation involves the binding of small molecules to enzymes, changing their activity. Glutamine synthetase, for example, is regulated allosterically by several metabolites, including alanine, glycine, and AMP, which signal that nitrogen is abundant and glutamine synthesis should be downregulated.

25.3.3 Gene Expression Control

The synthesis of enzymes for amino acid biosynthesis can also be regulated at the genetic level. When amino acids are in short supply, transcription of biosynthetic enzyme genes is upregulated. Conversely, when amino acids are abundant, transcription is downregulated. This mechanism allows cells to conserve energy by synthesizing amino acids only when needed.

25.4 PHYSIOLOGICAL SIGNIFICANCE OF AMINO ACID BIOSYNTHESIS

Amino acid biosynthesis is essential for numerous physiological functions, including protein synthesis, cellular signaling, and the production of neurotransmitters, hormones, and nucleotides. Adequate amino acid production or dietary intake is critical for growth, immune function, and tissue repair. Additionally, amino acids serve as precursors for other biomolecules, such as heme, purines, pyrimidines, and nitric oxide.

The capacity for amino acid biosynthesis also impacts health and disease. For example, disruptions in amino acid synthesis can lead to metabolic disorders, and deficiencies in essential amino acids can lead to muscle wasting, immune deficiencies, and impaired cognitive function.

25.5. METABOLIC DISORDERS ASSOCIATED WITH AMINO ACID BIOSYNTHESIS

Genetic defects in amino acid biosynthetic pathways can result in metabolic disorders. For instance:

- **Phenylketonuria (PKU):** A mutation in the gene for phenylalanine hydroxylase prevents the conversion of phenylalanine to tyrosine, leading to toxic accumulation of phenylalanine, which can cause developmental delays and neurological impairment if untreated.
- **Homocystinuria:** This disorder results from a deficiency in enzymes involved in methionine and cysteine metabolism, leading to elevated homocysteine levels, which is associated with cardiovascular and connective tissue disorders.

Understanding these pathways has led to treatments and dietary interventions for affected individuals, highlighting the medical relevance of amino acid biosynthesis.

25.6 CONCLUSION

The biosynthesis of amino acids is a complex and highly regulated process, involving interconnected metabolic pathways and essential precursor molecules. Each amino acid synthesis pathway is adapted to maintain a balance, with feedback mechanisms ensuring appropriate amino acid levels for cellular needs. This regulatory integration underscores the role of amino acids not only in protein synthesis but also as crucial players in metabolic and physiological processes. The study of amino acid biosynthesis provides insights into metabolic health and paves the way for therapeutic interventions in metabolic disorders.

CHAPTER 26
NUCLEOTIDE BIOSYNTHESIS

Nucleotide biosynthesis is a vital cellular process, providing the fundamental building blocks for DNA and RNA synthesis, energy transfer, signaling, and coenzyme function. Nucleotides are composed of a nitrogenous base (purine or pyrimidine), a sugar (ribose or deoxyribose), and one or more phosphate groups. Cells have developed intricate biosynthetic pathways to generate these molecules from simple precursors, with tight regulatory mechanisms ensuring balanced production.

26.1 UNDERSTANDING THE NUCLEOTIDE

Nucleotides are organic molecules composed of a nitrogenous base, a pentose sugar and a phosphate. They serve as monomeric units of the nucleic acid polymers - deoxyribonucleic acid (DNA) and ribonucleic acid (RNA), both of which are essential biomolecules within all life-forms on Earth. Nucleotides are obtained in the diet and are also synthesized from common nutrients by the liver.

Nucleotides are composed of three subunit molecules: a nucleobase, a five-carbon sugar (ribose or deoxyribose), and a phosphate group consisting of one to three phosphates. The four nucleobases in DNA are guanine, adenine, cytosine, and thymine; in RNA, uracil is used in place of thymine.

Nucleotides also play a central role in metabolism at a fundamental, cellular level. They provide chemical energy—in the form of the nucleoside triphosphates, adenosine triphosphate (ATP), guanosine triphosphate (GTP), cytidine triphosphate (CTP), and uridine triphosphate (UTP)—throughout the cell for the many cellular functions that demand energy, including: amino acid, protein and cell membrane synthesis, moving the cell and cell parts (both internally and intercellularly), cell division, etc..

In addition, nucleotides participate in cell signaling (cyclic guanosine monophosphate or cGMP and cyclic adenosine monophosphate or cAMP) and are incorporated into important cofactors of enzymatic reactions (e.g., coenzyme A, FAD, FMN, NAD, and $NADP^+$).

In experimental biochemistry, nucleotides can be radiolabeled using radionuclides to yield radionucleotides. 5-nucleotides are also used in flavour enhancers as food additive to enhance the umami taste, often in the form of a yeast extract.

26.1.1 Structure

A nucleotide is composed of three distinctive chemical sub-units: a five-carbon sugar molecule, a nucleobase (the two of which together are called a nucleoside), and one phosphate group. With all three joined, a nucleotide is also termed a "nucleoside *mono*phosphate", "nucleoside *di*phosphate" or "nucleoside *tri*phosphate", depending on how many phosphates make up the phosphate group.

In nucleic acids, nucleotides contain either a purine or a pyrimidine base—i.e., the nucleobase molecule, also known as a nitrogenous base—and are termed *ribo*nucleotides if the sugar is ribose, or *deoxyribo*nucleotides if the sugar is deoxyribose. Individual phosphate molecules repetitively connect the sugar-ring molecules in two adjacent nucleotide monomers, thereby connecting the nucleotide monomers of a nucleic acid end-to-end into a long chain. These chain-joins of sugar and phosphate molecules create a 'backbone' strand for a single- or double helix. In any one strand, the chemical orientation (directionality) of the chain-joins runs from the 5'-end to the 3'-end (*read*: 5 prime-end to 3 prime-end)—referring to the five carbon sites on sugar molecules in adjacent nucleotides. In a double helix, the two strands are oriented in opposite directions, which permits base pairing and complementarity between the base-pairs, all which is essential for replicating or transcribing the encoded information found in DNA.

Nucleic acids then are polymeric macromolecules assembled from nucleotides, the monomer-units of nucleic acids. The purine bases adenine and guanine and pyrimidine base cytosine occur in both DNA and RNA, while the pyrimidine bases thymine (in DNA) and uracil (in RNA) occur in just one.

Adenine forms a base pair with thymine with two hydrogen bonds, while guanine pairs with cytosine with three hydrogen bonds.

In addition to being building blocks for the construction of nucleic acid polymers, singular nucleotides play roles in cellular energy storage and provision, cellular signaling, as a source of phosphate groups used to modulate the activity of proteins and other signaling molecules, and as enzymatic cofactors, often carrying out redox reactions.

Signaling cyclic nucleotides are formed by binding the phosphate group twice to the same sugar molecule, bridging the 5'- and 3'- hydroxyl groups of the sugar. Some signaling nucleotides differ from the standard single-phosphate group configuration, in having multiple phosphate groups attached to different positions on the sugar. Nucleotide cofactors include a wider range of chemical groups attached to the sugar via the glycosidic bond, including nicotinamide and flavin, and in the latter case, the ribose sugar is linear rather than forming the ring seen in other nucleotides.

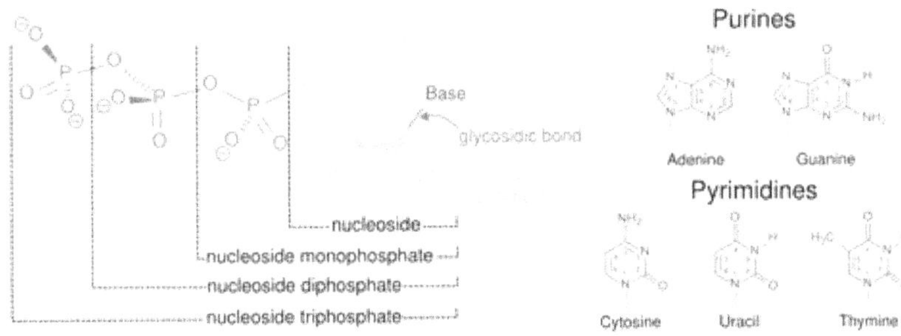

Fig 1: Structural Elements of Three Nucleotides—where One-, Two- or Three-Phosphates are Attached to the Nucleoside (in Yellow, Blue, Green) at Center: 1st, the Nucleotide Termed as a *Nucleoside Monophosphate* **is Formed by Adding a Phosphate (in Red); 2nd, Adding a Second Phosphate Forms a** *Nucleoside Diphosphate*; **3rd, Adding a Third Phosphate Results in a** *Nucleoside Triphosphate*. **+ the Nitrogenous Base (Nucleobase) is Indicated by "Base" and "Glycosidic Bond" (Sugar Bond). All Five Primary, or Canonical, Bases—the Purines and Pyrimidines—are Sketched at Right (in Blue)**

26.1.2 Synthesis

Nucleotides can be synthesized by a variety of means, both in vitro and in vivo. In vitro, protecting groups may be used during laboratory production of nucleotides. A purified nucleoside is protected to create a phosphoramidite, which can then be used to obtain analogues not found in nature and/or to synthesize an oligonucleotide. In vivo, nucleotides can be synthesized de novo or recycled through salvage pathways. The components used in de novo nucleotide synthesis are derived from biosynthetic precursors of carbohydrate and amino acid metabolism, and from ammonia and carbon dioxide. Recently it has been also demonstrated that cellular bicarbonate metabolism can be regulated by mTORC1 signaling. The liver is the major organ of de novo synthesis of all four nucleotides. De novo synthesis of pyrimidines and purines follows two different pathways. Pyrimidines are synthesized first from aspartate and carbamoyl-phosphate in the cytoplasm to the common precursor ring structure orotic acid, onto which a phosphorylated ribosyl unit is covalently linked. Purines, however, are first synthesized from the sugar template onto which the ring synthesis occurs. For reference, the syntheses of the purine and pyrimidine nucleotides are carried out by several enzymes in the cytoplasm of the cell, not within a specific organelle. Nucleotides undergo breakdown such that useful parts can be reused in synthesis reactions to create new nucleotides.

Pyrimidine Ribonucleotide Synthesis

The synthesis of the pyrimidines CTP and UTP occurs in the cytoplasm and starts with the formation of carbamoyl phosphate from glutamine and CO_2. Next, aspartate carbamoyltransferase catalyzes a condensation reaction between aspartate and carbamoyl phosphate to form carbamoyl aspartic acid, which is cyclized into 4,5-dihydroorotic acid by dihydroorotase. The latter is converted to orotate by dihydroorotate oxidase. The net reaction is:

$$(S)\text{-Dihydroorotate} + O_2 \rightarrow \text{Orotate} + H_2O_2$$

Orotate is covalently linked with a phosphorylated ribosyl unit. The covalent linkage between the ribose and pyrimidine occurs at position C_1 of the ribose unit, which contains a pyrophosphate, and N_1 of the pyrimidine ring. Orotate phosphoribosyltransferase (PRPP transferase) catalyzes the net reaction yielding orotidine monophosphate (OMP):

$$\text{Orotate} + \text{5-Phospho-}\alpha\text{-D-ribose 1-diphosphate (PRPP)}$$
$$\rightarrow \text{Orotidine 5'-phosphate} + \text{Pyrophosphate}$$

Orotidine 5'-monophosphate is decarboxylated by orotidine-5'-phosphate decarboxylase to form uridine monophosphate (UMP). PRPP transferase catalyzes both the ribosylation and decarboxylation reactions, forming UMP from orotic acid in the presence of PRPP. It is from UMP that other pyrimidine nucleotides are derived. UMP is phosphorylated by two kinases to uridine triphosphate (UTP) via two sequential reactions with ATP. First, the diphosphate from UDP is produced, which in turn is phosphorylated to UTP. Both steps are fueled by ATP hydrolysis:

$$\text{ATP} + \text{UMP} \rightarrow \text{ADP} + \text{UDP}$$

$$\text{UDP} + \text{ATP} \rightarrow \text{UTP} + \text{ADP}$$

CTP is subsequently formed by the amination of UTP by the catalytic activity of CTP synthetase. Glutamine is the NH_3 donor and the reaction is fueled by ATP hydrolysis, too:

$$\text{UTP} + \text{Glutamine} + \text{ATP} + H_2O \rightarrow \text{CTP} + \text{ADP} + P_i$$

Cytidine monophosphate (CMP) is derived from cytidine triphosphate (CTP) with subsequent loss of two phosphates.

Purine Ribonucleotide Synthesis

The de novo synthesis of purine nucleotides by which these precursors are incorporated into the purine ring proceeds by a 10-step pathway to the branch-point intermediate IMP, the nucleotide of the base hypoxanthine. AMP and GMP are subsequently synthesized from this intermediate via separate, two-step pathways.

Thus, purine moieties are initially formed as part of the ribonucleotides rather than as free bases. Six enzymes take part in IMP synthesis. Three of them are multifunctional:

- GART (reactions 2, 3, and 5).

- PAICS (reactions 6, and 7).
- ATIC (reactions 9, and 10).

The atoms that are used to build the purine nucleotides come from a variety of sources:

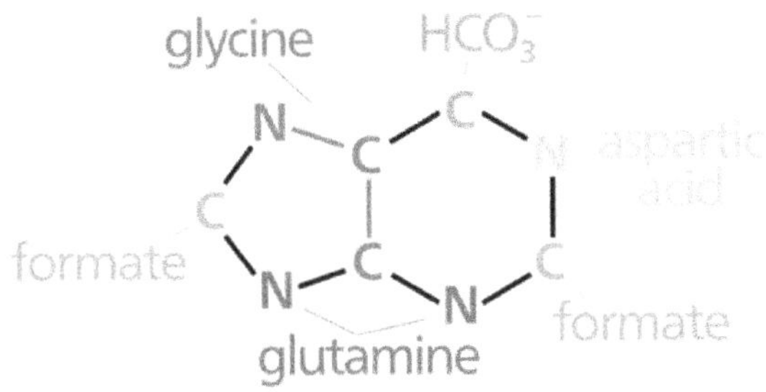

Fig 2: The Biosynthetic Origins of Purine Ring Atoms N_1 Arises from the Amine Group of Asp C_2 and C_8 Originate from Formate N_3 and N_9 are Contributed by the Amide Group of Gln C_4, C_5 and N_7 are Derived from Gly C_6 Comes from HCO_3^- (CO_2)

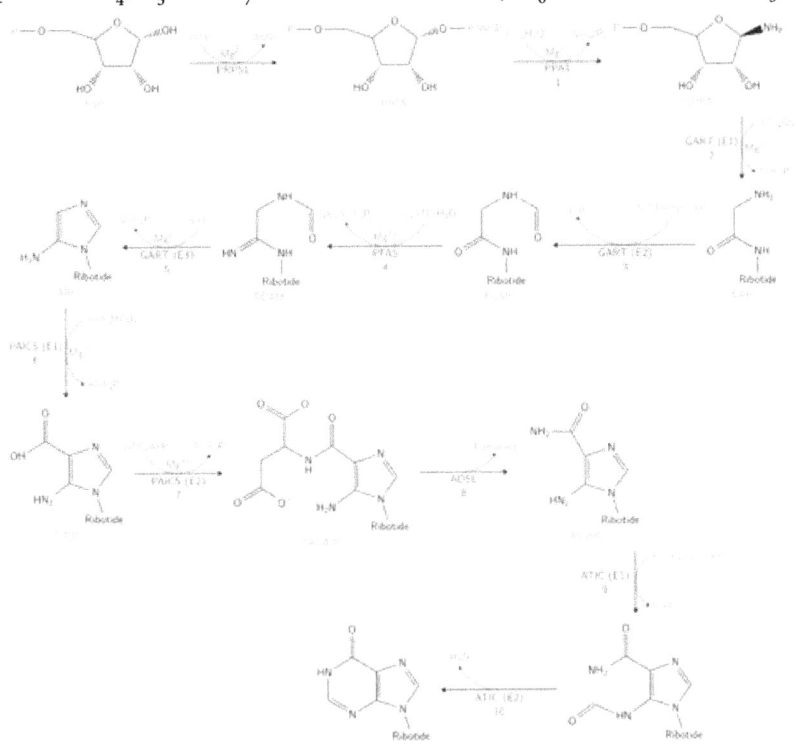

Fig 3: The Synthesis of IMP

The pathway starts with the formation of PRPP. PRPS1 is the enzyme that activates R5P, which is formed primarily by the pentose phosphate pathway, to PRPP by reacting it with ATP. The reaction is unusual in that a pyrophosphoryl group is directly transferred from ATP to C_1 of R5P and that the product has the

α configuration about C1. This reaction is also shared with the pathways for the synthesis of Trp, His, and the pyrimidine nucleotides. Being on a major metabolic crossroad and requiring much energy, this reaction is highly regulated.

In the first reaction unique to purine nucleotide biosynthesis, PPAT catalyzes the displacement of PRPP's pyrophosphate group (PP_i) by an amide nitrogen donated from either glutamine (N), glycine (N&C), aspartate (N), folic acid (C_1), or CO_2. This is the committed step in purine synthesis. The reaction occurs with the inversion of configuration about ribose C_1, thereby forming β-5-phosphorybosylamine (5-PRA) and establishing the anomeric form of the future nucleotide.

Next, a glycine is incorporated fueled by ATP hydrolysis, and the carboxyl group forms an amine bond to the NH_2 previously introduced. A one-carbon unit from folic acid coenzyme N_{10}-formyl-THF is then added to the amino group of the substituted glycine followed by the closure of the imidazole ring.

Next, a second NH_2 group is transferred from glutamine to the first carbon of the glycine unit. A carboxylation of the second carbon of the glycin unit is concomitantly added. This new carbon is modified by the addition of a third NH_2 unit, this time transferred from an aspartate residue. Finally, a second one-carbon unit from formyl-THF is added to the nitrogen group and the ring is covalently closed to form the common purine precursor inosine monophosphate (IMP).

Inosine monophosphate is converted to adenosine monophosphate in two steps. First, GTP hydrolysis fuels the addition of aspartate to IMP by adenylosuccinate synthase, substituting the carbonyl oxygen for a nitrogen and forming the intermediate adenylosuccinate.

Fumarate is then cleaved off forming adenosine monophosphate. This step is catalyzed by adenylosuccinate lyase.

Inosine monophosphate is converted to guanosine monophosphate by the oxidation of IMP forming xanthylate, followed by the insertion of an amino group at C_2. NAD^+ is the electron acceptor in the oxidation reaction. The amide group transfer from glutamine is fueled by ATP hydrolysis.

Pyrimidine and Purine Degradation

In humans, pyrimidine rings (C, T, U) can be degraded completely to CO_2 and NH_3 (urea excretion). That having been said, purine rings (G, A) cannot. Instead, they are degraded to the metabolically inert uric acid which is then excreted from the body. Uric acid is formed when GMP is split into the base guanine and ribose. Guanine is deaminated to xanthine which in turn is oxidized to uric acid. This last reaction is irreversible. Similarly, uric acid can be formed when AMP is deaminated to IMP from which the ribose unit is removed to form hypoxanthine. Hypoxanthine is oxidized to xanthine and finally to uric acid. Instead of uric acid secretion, guanine and IMP can be used for recycling purposes and nucleic acid synthesis in the presence of PRPP and aspartate (NH_3 donor).

26.1.3 Prebiotic Synthesis of Nucleotides

Theories about the origin of life require knowledge of chemical pathways that permit formation of life's key building blocks under plausible prebiotic conditions. The RNA world hypothesis holds that in the primordial soup there existed free-floating ribonucleotides, the fundamental molecules that combine in series to form RNA. Complex molecules like RNA must have arisen from small molecules whose reactivity was governed by physico-chemical processes. RNA is composed of purine and pyrimidine nucleotides, both of which are necessary for reliable information transfer, and thus Darwinian evolution. Becker et al. showed how pyrimidine nucleosides can be synthesized from small molecules and ribose, driven solely by wet-dry cycles. Purine nucleosides can be synthesized by a similar pathway. 5'-mono- and di-phosphates also form selectively from phosphate-containing minerals, allowing concurrent formation of polyribonucleotides with both the purine and pyrimidine bases. Thus a reaction network towards the purine and pyrimidine RNA building blocks can be established starting from simple atmospheric or volcanic molecules.

26.1.4 Unnatural Base Pair (UBP)

An unnatural base pair (UBP) is a designed subunit (or nucleobase) of DNA which is created in a laboratory and does not occur in nature. Examples include d5SICS and dNaM. These artificial nucleotides bearing hydrophobic nucleobases, feature two fused aromatic rings that form a (d5SICS–dNaM) complex or base pair in DNA. *E. coli* have been induced to replicate a plasmid containing UBPs through multiple generations. This is the first known example of a living organism passing along an expanded genetic code to subsequent generations.

26.1.5 Medical Applications of Synthetic Nucleotides

The applications of synthetic nucleotides vary widely and include disease diagnosis, treatment, or precision medicine.

- **Antiviral or Antiretroviral Agents:** Several nucleotide derivatives have been used in the treatment against infection with Hepatitis and HIV. Examples of direct nucleoside analog reverse-transcriptase inhibitors (NRTIs) include Tenofovir disoproxil, Tenofovir alafenamide, and Sofosbuvir. On the other hand, agents such as Mericitabine, Lamivudine, Entecavir and Telbivudine must first undergo metabolization via phosphorylation to become activated.
- **Antisense Oligonucleotides (ASO):** Synthetic oligonucleotides have

been used in the treatment of rare heritable diseases since they can bind specific RNA transcripts and ultimately modulate protein expression. Spinal muscular atrophy, amyotrophic lateral sclerosis, homozygous familial hypercholesterolemia, and primary hyperoxaluria type 1 are all amenable to ASO-based therapy. The application of oligonucleotides is a new frontier in precision medicine and management of conditions which are untreatable.

- **Synthetic Guide RNA (gRNA):** Synthetic nucleotides can be used to design gRNA which are essential for the proper function of gene-editing technologies such as CRISPR-Cas9.

26.1.6 Length Unit

Nucleotide (abbreviated "nt") is a common unit of length for single-stranded nucleic acids, similar to how base pair is a unit of length for double-stranded nucleic acids.

26.2 OVERVIEW OF NUCLEOTIDE BIOSYNTHESIS

The biosynthesis of nucleotides occurs through two primary pathways: *de novo synthesis* and *salvage pathways*. In de novo synthesis, nucleotides are synthesized from simpler molecules, while the salvage pathways recycle bases and nucleosides from cellular breakdown to conserve energy. Both pathways are essential for maintaining a steady supply of nucleotides, particularly during periods of cell division and growth.

26.2.1 The De Novo Pathways

The de novo pathways build nucleotides from small, simple molecules such as amino acids, carbon dioxide, and ribose-5-phosphate. These pathways are energy-intensive but are essential when there is a need for rapid nucleotide production, such as during DNA replication.

- **Purine Nucleotide Biosynthesis:** Purine nucleotides, such as adenine and guanine, are synthesized in a stepwise fashion, with the purine ring structure built directly onto a ribose-5-phosphate backbone, derived from the pentose phosphate pathway.
- **Pyrimidine Nucleotide Biosynthesis:** Pyrimidine nucleotides, such as cytosine, thymine, and uracil, are synthesized by constructing the ring structure separately and attaching it to ribose-5-phosphate.

26.2.2 Salvage Pathways

The salvage pathways recycle purine and pyrimidine bases from nucleic acid breakdown, conserving energy by avoiding the need to synthesize nucleotides from scratch. Enzymes like hypoxanthine-guanine phosphoribosyltransferase (HGPRT) play a critical role in reattaching recycled bases to ribose-5-phosphate, forming nucleotides like GMP and AMP. The salvage pathway is particularly important in cells with high turnover, such as immune cells and neurons.

26.3 PURINE BIOSYNTHESIS

Purine biosynthesis is a multi-step pathway involving several enzymes and coenzymes. The purine ring structure is constructed by adding atoms sequentially to a ribose-5-phosphate molecule, which is activated by conversion to phosphoribosyl pyrophosphate (PRPP).

26.3.1 Steps in Purine Biosynthesis

- **Formation of Inosine Monophosphate (IMP):** The initial steps involve the addition of nitrogen from glutamine and carbon from glycine, leading to the formation of IMP, which contains the basic purine ring structure. IMP serves as a precursor to both adenosine monophosphate (AMP) and guanosine monophosphate (GMP).
- **Synthesis of AMP and GMP:** AMP is synthesized from IMP by adding an amino group from aspartate, while GMP is formed by oxidizing IMP and adding an amino group from glutamine. These reactions require energy in the form of GTP and ATP, respectively, demonstrating the need for coordinated regulation to balance purine synthesis.

26.3.2 Regulation of Purine Biosynthesis

The purine biosynthetic pathway is tightly regulated through feedback inhibition. High levels of AMP and GMP inhibit the activity of PRPP amidotransferase, the enzyme catalyzing the first committed step in purine synthesis. Additionally, IMP conversion to AMP or GMP is regulated by the availability of ATP and GTP, ensuring balanced levels of these nucleotides for cellular functions.

26.4 PYRIMIDINE BIOSYNTHESIS

Pyrimidine biosynthesis differs from purine synthesis in that the ring structure is synthesized independently before being attached to a ribose-5-phosphate molecule. This pathway also uses simpler precursors, making it more straightforward than purine synthesis.

26.4.1 Steps in Pyrimidine Biosynthesis

- **Formation of Carbamoyl Phosphate:** The initial step involves combining bicarbonate, glutamine, and ATP to form carbamoyl phosphate, catalyzed by carbamoyl phosphate synthetase II. This enzyme plays a rate-limiting role and is a key regulatory point.

- **Synthesis of Orotate:** Carbamoyl phosphate combines with aspartate to form carbamoyl aspartate, which is converted to orotate. Orotate is then attached to PRPP, yielding orotidine monophosphate (OMP).
- **Conversion to UMP, CMP, and dTMP:** OMP is decarboxylated to form uridine monophosphate (UMP), a precursor for cytidine monophosphate (CMP) and thymidine monophosphate (dTMP). CMP is derived from UMP through amination, while dTMP is synthesized by methylating deoxyuridine monophosphate (dUMP), requiring folate as a methyl donor.

26.4.2 Regulation of Pyrimidine Biosynthesis

Similar to purine synthesis, pyrimidine biosynthesis is regulated by feedback inhibition. UTP, the end product of the pathway, inhibits carbamoyl phosphate synthetase II, the enzyme catalyzing the first step.

Additionally, ATP can stimulate the pathway, coordinating pyrimidine production with purine levels and ensuring balanced nucleotide pools.

26.5 CONVERSION TO DEOXYRIBONUCLEOTIDES

To produce deoxyribonucleotides required for DNA synthesis, ribonucleotide reductase reduces ribonucleotides to deoxyribonucleotides.

This enzyme requires a stable supply of ribonucleoside diphosphates and uses NADPH as a reducing agent. Regulation of ribonucleotide reductase is critical, as balanced deoxyribonucleotide pools are necessary to maintain DNA integrity during replication.

26.6 NUCLEOTIDE DEGRADATION AND RECYCLING

Nucleotide degradation involves breaking down purines and pyrimidines to smaller molecules. Purines are degraded to uric acid, a compound that can accumulate and cause gout if not excreted efficiently.

Pyrimidines, on the other hand, are degraded to simpler metabolites like beta-alanine, which are more easily disposed of by the body. Enzymes in the salvage pathways recycle bases for nucleotide synthesis, reducing the energy demands of de novo synthesis.

26.7 PHYSIOLOGICAL ROLES OF NUCLEOTIDES

Nucleotides serve as building blocks for nucleic acids, but their functions extend beyond DNA and RNA synthesis.

- **Energy Transfer:** ATP and GTP are essential energy carriers, driving endergonic reactions and cellular processes like muscle contraction, active transport, and biosynthesis.

- **Cellular Signaling:** Cyclic nucleotides like cAMP and cGMP act as second messengers in signal transduction pathways, transmitting signals from cell surface receptors to intracellular targets.
- **Cofactor Components:** Nucleotides are also precursors for coenzymes like NAD+, FAD, and CoA, which are central to redox reactions and metabolic processes.
- **Regulation of Enzyme Activity:** Nucleotides serve as allosteric regulators of enzymes, modulating metabolic pathways based on cellular needs.

26.8 METABOLIC DISORDERS ASSOCIATED WITH NUCLEOTIDE BIOSYNTHESIS

Defects in nucleotide metabolism can lead to metabolic disorders, with purine and pyrimidine metabolism disorders resulting in a range of clinical symptoms.

- **Lesch-Nyhan Syndrome:** This genetic disorder arises from a deficiency in HGPRT, leading to the inability to recycle purines efficiently. Patients experience high levels of uric acid, causing gout and neurological symptoms.
- **Gout:** Excessive purine degradation can lead to elevated uric acid levels, resulting in gout, a painful inflammatory condition characterized by crystal deposits in the joints. Managing purine intake through diet and medication can help mitigate symptoms.
- **Orotic Aciduria:** Caused by defects in pyrimidine biosynthesis, orotic aciduria leads to the accumulation of orotic acid. This disorder is treated by supplementing uridine, bypassing the defective enzyme in pyrimidine synthesis.

26.9 THERAPEUTIC TARGETING OF NUCLEOTIDE BIOSYNTHESIS

Given the central role of nucleotides in cell division, nucleotide biosynthesis pathways are often targeted in cancer treatment. Rapidly dividing cancer cells have high demands for nucleotides, making them vulnerable to drugs that inhibit nucleotide synthesis.

- **Methotrexate:** This drug inhibits dihydrofolate reductase, preventing the synthesis of thymidine, a nucleotide essential for DNA replication. By limiting dTMP synthesis, methotrexate impairs DNA synthesis and cell division in cancer cells.
- **5-Fluorouracil (5-FU):** This drug inhibits thymidylate synthase, an enzyme essential for dTMP production, effectively reducing thymidine pools and preventing DNA synthesis.

- **Nucleotide Analogues:** Drugs like cytarabine mimic natural nucleotides and get incorporated into DNA, causing chain termination and halting replication. These analogues are commonly used in treating leukemia and other cancers.

26.10 CONCLUSION

Nucleotide biosynthesis is a complex and highly regulated process, fundamental to cellular functions and critical for DNA and RNA synthesis, energy transfer, and cellular signaling. The pathways of purine and pyrimidine synthesis involve intricate regulatory mechanisms, balancing the need for nucleotides with cellular demands and availability of precursors. Disorders in nucleotide metabolism can lead to severe health issues, highlighting the importance of these pathways in maintaining cellular and physiological balance. Furthermore, understanding nucleotide biosynthesis has paved the way for therapeutic interventions, particularly in targeting cancer cell proliferation. In essence, nucleotide biosynthesis represents a cornerstone of molecular biology, integral to life and cellular function.

CHAPTER 27
BIOSYNTHESIS OF MEMBRANE LIPIDS AND STEROIDS

The biosynthesis of membrane lipids and steroids is a fundamental aspect of cellular biochemistry, playing vital roles in membrane structure, cellular signaling, and metabolic regulation. Membrane lipids, including phospholipids, glycolipids, and cholesterol, form the basis of cellular membranes, providing both structural integrity and functional fluidity. Steroids, derived from cholesterol, serve as crucial signaling molecules and precursors for hormones that regulate various physiological processes.

27.1 OVERVIEW OF MEMBRANE LIPIDS

Membrane lipids are a group of compounds (structurally similar to fats and oils) which form the lipid bilayer of the cell membrane. The three major classes of membrane lipids are phospholipids, glycolipids, and cholesterol. Lipids are amphiphilic: they have one end that is soluble in water ('polar') and an ending that is soluble in fat ('nonpolar'). By forming a double layer with the polar ends pointing outwards and the nonpolar ends pointing inwards membrane lipids can form a 'lipid bilayer' which keeps the watery interior of the cell separate from the watery exterior.

The arrangements of lipids and various proteins, acting as receptors and channel pores in the membrane, control the entry and exit of other molecules and ions as part of the cell's metabolism. In order to perform physiological functions, membrane proteins are facilitated to rotate and diffuse laterally in two dimensional expanse of lipid bilayer by the presence of a shell of lipids closely attached to protein surface, called annular lipid shell.

27.1.1 Biological Roles

The bilayer formed by membrane lipids serves as a containment unit of a living cell. Membrane lipids also form a matrix in which membrane proteins reside. Historically lipids were thought to merely serve a structural role. Functional roles of lipids are in fact many: They serve as regulatory agents in cell growth and adhesion. They participate in the biosynthesis of other biomolecules. They can serve to increase enzymatic activities of enzymes.

Non-bilayer forming lipid like monogalactosyl diglyceride (MGDG) predominates the bulk lipids in thylakoid membranes, which when hydrated alone, forms reverse hexagonal cylindrical phase. However, in combination with other lipids and carotenoids/chlorophylls of thylakoid membranes, they too conform together as lipid bilayers.

27.1.2 Major Classes

Phospholipids

Phospholipids and glycolipids consist of two long, nonpolar (hydrophobic) hydrocarbon chains linked to a hydrophilic head group.

The heads of phospholipids are phosphorylated and they consist of either:

- **Glycerol** (and hence the name phosphoglycerides given to this group of lipids), or
- **Sphingosine** (e.g., sphingomyelin and ceramide).

Glycerol dialkyl glycerol tetraether (GDGT) is helping to study ancient environmental factors.

Glycolipids

The heads of glycolipids (glyco- stands for sugar) contain a sphingosine with one or several sugar units attached to it. The hydrophobic chains belong either to:

- Two fatty acids (FA) – in the case of the phosphoglycerides, or
- One FA and the hydrocarbon tail of sphingosine – in the case of sphingomyelin and the glycolipids.

Galactolipids – monogalactosyl diglyceride (MGDG) and digalactosyl diglycreride (DGDG) form the predominant lipids in higher plant chloroplast thylakoid membranes; liposomal structures formed by total lipid extract of thylakoid membranes have been found sensitive to sucrose as it turns bilayers into micellar structures.

Fatty Acids

The fatty acids in phospho- and glycolipids usually contain an even number, typically between 14 and 24, of carbon atoms, with 16- and 18-carbon being the most common. FAs may be saturated or unsaturated, with the configuration of the double bonds nearly always *cis*. The length and the degree of unsaturation of FAs chains have a profound effect on membranes' fluidity. Plant *thylakoid membranes* maintain high fluidity, even at relatively cold environmental temperatures, due to the abundance of 18-carbon fatty acyl chains with three double bonds, *linolenic acid*, as has been revealed by 13-C NMR studies.

Phosphoglycerides

In phosphoglycerides, the hydroxyl groups at C-1 and C-2 of glycerol are esterified to the carboxyl groups of the FAs. The C-3 hydroxyl group is esterified to phosphoric acid. The resulting compound, called phosphatidate, is the simplest phosphoglycerate. Only small amounts of phosphatidate are present in membranes. However, it is a key intermediate in the biosynthesis of the other phosphoglycerides.

Sphingolipids

Sphingosine is an amino alcohol that contains a long, unsaturated hydrocarbon chain. In sphingomyelin and glycolipids, the amino group of sphingosine is linked to FAs by an amide bond. In sphingomyelin the primary hydroxyl group of sphingosine is esterified to phosphoryl choline.

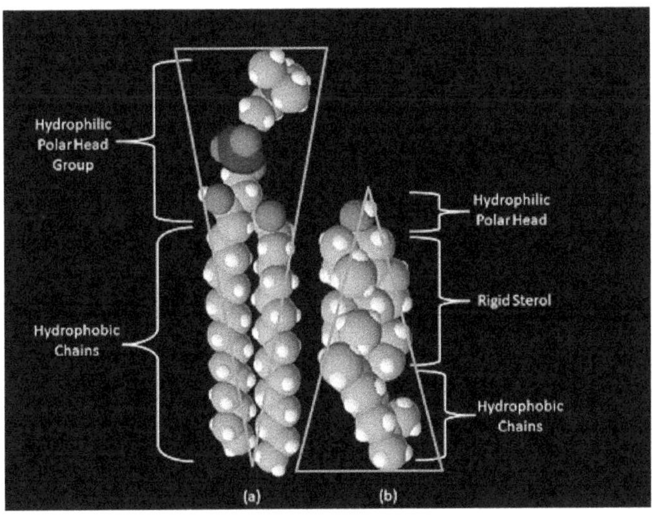

Fig 1: Space-Filling Models of (a) Sphingomyelin and (b) Cholesterol

In glycolipids, the sugar component is attached to this group. The simplest glycolipid is cerebroside, in which there is only one sugar residue, either Glc or Gal. More complex glycolipids, such as gangliosides, contain a branched chain of as many as seven sugar residues.

Sterols

The best known sterol is cholesterol, which is found in humans. Cholesterol also occurs naturally in other eukaryote cell membranes. Sterols have a hydrophobic four-membered fused ring rigid structure, and a small polar head group.

Cholesterol is bio-synthesised from mevalonate via a squalene cyclisation of terpenoids. Cell membranes require high levels of cholesterol – typically an average of 20% cholesterol in the whole membrane, increasing locally in raft areas up to 50% cholesterol (- % is molecular ratio). It associates preferentially with sphingolipids in cholesterol-rich lipid rafts areas of the membranes in eukaryotic cells. Formation of lipid rafts promotes aggregation of peripheral and transmembrane proteins including docking of SNARE and VAMP proteins. Phytosterols, such as sitosterol and stigmasterol, and hopanoids serve a similar function in plants and prokaryotes.

27.2 OVERVIEW OF STEROIDS

A steroid is an organic compound with four fused rings (designated A, B, C, and D) arranged in a specific molecular configuration. Steroids have two principal biological functions: as important components of cell membranes that alter membrane fluidity; and as signaling molecules.

Examples include the lipid cholesterol, sex hormones estradiol and testosterone, anabolic steroids, and the anti-inflammatory corticosteroid drug dexamethasone. Hundreds of steroids are found in fungi, plants, and animals. All steroids are manufactured in cells from the sterols lanosterol (opisthokonts) or cycloartenol (plants). Lanosterol and cycloartenol are derived from the cyclization of the triterpene squalene.

Steroids are named after the steroid cholesterol which was first described in gall stones from Ancient Greek *chole-* 'bile' and *stereos* 'solid'.

The steroid nucleus (core structure) is called gonane (cyclopentano-perhydrophenanthrene). It is typically composed of seventeen carbon atoms, bonded in four fused rings: three six-member cyclohexane rings (rings A, B and C in the first illustration) and one five-member cyclopentane ring (the D ring). Steroids vary by the functional groups attached to this four-ring core and by the oxidation state of the rings.

Sterols are forms of steroids with a hydroxy group at position three and a skeleton derived from cholestane. Steroids can also be more radically modified, such as by changes to the ring structure, for example, cutting one of the rings. Cutting Ring B produces secosteroids one of which is vitamin D_3.

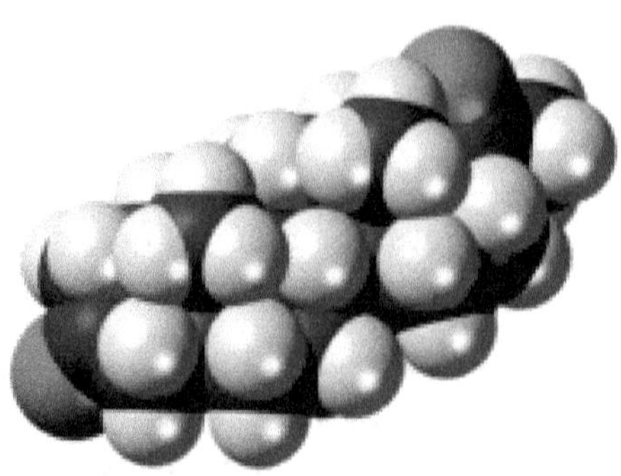

Fig 2: Space-Filling Representation

5α-dihydroprogesterone (5α-DHP), a steroid. The shape of the four rings of most steroids is illustrated (carbon atoms in black, oxygens in red and hydrogens in grey). The nonpolar "slab" of hydrocarbon in the middle (grey, black) and the polar groups at opposing ends (red) are common features of natural steroids. 5α-DHP is an endogenous steroid hormone and a biosynthetic intermediate.

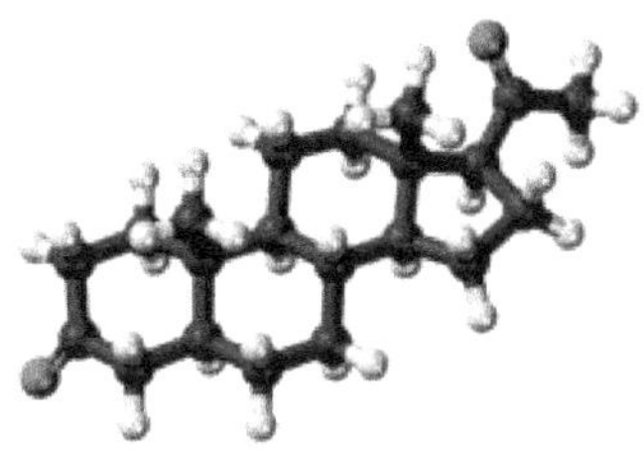

Fig 3: Ball-and-Stick Representation

27.2.1 Nomenclature

Rings and Functional Groups

Gonane, also known as steran or cyclopentanoperhydrophenanthrene, the simplest steroid and the nucleus of all steroids and sterols, is composed of seventeen carbon atoms in carbon-carbon bonds forming four fused rings in a three-dimensional shape. The three cyclohexane rings (A, B, and C in the first illustration) form the skeleton of a perhydro derivative of phenanthrene. The D ring has a cyclopentane structure. When the two methyl groups and eight carbon side chains (at C-17, as shown for cholesterol) are present, the steroid is said to have a cholestane framework. The two common 5α and 5β stereoisomeric forms of steroids exist because of differences in the side of the largely planar ring system where the hydrogen (H) atom at carbon-5 is attached, which results in a change in steroid A-ring conformation. Isomerisation at the C-21 side chain produces a parallel series of compounds, referred to as isosteroids.

Examples of steroid structures are:

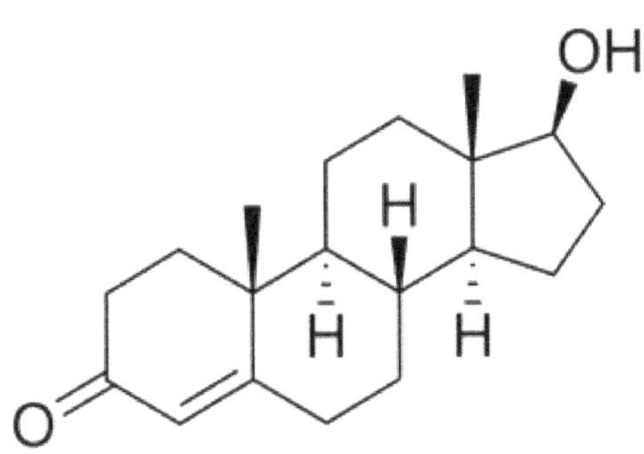

Fig 4: Testosterone, the Principal Male Sex Hormone and an Anabolic Steroid

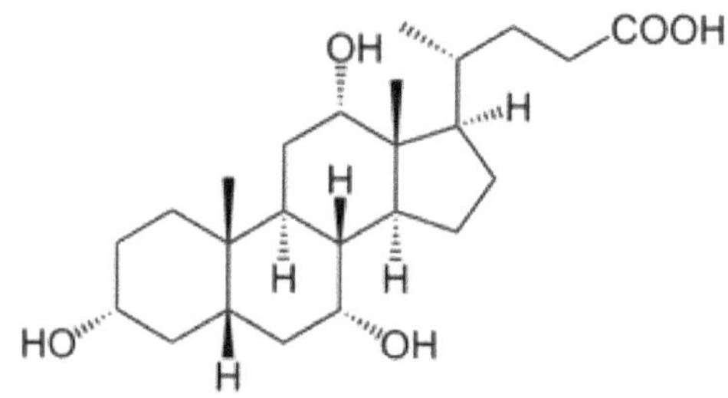

Fig 5: Cholic Acid, a Bile Acid, Showing the Carboxylic Acid and Additional Hydroxyl Groups often Present

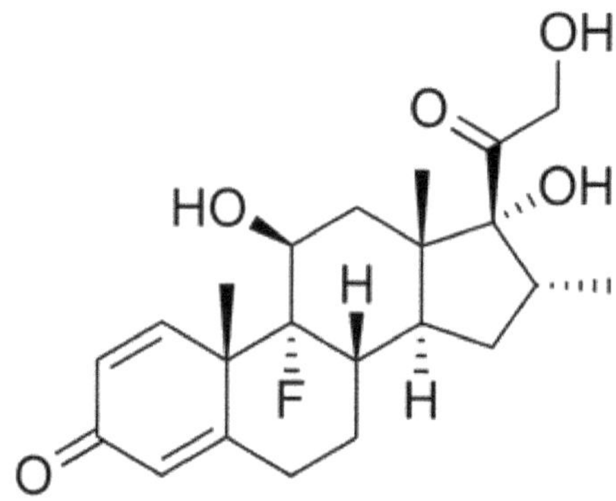

Fig 6: Dexamethasone, a Synthetic Corticosteroid Drug

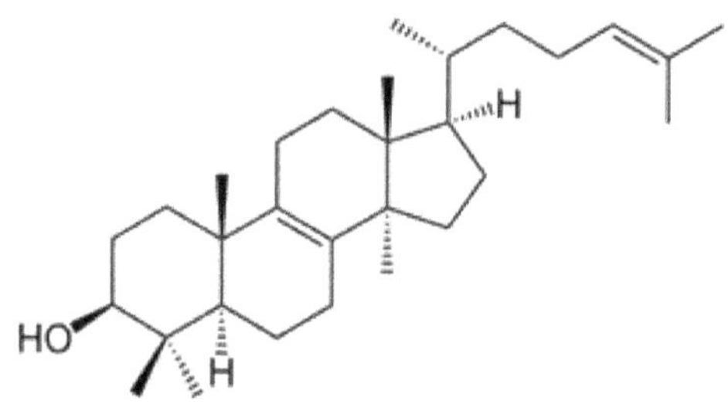

Fig 7: Lanosterol, the Biosynthetic Precursor to Animal Steroids. The Number of Carbons (30) Indicates its Triterpenoid Classification

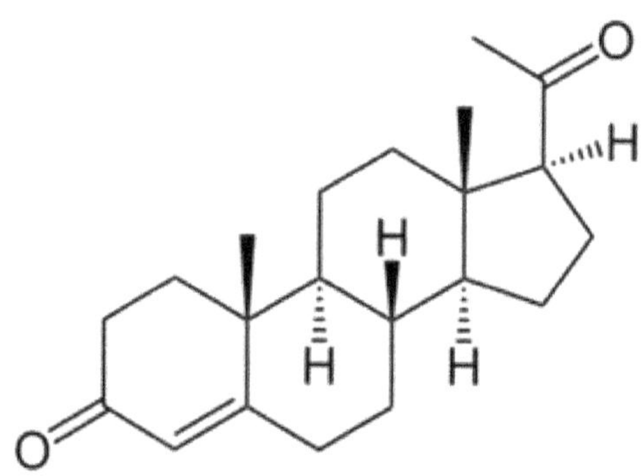

Fig 8: Progesterone, a Steroid Hormone Involved in the Female Menstrual Cycle, Pregnancy and Embryogenesis

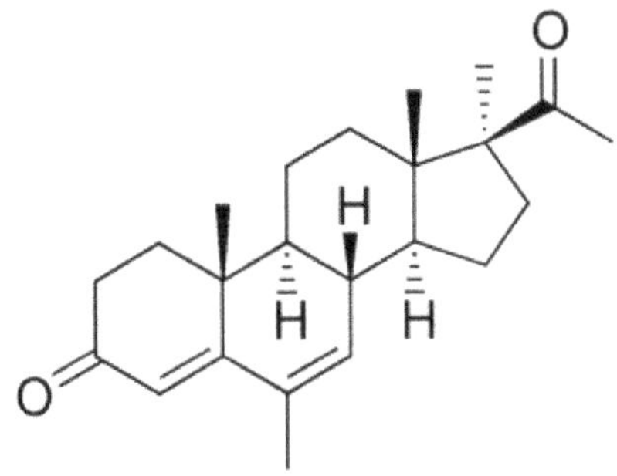

Fig 9: Medrogestone, a Synthetic Drug with Effects Similar to Progesterone

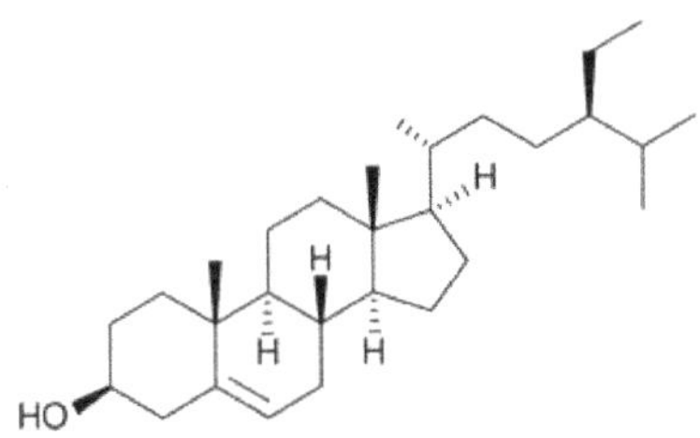

Fig 10: β-Sitosterol, a Plant or Phytosterol, with a Fully Branched Hydrocarbon Side Chain at C-17 and an Hydroxyl Group at C-3

In addition to the ring scissions (cleavages), expansions and contractions (cleavage and reclosing to a larger or smaller rings)—all variations in the carbon-carbon bond framework—steroids can also vary:

- in the bond orders within the rings,
- in the number of methyl groups attached to the ring (and, when present, on the prominent side chain at C17),
- in the functional groups attached to the rings and side chain, and
- in the configuration of groups attached to the rings and chain.

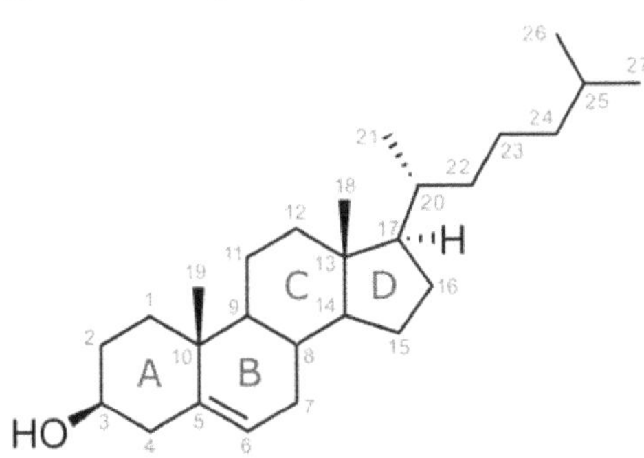

Fig 11: Cholesterol, a Prototypical Animal Sterol. This Structural Lipid and Key Steroid Biosynthetic

For instance, sterols such as cholesterol and lanosterol have a hydroxyl group attached at position C-3, while testosterone and progesterone have a carbonyl (oxo substituent) at C-3. Among these compounds, only lanosterol has two methyl groups at C-4. Cholesterol which has a C-5 to C-6 double bond, differs from testosterone and progesterone which have a C-4 to C-5 double bond.

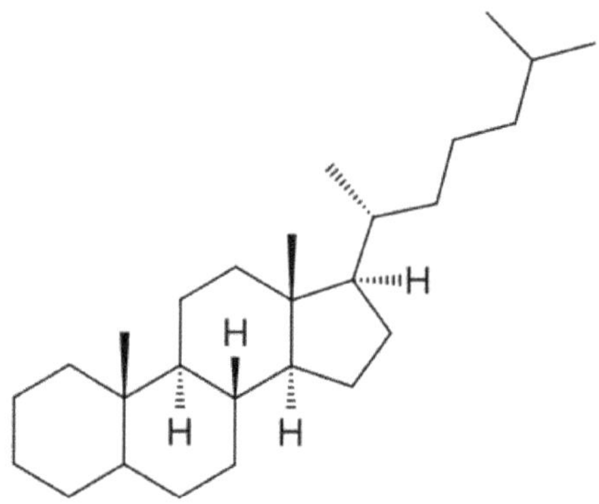

Fig 12: α-Cholestane, a Common Steroid Core

Naming Convention

Almost all biologically relevant steroids can be presented as a derivative of a parent cholesterol-like hydrocarbon structure that serves as a skeleton. These parent structures have specific names, such as pregnane, androstane, etc. The derivatives carry various functional groups called suffixes or prefixes after the respective numbers, indicating their position in the steroid nucleus. There are widely used trivial steroid names of natural origin with significant biologic activity, such as progesterone, testosterone or cortisol. Some of these names are defined in The Nomenclature of Steroids.

These trivial names can also be used as a base to derive new names, however, by adding prefixes only rather than suffixes, e.g., the steroid 17α-hydroxyprogesterone has a hydroxy group (-OH) at position 17 of the steroid nucleus comparing to progesterone.

The letters α and β denote absolute stereochemistry at chiral centers—a specific nomenclature distinct from the R/S convention of organic chemistry to denote absolute configuration of functional groups, known as Cahn–Ingold–Prelog priority rules. The R/S convention assigns priorities to substituents on a chiral center based on their atomic number.

The highest priority group is assigned to the atom with the highest atomic number, and the lowest priority group is assigned to the atom with the lowest atomic number. The molecule is then oriented so that the lowest priority group points away from the viewer, and the remaining three groups are arranged in order of decreasing priority around the chiral center. If this arrangement is clockwise, it is assigned an R configuration; if it is counterclockwise, it is assigned an S configuration. In contrast, steroid nomenclature uses α and β to denote stereochemistry at chiral centers.

The α and β designations are based on the orientation of substituents relative to each other in a specific ring system. In general, α refers to a substituent that is oriented towards the plane of the ring system, while β refers to a substituent that is oriented away from the plane of the ring system. In steroids drawn from the standard perspective used in this paper, α-bonds are depicted on figures as dashed wedges and β-bonds as solid wedges.

The name "11-deoxycortisol" is an example of a derived name that uses cortisol as a parent structure without an oxygen atom (hence "deoxy") attached to position 11 (as a part of a hydroxy group). The numbering of positions of carbon atoms in the steroid nucleus is set in a template found in the Nomenclature of Steroids that is used regardless of whether an atom is present in the steroid in question.

Unsaturated carbons (generally, ones that are part of a double bond) in the steroid nucleus are indicated by changing -ane to -ene. This change was traditionally done in the parent name, adding a prefix to denote the position, with or without Δ (Greek capital delta) which designates unsaturation, for example, 4-pregnene-11β,17α-diol-3,20-dione (also Δ^4-pregnene-11β,17α-diol-3,20-dione) or 4-androstene-3,11,17-trione (also Δ^4-androstene-3,11,17-trione). However, the Nomenclature of Steroids recommends the locant of a double bond to be always adjacent to the syllable designating the unsaturation, therefore, having it as a suffix rather than a prefix, and without the use of the Δ character, i.e., pregn-4-ene-11β,17α-diol-3,20-dione or androst-4-ene-3,11,17-trione. The double bond is designated by the lower-numbered carbon atom, i.e. "Δ^4-" or "4-ene" means the double bond between positions 4 and 5. The saturation of carbons of a parent steroid can be done by adding "dihydro-" prefix, i.e., a saturation of carbons 4 and 5 of testosterone with two hydrogen atoms is 4,5α-dihydrotestosterone or 4,5β-dihydrotestosterone. Generally, when there is no ambiguity, one number of a hydrogen position from a steroid with a saturated bond may be omitted, leaving only the position of the second hydrogen atom, e.g., 5α-dihydrotestosterone or 5β-dihydrotestosterone. The Δ^5-steroids are those with a double bond between carbons 5 and 6 and the Δ^4 steroids are those with a double bond between carbons 4 and 5.

The abbreviations like "P4" for progesterone and "A4" for androstenedione for refer to Δ^4-steroids, while "P5" for pregnenolone and "A5" for androstenediol refer to Δ^5-steroids.

The suffix -ol denotes a hydroxy group, while the suffix -one denotes an oxo group. When two or three identical groups are attached to the base structure at different positions, the suffix is indicated as -diol or -triol for hydroxy, and -dione or -trione for oxo groups, respectively. For example, 5α-pregnane-3α,17α-diol-20-one has a hydrogen atom at the 5α position (hence the "5α-" prefix), two hydroxy groups (-OH) at the 3α and 17α positions (hence "3α,17α-diol" suffix) and an oxo group (=O) at the position 20 (hence the "20-one" suffix). However, erroneous use of suffixes can be found, e.g., "5α-pregnan-17α-diol-3,11,20-trione" [*sic*] — since it has just one hydroxy group (at 17α) rather than two, then the suffix should be -ol, rather than -diol, so that the correct name to be "5α-pregnan-17α-ol-3,11,20-trione".

According to the rule set in the Nomenclature of Steroids, the terminal "e" in the parent structure name should be elided before the vowel (the presence or absence of a number does not affect such elision). This means, for instance, that if the suffix immediately appended to the parent structure name begins with a vowel, the trailing "e" is removed from that name. An example of such removal is "5α-pregnan-17α-ol-3,20-dione", where the last "e" of "pregnane" is dropped due to the vowel ("o") at the beginning of the suffix -ol. Some authors incorrectly use this rule, eliding the terminal "e" where it should be kept, or vice versa.

The term "11-oxygenated" refers to the presence of an oxygen atom as an oxo (=O) or hydroxy (-OH) substituent at carbon 11. "Oxygenated" is consistently used within the chemistry of the steroids since the 1950s. Some studies use the term "11-oxyandrogens" as an abbreviation for 11-oxygenated androgens, to emphasize that they all have an oxygen atom attached to carbon at position 11. However, in chemical nomenclature, the prefix "oxy" is associated with ether functional groups, i.e., a compound with an oxygen atom connected to two alkyl or aryl groups (R-O-R), therefore, using "oxy" within the name of a steroid class may be misleading. One can find clear examples of "oxygenated" to refer to a broad class of organic molecules containing a variety of oxygen containing functional groups in other domains of organic chemistry, and it is appropriate to use this convention.

Even though "keto" is a standard prefix in organic chemistry, the 1989 recommendations of the Joint Commission on Biochemical Nomenclature discourage the application of the prefix "keto" for steroid names, and favor the prefix "oxo" (e.g., 11-oxo steroids rather than 11-keto steroids), because "keto" includes the carbon that is part of the steroid nucleus and the same carbon atom should not be specified twice.

27.2.2 Species Distribution

Steroids are found in all domains of life including bacteria, archaea, and eukaryotes. In eukaryotes, steroids are found in fungi, plants, and animals.

Eukaryotic and Prokaryotic

- Eukaryotic cells, which include animals, plants, fungi, and protists, have complex cellular structures with a true nucleus and membrane-bound organelles.

- Steroids are integral to eukaryotic cellular membranes, where they help maintain membrane integrity and function.
- During eukaryogenesis (the emergence of modern eukaryotic cells), steroids likely played a role in the acquisition of mitochondria via endocytosis.
- In prokaryotes, biosynthetic pathways exist for the tetracyclic steroid framework (e.g. in myxobacteria) – where its origin from eukaryotes is conjectured – and the more-common pentacyclic triterpinoid hopanoid framework.

Fungal

Fungal steroids include the ergosterols, which are involved in maintaining the integrity of the fungal cellular membrane. Various antifungal drugs, such as amphotericin B and azole antifungals, utilize this information to kill pathogenic fungi. Fungi can alter their ergosterol content (e.g. through loss of function mutations in the enzymes ERG3 or ERG6, inducing depletion of ergosterol, or mutations that decrease the ergosterol content) to develop resistance to drugs that target ergosterol.

Ergosterol is analogous to the cholesterol found in the cellular membranes of animals (including humans), or the phytosterols found in the cellular membranes of plants. All mushrooms contain large quantities of ergosterol, in the range of tens to hundreds of milligrams per 100 grams of dry weight. Oxygen is necessary for the synthesis of ergosterol in fungi.

Ergosterol is responsible for the vitamin D content found in mushrooms; ergosterol is chemically converted into provitamin D2 by exposure to ultraviolet light. Provitamin D2 spontaneously forms vitamin D2. However, not all fungi utilize ergosterol in their cellular membranes; for example, the pathogenic fungal species *Pneumocystis jirovecii* does not, which has important clinical implications (given the mechanism of action of many antifungal drugs). Using the fungus *Saccharomyces cerevisiae* as an example, other major steroids include ergosta-5,7,22,24(28)-tetraen-3β-ol, zymosterol, and lanosterol. *S. cerevisiae* utilizes 5,6-dihydroergosterol in place of ergosterol in its cell membrane.

Plant

Plant steroids include steroidal alkaloids found in Solanaceae and Melanthiaceae (specially the genus Veratrum), cardiac glycosides, the phytosterols and the brassinosteroids (which include several plant hormones).

Animal

Animal steroids include compounds of vertebrate and insect origin, the latter including ecdysteroids such as ecdysterone (controlling molting in some species). Vertebrate examples include the steroid hormones and cholesterol; the latter is a structural component of cell membranes that helps determine the fluidity of cell membranes and is a principal constituent of plaque (implicated in atherosclerosis). Steroid hormones include:

- **Sex hormones**, which influence sex differences and support reproduction. These include androgens, estrogens, and progestogens.
- **Corticosteroids**, including most synthetic steroid drugs, with natural product classes the glucocorticoids (which regulate many aspects of metabolism and immune function) and the mineralocorticoids (which help maintain blood volume and control renal excretion of electrolytes)
- **Anabolic steroids**, natural and synthetic, which interact with androgen receptors to increase muscle and bone synthesis. In popular use, the term "steroids" often refers to anabolic steroids.

27.2.3 Types

By Function

The major classes of steroid hormones, with prominent members and examples of related functions, are:

- **Corticosteroids:**
  - *Glucocorticoids:* Cortisol, a glucocorticoid whose functions include immunosuppression.
  - *Mineralocorticoids:* Aldosterone, a mineralocorticoid that helps regulate blood pressure through water and electrolyte balance.
- **Sex Steroids:**
  - *Progestogens:* Progesterone, which regulates cyclical changes in the endometrium of the uterus and maintains a pregnancy.
  - *Androgens:* Testosterone, which contributes to the development and maintenance of male secondary sex characteristics.
  - *Estrogens:* Estradiol, which contributes to the development and maintenance of female secondary sex characteristics.

Additional classes of steroids include:

- Neurosteroids such as DHEATooltip dehydroepiandrosterone and allopregnanolone.
- Bile acids such as taurocholic acid.

- Aminosteroid neuromuscular blocking agents (mainly synthetic) such as pancuronium bromide.
- Steroidal antiandrogens (mainly synthetic) such as cyproterone acetate.
- Steroidogenesis inhibitors (mainly exogenous) such as alfatradiol.
- Membrane sterols such as cholesterol, ergosterol, and various phytosterols.
- Toxins such as steroidal saponins and cardenolides/cardiac glycosides.

As well as the following class of secosteroids (open-ring steroids):

- Vitamin D forms such as ergocalciferol, cholecalciferol, and calcitriol.

By Structure

Intact Ring System

Steroids can be classified based on their chemical composition. Examples of this classification include:

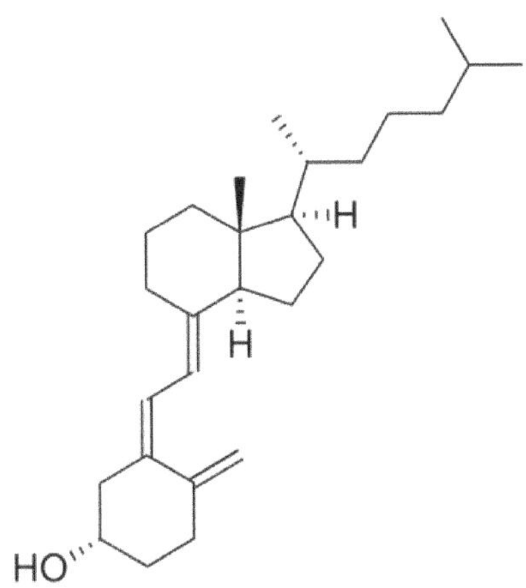

Fig 13: Cholecalciferol (Vitamin D_3), an Example of a 9,10-Secosteroid

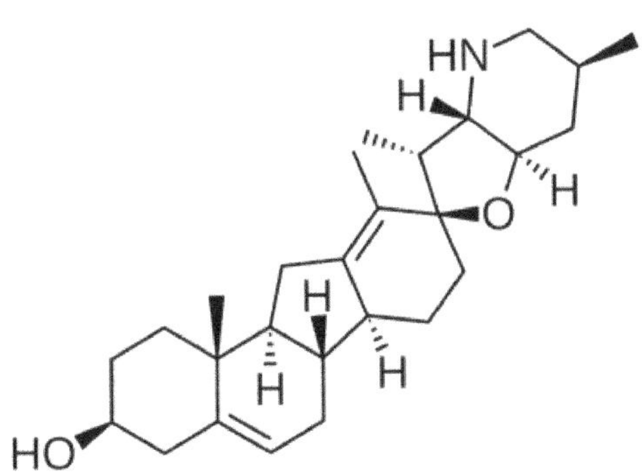

Fig 14: Cyclopamine, an example of a complex C-nor-D-homosteroid

Class	Example	Number of Carbon Atoms
Cholestanes	Cholesterol	27
Cholanes	Cholic acid	24
Pregnanes	Progesterone	21
Androstanes	Testosterone	19
Estranes	Estradiol	18

In biology, it is common to name the above steroid classes by the number of carbon atoms present when referring to hormones: C_{18}-steroids for the estranes (mostly estrogens), C_{19}-steroids for the androstanes (mostly androgens), and C_{21}-steroids for the pregnanes (mostly corticosteroids). The classification "17-ketosteroid" is also important in medicine.

The gonane (steroid nucleus) is the parent 17-carbon tetracyclic hydrocarbon molecule with no alkyl sidechains.

Cleaved, Contracted, and Expanded Rings

Secosteroids (Latin *seco*, "to cut") are a subclass of steroidal compounds resulting, biosynthetically or conceptually, from scission (cleavage) of parent steroid rings (generally one of the four). Major secosteroid subclasses are defined by the steroid carbon atoms where this scission has taken place. For instance, the prototypical secosteroid cholecalciferol, vitamin D_3 (shown), is in the 9,10-secosteroid subclass and derives from the cleavage of carbon atoms C-9 and C-10 of the steroid B-ring; 5,6-secosteroids and 13,14-steroids are similar.

Norsteroids (nor-, L. *norma*; "normal" in chemistry, indicating carbon removal) and homosteroids (homo-, Greek *homos*; "same", indicating carbon addition) are structural subclasses of steroids formed from biosynthetic steps. The former involves enzymic ring expansion-contraction reactions, and the latter is accomplished (biomimetically) or (more frequently) through ring closures of acyclic precursors with more (or fewer) ring atoms than the parent steroid framework.

Combinations of these ring alterations are known in nature. For instance, ewes who graze on corn lily ingest cyclopamine (shown) and veratramine, two of a sub-family of steroids where the C- and D-rings are contracted and expanded respectively via a biosynthetic migration of the original C-13 atom. Ingestion of these C-nor-D-homosteroids results in birth defects in lambs: cyclopia from cyclopamine and leg deformity from veratramine. A further C-nor-D-homosteroid (nakiterpiosin) is excreted by Okinawan cyanobacteriosponges. e.g., *Terpios hoshinota*, leading to coral mortality from black coral disease. Nakiterpiosin-type steroids are active against the signaling pathway involving the smoothened and hedgehog proteins, a pathway which is hyperactive in a number of cancers.

27.2.4 Biological Significance

Steroids and their metabolites often function as signalling molecules (the most notable examples are steroid hormones), and steroids and phospholipids are components of cell membranes. Steroids such as cholesterol decrease membrane fluidity. Similar to lipids, steroids are highly concentrated energy stores. However, they are not typically sources of energy; in mammals, they are normally metabolized and excreted. Steroids play critical roles in a number of disorders, including malignancies like prostate cancer, where steroid production inside and outside the tumour promotes cancer cell aggressiveness.

27.2.5 Biosynthesis and Metabolism

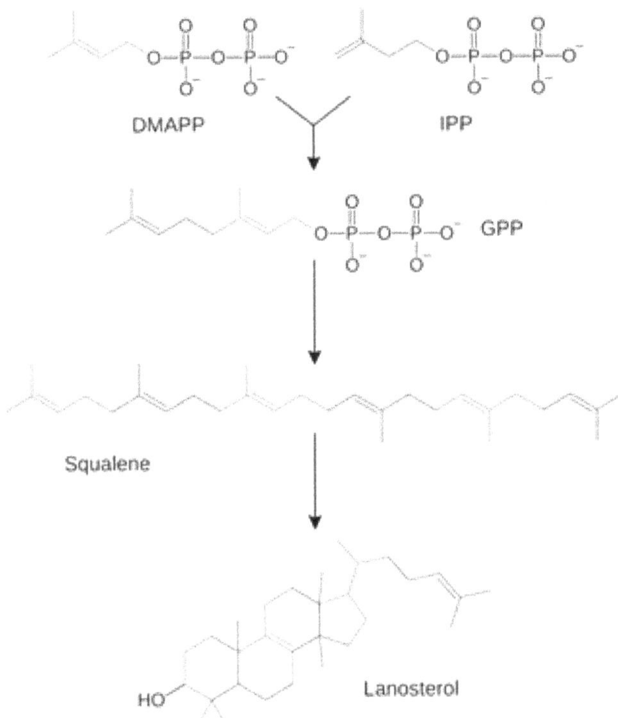

Fig 15: Simplification of the End of the Steroid Synthesis Pathway, where the Intermediates Isopentenyl Pyrophosphate (PP or IPP) and Dimethylallyl Pyrophosphate (DMAPP) form Geranyl Pyrophosphate (GPP), Squalene and Lanosterol (the First Steroid in the Pathway)

The hundreds of steroids found in animals, fungi, and plants are made from lanosterol or cycloartenol (in other eukaryotes). Both lanosterol and cycloartenol derive from cyclization of the triterpenoid squalene. Lanosterol and cycloartenol are sometimes called protosterols because they serve as the starting compounds for all other steroids.

Steroid biosynthesis is an anabolic pathway which produces steroids from simple precursors. A unique biosynthetic pathway is followed in animals (compared to many other organisms), making the pathway a common target for antibiotics and other anti-infection drugs. Steroid metabolism in humans is also the target of cholesterol-lowering drugs, such as statins. In humans and other animals the biosynthesis of steroids follows the mevalonate pathway, which uses acetyl-CoA as building blocks for dimethylallyl diphosphate (DMAPP) and isopentenyl diphosphate (IPP).

In subsequent steps DMAPP and IPP conjugate to form farnesyl diphosphate (FPP), which further conjugates with each other to form the linear triterpenoid squalene. Squalene biosynthesis is catalyzed by squalene synthase, which belongs to the squalene/phytoene synthase family. Subsequent epoxidation and cyclization of squalene generate lanosterol, which is the starting point for additional modifications into other steroids (steroidogenesis). In other eukaryotes, the cyclization product of epoxidized squalene (oxidosqualene) is cycloartenol.

Mevalonate Pathway

The mevalonate pathway begins with acetyl-CoA and ends with dimethylallyl diphosphate (DMAPP) and isopentenyl diphosphate (IPP). DMAPP and IPP donate isoprene units, which are assembled and modified to form terpenes and isoprenoids (a large class of lipids, which include the carotenoids and form the largest class of plant natural products). Here, the isoprene units are joined to make squalene and folded into a set of rings to make lanosterol. Lanosterol can then be converted into other steroids, such as cholesterol and ergosterol.

Two classes of drugs target the mevalonate pathway: statins (like rosuvastatin), which are used to reduce elevated cholesterol levels, and bisphosphonates (like zoledronate), which are used to treat a number of bone-degenerative diseases.

Steroidogenesis

Steroidogenesis is the biological process by which steroids are generated from cholesterol and changed into other steroids. The pathways of steroidogenesis differ among species. The major classes of steroid hormones, as noted above (with their prominent members and functions), are the progestogens, corticosteroids (corticoids), androgens, and estrogens. Human steroidogenesis of these classes occurs in a number of locations:

- Progestogens are the precursors of all other human steroids, and all human tissues which produce steroids must first convert cholesterol to pregnenolone. This conversion is the rate-limiting step of steroid synthesis, which occurs inside the mitochondrion of the respective tissue.

- Cortisol, corticosterone, aldosterone are produced in the adrenal cortex.
- Estradiol, estrone and progesterone are made primarily in the ovary, estriol in placenta during pregnancy, and testosterone primarily in the testes (some testosterone may also be produced in the adrenal cortex).
- Estradiol is converted from testosterone directly (in males), or via the primary pathway DHEA - androstenedione - estrone and secondarily via testosterone (in females).
- Stromal cells have been shown to produce steroids in response to signaling produced by androgen-starved prostate cancer cells.
- Some neurons and glia in the central nervous system (CNS) express the enzymes required for the local synthesis of pregnenolone, progesterone, DHEA and DHEAS, *de novo* or from peripheral sources.

Alternative Pathways

In plants and bacteria, the non-mevalonate pathway (MEP pathway) uses pyruvate and glyceraldehyde 3-phosphate as substrates to produce IPP and DMAPP. During diseases pathways otherwise not significant in healthy humans can become utilized. For example, in one form of congenital adrenal hyperplasia a deficiency in the 21-hydroxylase enzymatic pathway leads to an excess of 17α-Hydroxyprogesterone (17-OHP) - this pathological excess of 17-OHP in turn may be converted to dihydrotestosterone (DHT, a potent androgen) through among others 17,20 Lyase (a member of the cytochrome P450 family of enzymes), 5α-Reductase and 3α-Hydroxysteroid dehydrogenase.

27.2.6 Catabolism and Excretion

Steroids are primarily oxidized by cytochrome P450 oxidase enzymes, such as CYP3A4. These reactions introduce oxygen into the steroid ring, allowing the cholesterol to be broken up by other enzymes into bile acids. These acids can then be eliminated by secretion from the liver in bile. The expression of the oxidase gene can be upregulated by the steroid sensor PXR when there is a high blood concentration of steroids. Steroid hormones, lacking the side chain of cholesterol and bile acids, are typically hydroxylated at various ring positions or oxidized at the 17 position, conjugated with sulfate or glucuronic acid and excreted in the urine.

27.2.7 Isolation, Structure Determination, and Methods of Analysis

Steroid *isolation*, depending on context, is the isolation of chemical matter required for chemical structure elucidation, derivitzation or degradation chemistry, biological testing, and other research needs (generally milligrams to grams, but often more or the isolation of "analytical quantities" of the substance of interest (where the focus is on identifying and quantifying the substance (for example, in biological tissue or fluid). The amount isolated depends on the analytical method, but is generally less than one microgram.

The methods of isolation to achieve the two scales of product are distinct, but include extraction, precipitation, adsorption, chromatography, and crystallization. In both cases, the isolated substance is purified to chemical homogeneity; combined separation and analytical methods, such as LC-MS, are chosen to be "orthogonal"—achieving their separations based on distinct modes of interaction between substance and isolating matrix—to detect a single species in the pure sample.

Structure determination refers to the methods to determine the chemical structure of an isolated pure steroid, using an evolving array of chemical and physical methods which have included NMR and small-molecule crystallography.

Methods of analysis overlap both of the above areas, emphasizing analytical methods to determining if a steroid is present in a mixture and determining its quantity.

27.2.8 Chemical Synthesis

Microbial catabolism of phytosterol side chains yields C-19 steroids, C-22 steroids, and 17-ketosteroids (i.e. precursors to adrenocortical hormones and contraceptives).

The addition and modification of functional groups is key when producing the wide variety of medications available within this chemical classification. These modifications are performed using conventional organic synthesis and/or biotransformation techniques.

Precursors

Semisynthesis

The semisynthesis of steroids often begins from precursors such as cholesterol, phytosterols, or sapogenins. The efforts of Syntex, a company involved in the Mexican barbasco trade, used *Dioscorea mexicana* to produce the sapogenin diosgenin in the early days of the synthetic steroid pharmaceutical industry.

Total Synthesis

Some steroidal hormones are economically obtained only by total synthesis from petrochemicals (e.g. 13-alkyl steroids). For example, the pharmaceutical Norgestrel begins from methoxy-1-tetralone, a petrochemical derived from phenol.

27.3 BIOSYNTHESIS OF PHOSPHOLIPIDS

Phospholipid biosynthesis occurs primarily in the endoplasmic reticulum (ER) and involves several enzymatic reactions. The primary starting point for phospholipid synthesis is glycerol-3-phosphate, which undergoes several modifications to yield various phospholipid species.

27.3.1 Steps in Phospholipid Biosynthesis

- **Formation of Phosphatidic Acid:** The biosynthesis begins with the acylation of glycerol-3-phosphate. Two fatty acids are esterified to the glycerol backbone, resulting in the formation of phosphatidic acid (PA). This reaction is catalyzed by enzymes known as acyltransferases, which facilitate the addition of fatty acyl-CoA molecules.
- **Conversion to Diacylglycerol:** Phosphatidic acid can be dephosphorylated by the enzyme phosphatidic acid phosphatase to yield diacylglycerol (DAG), a crucial intermediate in the biosynthetic pathway. DAG can then be converted into several important phospholipids.
- **Synthesis of Phosphatidylcholine and Phosphatidylethanolamine:** Diacylglycerol can be further modified to produce phosphatidylcholine (PC) and phosphatidylethanolamine (PE). The enzyme choline phosphotransferase transfers a phosphate group from CDP-choline to DAG, forming PC. Alternatively, PE is synthesized from DAG through the addition of an ethanolamine group.
- **Formation of Phosphatidylserine and Phosphatidylinositol:** Phosphatidylserine (PS) is generated by a base-exchange reaction in which serine is substituted for the ethanolamine in PE. Phosphatidylinositol (PI) is synthesized from the transfer of inositol to DAG, making it a key component in cellular signaling pathways.

27.3.2 Regulation of Phospholipid Biosynthesis

The regulation of phospholipid biosynthesis is complex and involves feedback mechanisms. Enzymes such as CDP-choline cytidylyltransferase and phosphatidic acid phosphatase are subject to regulation by various factors, including substrate availability and phosphorylation states. Additionally, the composition of fatty acids in membrane lipids is influenced by dietary intake and metabolic needs, allowing cells to adapt their membranes for optimal function under different conditions.

27.4 BIOSYNTHESIS OF GLYCOLIPIDS

Glycolipids, which are essential for cell recognition and communication, are synthesized in the ER and Golgi apparatus. Their biosynthesis begins with the formation of ceramide, a sphingolipid that acts as a precursor for various glycolipids.

27.4.1 Steps in Glycolipid Biosynthesis

- **Ceramide Formation:** Ceramide is synthesized by the condensation of serine and palmitoyl-CoA, catalyzed by the enzyme serine palmitoyltransferase. This reaction yields dihydrosphingosine, which is subsequently acylated to form ceramide.

- **Addition of Sugar Residues:** In the Golgi apparatus, glycosyltransferases catalyze the addition of sugar moieties to ceramide, producing various glycolipids. These enzymes determine the specific carbohydrate composition, leading to the formation of different glycolipid classes, including gangliosides and cerebrosides.
- **Diversity and Function:** The diversity of glycolipids in different tissues is essential for cellular functions such as recognition, signaling, and adhesion. For example, glycolipids are abundant in neuronal membranes, where they play critical roles in synaptic transmission and cell signaling.

27.4.2 Physiological Importance of Glycolipids

Glycolipids are key players in cell-cell interactions, serving as receptors for pathogens and mediating immune responses. They also contribute to the formation of lipid rafts, specialized membrane microdomains that facilitate signaling events and membrane protein organization.

27.5 BIOSYNTHESIS OF CHOLESTEROL

Cholesterol is synthesized in the liver and other tissues from acetyl-CoA through a series of complex reactions. The cholesterol biosynthetic pathway is a multistep process that produces various intermediates, each contributing to the final cholesterol molecule.

27.5.1 Steps in Cholesterol Biosynthesis

- **Formation of HMG-CoA:** The biosynthesis begins with the condensation of three acetyl-CoA molecules to form HMG-CoA (3-hydroxy-3-methylglutaryl-CoA). This reaction is catalyzed by the enzyme HMG-CoA synthase.
- **Reduction to Mevalonate:** HMG-CoA is then reduced to mevalonate by HMG-CoA reductase, which is the rate-limiting step in cholesterol biosynthesis. This enzyme is a critical regulatory point and is the target of statins, which are widely used to lower cholesterol levels in the body.
- **Conversion to Isoprenoid Units:** Mevalonate is phosphorylated and decarboxylated to produce isoprenoid units, which are essential for the further synthesis of cholesterol.
- **Formation of Squalene and Cholesterol:** The isoprenoid units condense to form squalene, which undergoes cyclization and other

modifications to yield cholesterol. This final step involves a series of enzymatic reactions that transform the linear squalene structure into the four-ring steroid structure characteristic of cholesterol.

27.5.2 Regulation of Cholesterol Biosynthesis

Cholesterol synthesis is tightly regulated at multiple levels, primarily through the activity of HMG-CoA reductase. High cellular cholesterol levels suppress the transcription of genes encoding this enzyme, while low levels stimulate its expression.

Additionally, the activity of SREBP (sterol regulatory element-binding protein) transcription factors plays a crucial role in regulating cholesterol and fatty acid biosynthesis in response to cellular cholesterol levels.

27.6 BIOSYNTHESIS OF STEROID HORMONES

Steroid hormones are synthesized from cholesterol in steroidogenic tissues, including the adrenal glands and gonads.

These hormones are vital for numerous physiological processes, including metabolism, immune response, and reproductive function.

27.6.1 Steps in Steroid Hormone Biosynthesis

- **Conversion to Pregnenolone**: Cholesterol is transported into the mitochondria, where it is converted into pregnenolone by the enzyme cholesterol side-chain cleavage enzyme. Pregnenolone serves as the precursor for all steroid hormones.
- **Formation of Specific Steroids**: Pregnenolone can be further converted into various steroid hormones through different pathways. For example, it can be transformed into progesterone, which serves as a precursor for cortisol (a glucocorticoid) and aldosterone (a mineralocorticoid). Pregnenolone can also give rise to testosterone and estrogens, which are critical for reproductive health.

27.6.2 Regulation of Steroid Hormone Biosynthesis

The biosynthesis of steroid hormones is regulated by a variety of factors, including hormonal signals and feedback mechanisms.

For example, adrenocorticotropic hormone (ACTH) stimulates cortisol production in the adrenal cortex, while luteinizing hormone (LH) and follicle-stimulating hormone (FSH) regulate sex hormone synthesis in the gonads. The intricate regulation of steroid biosynthesis ensures that hormone levels remain balanced and responsive to physiological needs.

27.7 PHYSIOLOGICAL ROLES OF MEMBRANE LIPIDS AND STEROIDS

The biosynthesis of membrane lipids and steroids serves numerous physiological roles essential for maintaining cellular and systemic functions.

- **Cell Membrane Structure and Function:** Membrane lipids are crucial for maintaining the integrity and fluidity of cellular membranes. The lipid bilayer created by phospholipids provides a barrier to protect cellular contents and facilitates the compartmentalization of cellular processes.
- **Signaling and Communication:** Membrane lipids, particularly phosphoinositides, play vital roles in intracellular signaling pathways. They act as second messengers that relay signals from membrane receptors to intracellular targets, influencing processes such as cell growth, differentiation, and metabolism.
- **Hormonal Regulation and Metabolism:** Steroid hormones, derived from cholesterol, regulate a wide array of physiological processes, including metabolism, immune response, and reproductive functions. For instance, glucocorticoids such as cortisol are critical for stress response, while sex hormones regulate reproductive cycles and secondary sexual characteristics.
- **Membrane Fluidity and Dynamics:** Cholesterol is integral to maintaining membrane fluidity across various temperatures, allowing for the proper functioning of membrane proteins and lipid raft formation, which are important for cell signaling and protein interactions.

CHAPTER 28

DNA REPLICATION, REPAIR AND RECOMBINATION

DNA replication, repair, and recombination are fundamental biological processes that ensure the integrity and stability of the genetic material in all living organisms. These processes are crucial for cellular function, growth, and reproduction, and they play a significant role in evolution and genetic diversity.

28.1 DNA REPLICATION

In molecular biology, DNA replication is the biological process of producing two identical replicas of DNA from one original DNA molecule. DNA replication occurs in all living organisms acting as the most essential part of biological inheritance. This is essential for cell division during growth and repair of damaged tissues, while it also ensures that each of the new cells receives its own copy of the DNA. The cell possesses the distinctive property of division, which makes replication of DNA essential.

DNA is made up of a double helix of two complementary strands. The double helix describes the appearance of a double-stranded DNA which is thus composed of two linear strands that run opposite to each other and twist together to form. During replication, these strands are separated.

Each strand of the original DNA molecule then serves as a template for the production of its counterpart, a process referred to as semiconservative replication. As a result of semi-conservative replication, the new helix will be composed of an original DNA strand as well as a newly synthesized strand. Cellular proofreading and error-checking mechanisms ensure near perfect fidelity for DNA replication.

In a cell, DNA replication begins at specific locations, or origins of replication, in the genome which contains the genetic material of an organism. Unwinding of DNA at the origin and synthesis of new strands, accommodated by an enzyme known as helicase, results in replication forks growing bi-directionally from the origin. A number of proteins are associated with the replication fork to help in the initiation and continuation of DNA synthesis. Most prominently, DNA polymerase synthesizes the new strands by adding nucleotides that complement each (template) strand. DNA replication occurs during the S-stage of interphase.

DNA replication (DNA amplification) can also be performed *in vitro* (artificially, outside a cell). DNA polymerases isolated from cells and artificial DNA primers can be used to start DNA synthesis at known sequences in a template DNA molecule. Polymerase chain reaction (PCR), ligase chain reaction (LCR), and transcription-mediated amplification (TMA) are examples. In March 2021, researchers reported evidence suggesting that a preliminary form of transfer RNA, a necessary component of translation, the biological synthesis of new proteins in accordance with the genetic code, could have been a replicator molecule itself in the very early development of life, or abiogenesis.

28.1.1 DNA Structure

DNA exists as a double-stranded structure, with both strands coiled together to form the characteristic double helix. Each single strand of DNA is a chain of four types of nucleotides. Nucleotides in DNA contain a deoxyribose sugar, a phosphate, and a nucleobase. The four types of nucleotide correspond to the four nucleobases adenine, cytosine, guanine, and thymine, commonly abbreviated as A, C, G, and T. Adenine and guanine are purine bases, while cytosine and thymine are pyrimidines. These nucleotides form phosphodiester bonds, creating the phosphate-deoxyribose backbone of the DNA double helix with the nucleobases pointing inward (i.e., toward the opposing strand). Nucleobases are matched between strands through hydrogen bonds to form base pairs. Adenine pairs with thymine (two hydrogen bonds), and guanine pairs with cytosine (three hydrogen bonds).

DNA strands have a directionality, and the different ends of a single strand are called the "3′ (three-prime) end" and the "5′ (five-prime) end". By convention, if the base sequence of a single strand of DNA is given, the left end of the sequence is the 5′ end, while the right end of the sequence is the 3′ end. The strands of the double helix are anti-parallel, with one being 5′ to 3′, and the opposite strand 3′ to 5′. These terms refer to the carbon atom in deoxyribose to which the next phosphate in the chain attaches. Directionality has consequences in DNA synthesis, because DNA polymerase can synthesize DNA in only one direction by adding nucleotides to the 3′ end of a DNA strand.

The pairing of complementary bases in DNA (through hydrogen bonding) means that the information contained within each strand is redundant. Phosphodiester (intra-strand) bonds are stronger than hydrogen (inter-strand) bonds. The actual job of the phosphodiester bonds is where in DNA polymers connect the 5' carbon atom of one nucleotide to the 3' carbon atom of another nucleotide, while the hydrogen bonds stabilize DNA double helices across the helix axis but not in the direction of the axis. This makes it possible to separate the strands from one another. The nucleotides on a single strand can therefore be used to reconstruct nucleotides on a newly synthesized partner strand.

28.1.2 DNA Polymerase

DNA polymerases are a family of enzymes that carry out all forms of DNA replication. DNA polymerases in general cannot initiate synthesis of new strands but can only extend an existing DNA or RNA strand paired with a template strand. To begin synthesis, a short fragment of RNA, called a primer, must be created and paired with the template DNA strand.

DNA polymerase adds a new strand of DNA by extending the 3′ end of an existing nucleotide chain, adding new nucleotides matched to the template strand, one at a time, via the creation of phosphodiester bonds. The energy for this process of DNA polymerization comes from hydrolysis of the high-energy phosphate (phosphoanhydride) bonds between the three phosphates attached to each unincorporated base.

Free bases with their attached phosphate groups are called nucleotides; in particular, bases with three attached phosphate groups are called nucleoside triphosphates. When a nucleotide is being added to a growing DNA strand, the formation of a phosphodiester bond between the proximal phosphate of the nucleotide to the growing chain is accompanied by hydrolysis of a high-energy phosphate bond with release of the two distal phosphate groups as a pyrophosphate. Enzymatic hydrolysis of the resulting pyrophosphate into inorganic phosphate consumes a second high-energy phosphate bond and renders the reaction effectively irreversible.

In general, DNA polymerases are highly accurate, with an intrinsic error rate of less than one mistake for every 10^7 nucleotides added. Some DNA polymerases can also delete nucleotides from the end of a developing strand in order to fix mismatched bases.

This is known as proofreading. Finally, post-replication mismatch repair mechanisms monitor the DNA for errors, being capable of distinguishing mismatches in the newly synthesized DNA Strand from the original strand sequence. Together, these three discrimination steps enable replication fidelity of less than one mistake for every 10^9 nucleotides added.

The rate of DNA replication in a living cell was first measured as the rate of phage T4 DNA elongation in phage-infected *E. coli.*

During the period of exponential DNA increase at 37 °C, the rate was 749 nucleotides per second. The mutation rate per base pair per replication during phage T4 DNA synthesis is 1.7 per 10^8.

28.1.3 Replication Process

DNA replication, like all biological polymerization processes, proceeds in three enzymatically catalyzed and coordinated steps: initiation, elongation and termination.

Initiation

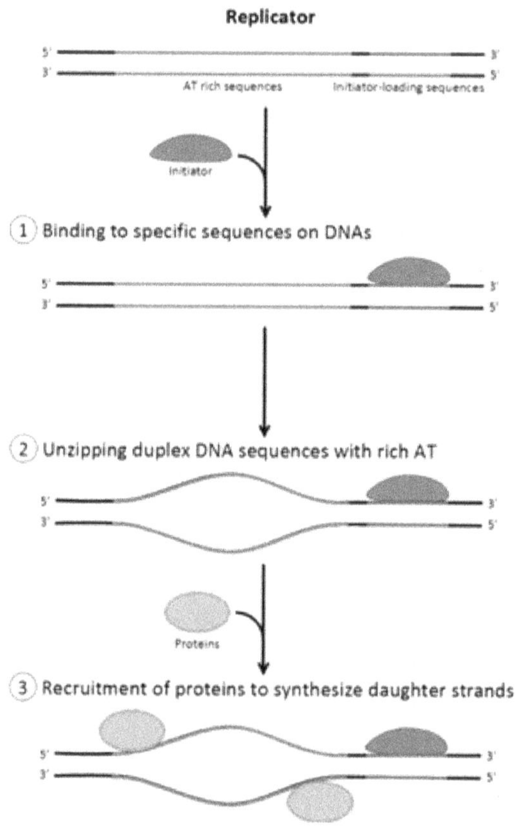

Fig: Role of Initiators for Initiation of DNA Replication

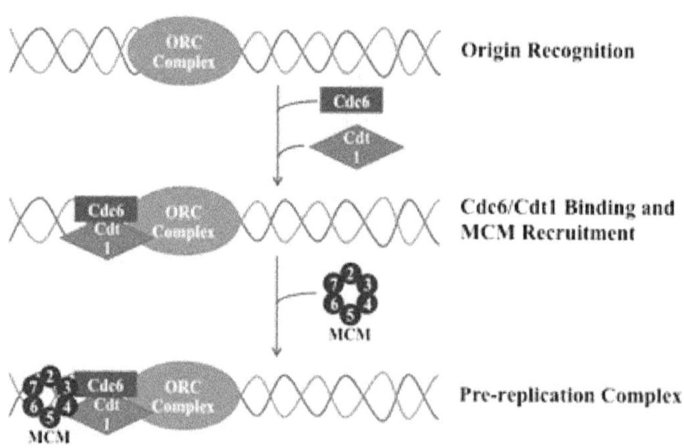

Fig 1: Formation of Pre-Replication Complex

For a cell to divide, it must first replicate its DNA. DNA replication is an all-or-none process; once replication begins, it proceeds to completion. Once replication is complete, it does not occur again in the same cell cycle. This is made possible by the division of initiation of the pre-replication complex.

Pre-Replication Complex

In late mitosis and early G1 phase, a large complex of initiator proteins assembles into the pre-replication complex at particular points in the DNA, known as "origins". In *E. coli* the primary initiator protein is Dna A; in yeast, this is the origin recognition complex. Sequences used by initiator proteins tend to be "AT-rich" (rich in adenine and thymine bases), because A-T base pairs have two hydrogen bonds (rather than the three formed in a C-G pair) and thus are easier to strand-separate.

In eukaryotes, the origin recognition complex catalyzes the assembly of initiator proteins into the pre-replication complex. In addition, a recent report suggests that budding yeast ORC dimerizes in a cell cycle dependent manner to control licensing. In turn, the process of ORC dimerization is mediated by a cell cycle-dependent Noc3p dimerization cycle in vivo, and this role of Noc3p is separable from its role in ribosome biogenesis.

An essential Noc3p dimerization cycle mediates ORC double-hexamer formation in replication licensing ORC and Noc3p are continuously bound to the chromatin throughout the cell cycle. Cdc6 and Cdt1 then associate with the bound origin recognition complex at the origin in order to form a larger complex necessary to load the Mcm complex onto the DNA. In eukaryotes, the Mcm complex is the helicase that will split the DNA helix at the replication forks and origins. The Mcm complex is recruited at late G1 phase and loaded by the ORC-Cdc6-Cdt1 complex onto the DNA via ATP-dependent protein remodeling. The loading of the Mcm complex onto the origin DNA marks the completion of pre-replication complex formation.

If environmental conditions are right in late G1 phase, the G1 and G1/S cyclin-Cdk complexes are activated, which stimulate expression of genes that encode components of the DNA synthetic machinery. G1/S-Cdk activation also promotes the expression and activation of S-Cdk complexes, which may play a role in activating replication origins depending on species and cell type. Control of these Cdks vary depending on cell type and stage of development. This regulation is best understood in budding yeast, where the S cyclins Clb5 and Clb6 are primarily responsible for DNA replication. Clb5,6-Cdk1 complexes directly trigger the activation of replication origins and are therefore required throughout S phase to directly activate each origin.

In a similar manner, Cdc7 is also required through S phase to activate replication origins. Cdc7 is not active throughout the cell cycle, and its activation is strictly timed to avoid premature initiation of DNA replication. In late G1, Cdc7 activity rises abruptly as a result of association with the regulatory subunit DBF4, which binds Cdc7 directly and promotes its protein kinase activity. Cdc7 has been found to be a rate-limiting regulator of origin activity. Together, the G1/S-Cdks and/or S-Cdks and Cdc7 collaborate to directly activate the replication origins, leading to initiation of DNA synthesis.

Preinitiation Complex

In early S phase, S-Cdk and Cdc7 activation lead to the assembly of the preinitiation complex, a massive protein complex formed at the origin. Formation of the preinitiation complex displaces Cdc6 and Cdt1 from the origin replication complex, inactivating and disassembling the pre-replication complex. Loading the preinitiation complex onto the origin activates the Mcm helicase, causing unwinding of the DNA helix. The preinitiation complex also loads α-primase and other DNA polymerases onto the DNA.

After α-primase synthesizes the first primers, the primer-template junctions interact with the clamp loader, which loads the sliding clamp onto the DNA to begin DNA synthesis. The components of the preinitiation complex remain associated with replication forks as they move out from the origin.

Elongation

DNA polymerase has 5′–3′ activity. All known DNA replication systems require a free 3′ hydroxyl group before synthesis can be initiated (note: the DNA template is read in 3′ to 5′ direction whereas a new strand is synthesized in the 5′ to 3′ direction—this is often confused). Four distinct mechanisms for DNA synthesis are recognized:

- All cellular life forms and many DNA viruses, phages and plasmids use a primase to synthesize a short RNA primer with a free 3′ OH group which is subsequently elongated by a DNA polymerase.
- The retroelements (including retroviruses) employ a transfer RNA that primes DNA replication by providing a free 3′ OH that is used for elongation by the reverse transcriptase.
- In the adenoviruses and the ϕ29 family of bacteriophages, the 3′ OH group is provided by the side chain of an amino acid of the genome attached protein (the terminal protein) to which nucleotides are added by the DNA polymerase to form a new strand.
- In the single stranded DNA viruses—a group that includes the circoviruses, the geminiviruses, the parvoviruses and others—and also the many phages and plasmids that use the rolling circle replication

(RCR) mechanism, the RCR endonuclease creates a nick in the genome strand (single stranded viruses) or one of the DNA strands (plasmids). The 5′ end of the nicked strand is transferred to a tyrosine residue on the nuclease and the free 3′ OH group is then used by the DNA polymerase to synthesize the new strand.

Cellular organisms use the first of these pathways since it is the most well-known. In this mechanism, once the two strands are separated, primase adds RNA primers to the template strands. The leading strand receives one RNA primer while the lagging strand receives several. The leading strand is continuously extended from the primer by a DNA polymerase with high processivity, while the lagging strand is extended discontinuously from each primer forming Okazaki fragments. RNase removes the primer RNA fragments, and a low processivity DNA polymerase distinct from the replicative polymerase enters to fill the gaps. When this is complete, a single nick on the leading strand and several nicks on the lagging strand can be found. Ligase works to fill these nicks in, thus completing the newly replicated DNA molecule.

The primase used in this process differs significantly between bacteria and archaea/eukaryotes. Bacteria use a primase belonging to the DnaG protein superfamily which contains a catalytic domain of the TOPRIM fold type. The TOPRIM fold contains an α/β core with four conserved strands in a Rossmann-like topology. This structure is also found in the catalytic domains of topoisomerase Ia, topoisomerase II, the OLD-family nucleases and DNA repair proteins related to the RecR protein.

The primase used by archaea and eukaryotes, in contrast, contains a highly derived version of the RNA recognition motif (RRM). This primase is structurally similar to many viral RNA-dependent RNA polymerases, reverse transcriptases, cyclic nucleotide generating cyclases and DNA polymerases of the A/B/Y families that are involved in DNA replication and repair. In eukaryotic replication, the primase forms a complex with Pol α.

Multiple DNA polymerases take on different roles in the DNA replication process. In *E. coli*, DNA Pol III is the polymerase enzyme primarily responsible for DNA replication. It assembles into a replication complex at the replication fork that exhibits extremely high processivity, remaining intact for the entire replication cycle.

In contrast, DNA Pol I is the enzyme responsible for replacing RNA primers with DNA. DNA Pol I has a 5′ to 3′ exonuclease activity in addition to its polymerase activity, and uses its exonuclease activity to degrade the RNA primers ahead of it as it extends the DNA strand behind it, in a process called nick translation. Pol I is much less processive than Pol III because its primary function in DNA replication is to create many short DNA regions rather than a few very long regions.

In eukaryotes, the low-processivity enzyme, Pol α, helps to initiate replication because it forms a complex with primase. In eukaryotes, leading strand synthesis is thought to be conducted by Pol ε; however, this view has recently been challenged, suggesting a role for Pol δ. Primer removal is completed Pol δ while repair of DNA during replication is completed by Pol ε.

As DNA synthesis continues, the original DNA strands continue to unwind on each side of the bubble, forming a replication fork with two prongs. In bacteria, which have a single origin of replication on their circular chromosome, this process creates a "theta structure" (resembling the Greek letter theta: θ). In contrast, eukaryotes have longer linear chromosomes and initiate replication at multiple origins within these.

Replication Fork

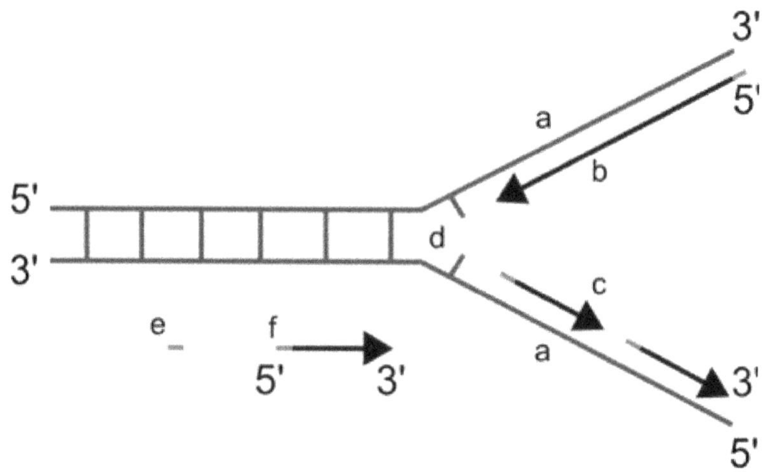

Fig 2: Scheme of the Replication Fork. A: Template, B: Leading Strand, C: Lagging Strand, D: Replication Fork, E: Primer, F: Okazaki Fragments

The replication fork is a structure that forms within the long helical DNA during DNA replication. It is produced by enzymes called helicases that break the hydrogen bonds that hold the DNA strands together in a helix. The resulting structure has two branching "prongs", each one made up of a single strand of DNA. These two strands serve as the template for the leading and lagging strands, which will be created as DNA polymerase matches complementary nucleotides to the templates; the templates may be properly referred to as the leading strand template and the lagging strand template.

DNA is read by DNA polymerase in the 3′ to 5′ direction, meaning the new strand is synthesized in the 5' to 3' direction. Since the leading and lagging strand templates are oriented in opposite directions at the replication fork, a major issue is how to achieve synthesis of new lagging strand DNA, whose direction of synthesis is opposite to the direction of the growing replication fork.

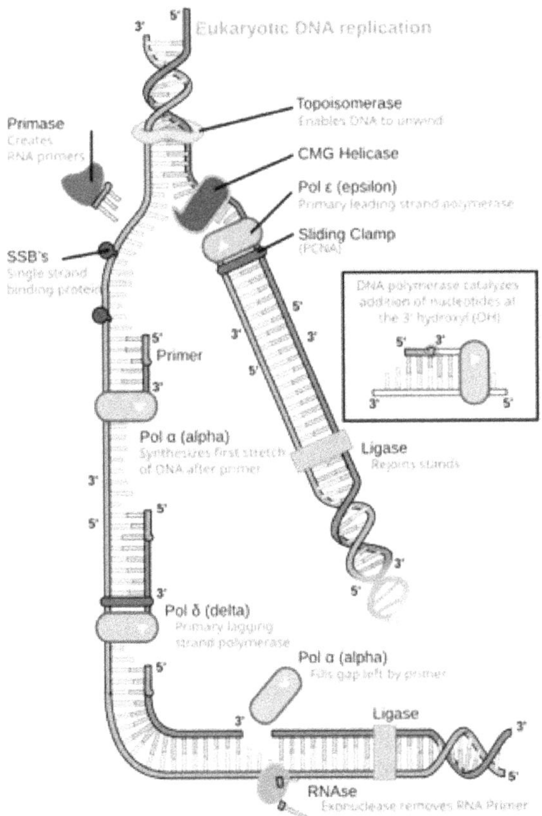

Fig 3: Many Enzymes are Involved in the DNA Replication Fork

Leading Strand

The leading strand is the strand of new DNA which is synthesized in the same direction as the growing replication fork. This sort of DNA replication is continuous.

Lagging Strand

The lagging strand is the strand of new DNA whose direction of synthesis is opposite to the direction of the growing replication fork. Because of its orientation, replication of the lagging strand is more complicated as compared to that of the leading strand. As a consequence, the DNA polymerase on this strand is seen to "lag behind" the other strand. The lagging strand is synthesized in short, separated segments. On the lagging strand *template*, a primase "reads" the template DNA and initiates synthesis of a short complementary RNA primer. A DNA polymerase extends the primed segments, forming Okazaki fragments. The RNA primers are then removed and replaced with DNA, and the fragments of DNA are joined by DNA ligase.

Dynamics at the Replication Fork

In all cases the helicase is composed of six polypeptides that wrap around only one strand of the DNA being replicated. The two polymerases are bound to the helicase hexamer. In eukaryotes the helicase wraps around the leading strand, and in prokaryotes it wraps around the lagging strand.

Fig 4: The Assembled Human DNA Clamp, a Trimer of the Protein PCNA

As helicase unwinds DNA at the replication fork, the DNA ahead is forced to rotate. This process results in a build-up of twists in the DNA ahead. This build-up creates a torsional load that would eventually stop the replication fork. Topoisomerases are enzymes that temporarily break the strands of DNA, relieving the tension caused by unwinding the two strands of the DNA helix; topoisomerases (including DNA gyrase) achieve this by adding negative supercoils to the DNA helix.

Bare single-stranded DNA tends to fold back on itself forming secondary structures; these structures can interfere with the movement of DNA polymerase. To prevent this, single-strand binding proteins bind to the DNA until a second strand is synthesized, preventing secondary structure formation.

Double-stranded DNA is coiled around histones that play an important role in regulating gene expression so the replicated DNA must be coiled around histones at the same places as the original DNA. To ensure this, histone chaperones disassemble the chromatin before it is replicated and replace the histones in the correct place. Some steps in this reassembly are somewhat speculative.

Clamp proteins act as a sliding clamp on DNA, allowing the DNA polymerase to bind to its template and aid in processivity. The inner face of the clamp enables DNA to be threaded through it. Once the polymerase reaches the end of the template or detects double-stranded DNA, the sliding clamp undergoes a conformational change that releases the DNA polymerase. Clamp-loading proteins are used to initially load the clamp, recognizing the junction between template and RNA primers.

DNA Replication Proteins

At the replication fork, many replication enzymes assemble on the DNA into a complex molecular machine called the replisome. The following is a list of major DNA replication enzymes that participate in the replisome:

Enzyme	Function in DNA Replication
DNA helicase	Also known as helix destabilizing enzyme. Helicase separates the two strands of DNA at the Replication Fork behind the topoisomerase.
DNA polymerase	The enzyme responsible for catalyzing the addition of nucleotide substrates to DNA in the 5? to 3? direction during DNA replication. Also performs proof-reading and error correction. There exist many different types of DNA Polymerase, each of which perform different functions in different types of cells.
DNA clamp	A protein which prevents elongating DNA polymerases from dissociating from the DNA parent strand.
Single-strand DNA-binding protein	Bind to ssDNA and prevent the DNA double helix from re-annealing after DNA helicase unwinds it, thus maintaining the strand separation, and facilitating the synthesis of the new strand.
Topoisomerase	Relaxes the DNA from its super-coiled nature.
DNA gyrase	Relieves strain of unwinding by DNA helicase; this is a specific type of topoisomerase
DNA ligase	Re-anneals the semi-conservative strands and joins Okazaki Fragments of the lagging strand.
Primase	Provides a starting point of RNA (or DNA) for DNA polymerase to begin synthesis of the new DNA strand.
Telomerase	Lengthens telomeric DNA by adding repetitive nucleotide sequences to the ends of eukaryotic chromosomes. This allows germ cells and stem cells to avoid the Hayflick limit on cell division.

In vitro single-molecule experiments (using optical tweezers and magnetic tweezers) have found synergetic interactions between the replisome enzymes (helicase, polymerase, and Single-strand DNA-binding protein) and with the DNA replication fork enhancing DNA-unwinding and DNA-replication.

These results lead to the development of kinetic models accounting for the synergetic interactions and their stability.

Replication Machinery

Replication machineries consist of factors involved in DNA replication and appearing on template ssDNAs. Replication machineries include primosotors are replication enzymes; DNA polymerase, DNA helicases, DNA clamps and DNA topoisomerases, and replication proteins; e.g. single-stranded DNA binding proteins (SSB). In the replication machineries these components coordinate.

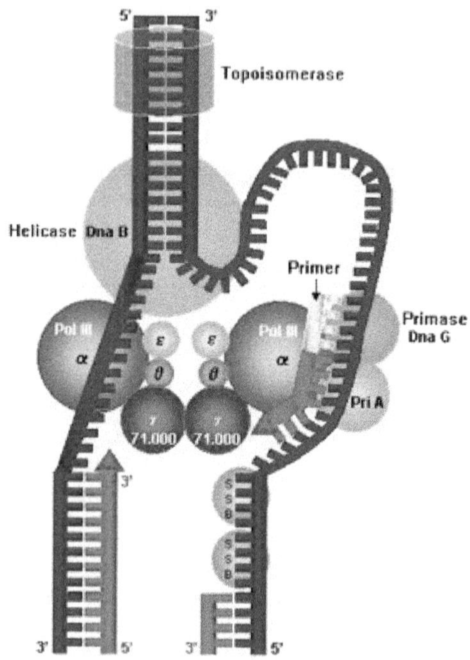

Fig 5: E. coli Replisome

In most of the bacteria, all of the factors involved in DNA replication are located on replication forks and the complexes stay on the forks during DNA replication. Replication machineries are also referred to as replisomes, or DNA replication systems. These terms are generic terms for proteins located on replication forks. In eukaryotic and some bacterial cells the replisomes are not formed.

- DNA factories are similar to projectors and DNAs are like as cinematic films passing constantly into the projectors. In the replication factory model, after both DNA helicases for leading strands and lagging strands are loaded on the template DNAs, the helicases run along the DNAs into each other. The helicases remain associated for the remainder of replication process. Peter Meister et al. observed directly replication sites in budding yeast by monitoring green fluorescent protein (GFP)-tagged DNA polymerases α. They detected DNA replication of pairs of the tagged loci spaced apart symmetrically from a replication origin and found that the distance between the pairs decreased markedly by time. This finding suggests that the mechanism of DNA replication goes with DNA factories. That is, couples of replication factories are loaded on replication origins and the factories associated with each other. Also, template DNAs move into the factories, which bring extrusion of the template ssDNAs and new DNAs. Meister's finding is the first direct evidence of replication factory model. Subsequent research has shown that DNA helicases form dimers in many eukaryotic cells and bacterial replication machineries stay in single intranuclear location during DNA synthesis.

Replication Factories Disentangle Sister Chromatids. The disentanglement is essential for distributing the chromatids into daughter cells after DNA replication. Because sister chromatids after DNA replication hold each other by Cohesin rings, there is the only chance for the disentanglement in DNA replication. Fixing of replication machineries as replication factories can improve the success rate of DNA replication. If replication forks move freely in chromosomes, catenation of nuclei is aggravated and impedes mitotic segregation.

Termination

Eukaryotes initiate DNA replication at multiple points in the chromosome, so replication forks meet and terminate at many points in the chromosome. Because eukaryotes have linear chromosomes, DNA replication is unable to reach the very end of the chromosomes.

Due to this problem, DNA is lost in each replication cycle from the end of the chromosome. Telomeres are regions of repetitive DNA close to the ends and help prevent loss of genes due to this shortening. Shortening of the telomeres is a normal process in somatic cells.

This shortens the telomeres of the daughter DNA chromosome. As a result, cells can only divide a certain number of times before the DNA loss prevents further division. (This is known as the Hayflick limit.) Within the germ cell line, which passes DNA to the next generation, telomerase extends the repetitive sequences of the telomere region to prevent degradation.

Telomerase can become mistakenly active in somatic cells, sometimes leading to cancer formation. Increased telomerase activity is one of the hallmarks of cancer.

Termination requires that the progress of the DNA replication fork must stop or be blocked. Termination at a specific locus, when it occurs, involves the interaction between two components: (1) a termination site sequence in the DNA, and (2) a protein which binds to this sequence to physically stop DNA replication. In various bacterial species, this is named the DNA replication terminus site-binding protein, or Ter protein.

Because bacteria have circular chromosomes, termination of replication occurs when the two replication forks meet each other on the opposite end of the parental chromosome. *E. coli* regulates this process through the use of termination sequences that, when bound by the Tus protein, enable only one direction of replication fork to pass through. As a result, the replication forks are constrained to always meet within the termination region of the chromosome.

28.1.4 Regulation

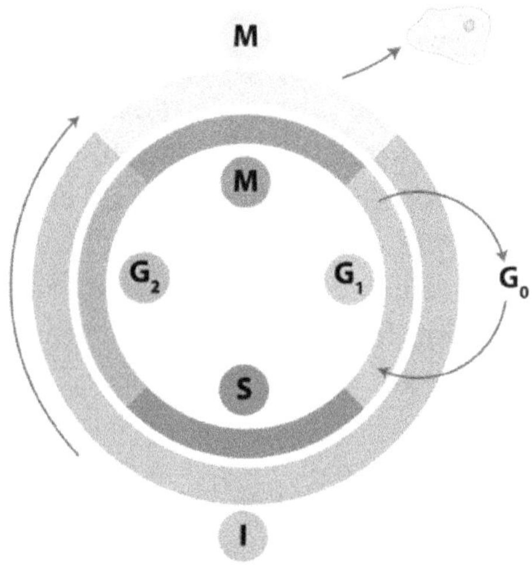

Fig 6: The Cell Cycle of Eukaryotic Cells

Eukaryotes

Within eukaryotes, DNA replication is controlled within the context of the cell cycle. As the cell grows and divides, it progresses through stages in the cell cycle; DNA replication takes place during the S phase (synthesis phase). The progress of the eukaryotic cell through the cycle is controlled by cell cycle checkpoints. Progression through checkpoints is controlled through complex interactions between various proteins, including cyclins and cyclin-dependent kinases. Unlike bacteria, eukaryotic DNA replicates in the confines of the nucleus.

The G1/S checkpoint (restriction checkpoint) regulates whether eukaryotic cells enter the process of DNA replication and subsequent division. Cells that do not proceed through this checkpoint remain in the G0 stage and do not replicate their DNA.

Once the DNA has gone through the "G1/S" test, it can only be copied once in every cell cycle. When the Mcm complex moves away from the origin, the pre-replication complex is dismantled. Because a new Mcm complex cannot be loaded at an origin until the pre-replication subunits are reactivated, one origin of replication can not be used twice in the same cell cycle.

Activation of S-Cdks in early S phase promotes the destruction or inhibition of individual pre-replication complex components, preventing immediate reassembly. S and M-Cdks continue to block pre-replication complex assembly even after S phase is complete, ensuring that assembly cannot occur again until all Cdk activity is reduced in late mitosis. In budding yeast, inhibition of assembly is caused by Cdk-dependent phosphorylation of pre-replication complex components. At the onset of S phase, phosphorylation of Cdc6 by Cdk1 causes the binding of Cdc6 to the SCF ubiquitin protein ligase, which causes proteolytic destruction of Cdc6. Cdk-dependent phosphorylation of Mcm proteins promotes their export out of the nucleus along with Cdt1 during S phase, preventing the loading of new Mcm complexes at origins during a single cell cycle. Cdk phosphorylation of the origin replication complex also inhibits pre-replication complex assembly. The individual presence of any of these three mechanisms is sufficient to inhibit pre-replication complex assembly. However, mutations of all three proteins in the same cell does trigger reinitiation at many origins of replication within one cell cycle.

In animal cells, the protein geminin is a key inhibitor of pre-replication complex assembly. Geminin binds Cdt1, preventing its binding to the origin recognition complex. In G1, levels of geminin are kept low by the APC, which ubiquitinates geminin to target it for degradation. When geminin is destroyed, Cdt1 is released, allowing it to function in pre-replication complex assembly. At the end of G1, the APC is inactivated, allowing geminin to accumulate and bind Cdt1.

Replication of chloroplast and mitochondrial genomes occurs independently of the cell cycle, through the process of D-loop replication.

Replication Focus

In vertebrate cells, replication sites concentrate into positions called replication foci. Replication sites can be detected by immunostaining daughter strands and replication enzymes and monitoring GFP-tagged replication factors. By these methods it is found that replication foci of varying size and positions appear in S phase of cell division and their number per nucleus is far smaller than the number of genomic replication forks.

P. Heun et al.,(2001) tracked GFP-tagged replication foci in budding yeast cells and revealed that replication origins move constantly in G1 and S phase and the dynamics decreased significantly in S phase. Traditionally, replication sites were fixed on spatial structure of chromosomes by nuclear matrix or lamins. The Heun's results denied the traditional concepts, budding yeasts do not have lamins, and support that replication origins self-assemble and form replication foci.

By firing of replication origins, controlled spatially and temporally, the formation of replication foci is regulated. D. A. Jackson et al.(1998) revealed that neighboring origins fire simultaneously in mammalian cells. Spatial juxtaposition of replication sites brings clustering of replication forks. The clustering do rescue of stalled replication forks and favors normal progress of replication forks. Progress of replication forks is inhibited by many factors; collision with proteins or with complexes binding strongly on DNA, deficiency of dNTPs, nicks on template DNAs and so on. If replication forks get stuck and the rest of the sequences from the stuck forks are not copied, then the daughter strands get nick nick unreplicated sites.

The un-replicated sites on one parent's strand hold the other strand together but not daughter strands. Therefore, the resulting sister chromatids cannot separate from each other and cannot divide into 2 daughter cells. When neighboring origins fire and a fork from one origin is stalled, fork from other origin access on an opposite direction of the stalled fork and duplicate the un-replicated sites. As other mechanism of the rescue there is application of dormant replication origins that excess origins do not fire in normal DNA replication.

Bacteria

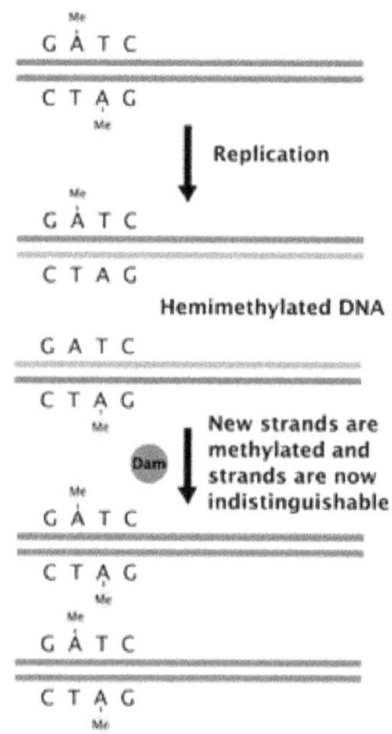

Fig 7: Dam Methylates Adenine of GATC Sites after Replication

Most bacteria do not go through a well-defined cell cycle but instead continuously copy their DNA; during rapid growth, this can result in the concurrent occurrence of multiple rounds of replication. In *E. coli*, the best-characterized bacteria, DNA replication is regulated through several mechanisms, including: the hemimethylation and sequestering of the origin sequence, the ratio of adenosine triphosphate (ATP) to adenosine diphosphate (ADP), and the levels of protein DnaA. All these control the binding of initiator proteins to the origin sequences.

Because *E. coli* methylates GATC DNA sequences, DNA synthesis results in hemimethylated sequences. This hemimethylated DNA is recognized by the protein SeqA, which binds and sequesters the origin sequence; in addition, DnaA (required for initiation of replication) binds less well to hemimethylated DNA. As a result, newly replicated origins are prevented from immediately initiating another round of DNA replication.

ATP builds up when the cell is in a rich medium, triggering DNA replication once the cell has reached a specific size. ATP competes with ADP to bind to DnaA, and the DnaA-ATP complex is able to initiate replication. A certain number of DnaA proteins are also required for DNA replication — each time the origin is copied, the number of binding sites for DnaA doubles, requiring the synthesis of more DnaA to enable another initiation of replication.

In fast-growing bacteria, such as *E. coli*, chromosome replication takes more time than dividing the cell. The bacteria solve this by initiating a new round of replication before the previous one has been terminated. The new round of replication will form the chromosome of the cell that is born two generations after the dividing cell. This mechanism creates overlapping replication cycles.

28.1.5 Problems with DNA Replication

There are many events that contribute to replication stress, including:

- Misincorporation of ribonucleotides.
- Unusual DNA structures.
- Conflicts between replication and transcription.
- Insufficiency of essential replication factors.
- Common fragile sites.
- Overexpression or constitutive activation of oncogenes.
- Chromatin inaccessibility.

28.1.6 Polymerase Chain Reaction

Researchers commonly replicate DNA *in vitro* using the polymerase chain reaction (PCR). PCR uses a pair of primers to span a target region in template DNA, and then polymerizes partner strands in each direction from these primers using a thermostable DNA polymerase. Repeating this process through multiple cycles amplifies the targeted DNA region. At the start of each cycle, the mixture of template and primers is heated, separating the newly synthesized molecule and template. Then, as the mixture cools, both of these become templates for annealing of new primers, and the polymerase extends from these. As a result, the number of copies of the target region doubles each round, increasing exponentially.

28.2 DNA REPAIR

DNA repair is a collection of processes by which a cell identifies and corrects damage to the DNA molecules that encode its genome. In human cells, both normal metabolic activities and environmental factors such as radiation can cause DNA damage, resulting in tens of thousands of individual molecular lesions per cell per day. Many of these lesions cause structural damage to the DNA molecule and can alter or eliminate the cell's ability to transcribe the gene that the affected DNA encodes. Other lesions induce potentially harmful mutations in the cell's genome, which affect the survival of its daughter cells after it undergoes mitosis. As a consequence, the DNA repair process is constantly active as it responds to damage in the DNA structure. When normal repair processes fail, and when cellular apoptosis does not occur, irreparable DNA damage may occur. This can eventually lead to malignant tumors, or cancer as per the two-hit hypothesis.

The rate of DNA repair depends on various factors, including the cell type, the age of the cell, and the extracellular environment. A cell that has accumulated a large amount of DNA damage or can no longer effectively repair its DNA may enter one of three possible states:

- An irreversible state of dormancy, known as senescence.
- Cell suicide, also known as apoptosis or programmed cell death.
- Unregulated cell division, which can lead to the formation of a tumor that is cancerous.

The DNA repair ability of a cell is vital to the integrity of its genome and thus to the normal functionality of that organism. Many genes that were initially shown to influence life span have turned out to be involved in DNA damage repair and protection.

The 2015 Nobel Prize in Chemistry was awarded to Tomas Lindahl, Paul Modrich, and Aziz Sancar for their work on the molecular mechanisms of DNA repair processes.

28.2.1 Evolution

The basic processes of DNA repair are highly conserved among both prokaryotes and eukaryotes and even among bacteriophages (viruses which infect bacteria); however, more complex organisms with more complex genomes have correspondingly more complex repair mechanisms. The ability of a large number of protein structural motifs to catalyze relevant chemical reactions has played a significant role in the elaboration of repair mechanisms during evolution.

The fossil record indicates that single-cell life began to proliferate on the planet at some point during the Precambrian period, although exactly when recognizably modern life first emerged is unclear. Nucleic acids became the sole and universal means of encoding genetic information, requiring DNA repair mechanisms that in their basic form have been inherited by all extant life forms from their common ancestor. The emergence of Earth's oxygen-rich atmosphere (known as the "oxygen catastrophe") due to photosynthetic organisms, as well as the presence of potentially damaging free radicals in the cell due to oxidative phosphorylation, necessitated the evolution of DNA repair mechanisms that act specifically to counter the types of damage induced by oxidative stress. The mechanism by which this came about, however, is unclear.

Rate of Evolutionary Change

On some occasions, DNA damage is not repaired or is repaired by an error-prone mechanism that results in a change from the original sequence. When this occurs, mutations may propagate into the genomes of the cell's progeny. Should such an event occur in a germ line cell that will eventually produce a gamete, the mutation has the potential to be passed on to the organism's offspring. The rate of evolution in a particular species (or, in a particular gene) is a function of the rate of mutation. As a consequence, the rate and accuracy of DNA repair mechanisms have an influence over the process of evolutionary change. DNA damage protection and repair does not influence the rate of adaptation by gene regulation and by recombination and selection of alleles. On the other hand, DNA damage repair and protection does influence the rate of accumulation of irreparable, advantageous, code expanding, inheritable mutations, and slows down the evolutionary mechanism for expansion of the genome of organisms with new functionalities. The tension between evolvability and mutation repair and protection needs further investigation.

28.2.2 Technology

A technology named clustered regularly interspaced short palindromic repeat (shortened to CRISPR-Cas9) was discovered in 2012. The new technology allows anyone with molecular biology training to alter the genes of any species with precision, by inducing DNA damage at a specific point and then altering DNA repair mechanisms to insert new genes. It is cheaper, more efficient, and more precise than other technologies. With the help of CRISPR–Cas9, parts of a genome can be edited by scientists by removing, adding, or altering parts in a DNA sequence.

28.2.3 DNA Damage

DNA damage, due to environmental factors and normal metabolic processes inside the cell, occurs at a rate of 10,000 to 1,000,000 molecular lesions per cell per day. While this constitutes at most only 0.0003125% of the human genome's approximately 3.2 billion bases, unrepaired lesions in critical genes (such as tumor suppressor genes) can impede a cell's ability to carry out its function and appreciably increase the likelihood of tumor formation and contribute to tumor heterogeneity.

The vast majority of DNA damage affects the primary structure of the double helix; that is, the bases themselves are chemically modified. These modifications can in turn disrupt the molecules' regular helical structure by introducing non-native chemical bonds or bulky adducts that do not fit in the standard double helix. Unlike proteins and RNA, DNA usually lacks tertiary structure and therefore damage or disturbance does not occur at that level. DNA is, however, supercoiled and wound around "packaging" proteins called histones (in eukaryotes), and both superstructures are vulnerable to the effects of DNA damage.

Sources

DNA damage can be subdivided into two main types:

- Endogenous damage such as attack by reactive oxygen species produced from normal metabolic byproducts (spontaneous mutation), especially the process of oxidative deamination:
  - Also includes replication errors.
- Exogenous Damage Caused by External Agents such as:
  - Ultraviolet (UV) radiation (200–400 nm) from the sun or other artificial light sources.
  - Other radiation frequencies, including x-rays and gamma rays.
  - Hydrolysis or thermal disruption.
  - Certain plant toxins.
  - Human-made mutagenic chemicals, especially aromatic compounds that act as DNA intercalating agents.
  - Viruses.

The replication of damaged DNA before cell division can lead to the incorporation of wrong bases opposite damaged ones. Daughter cells that inherit these wrong bases carry mutations from which the original DNA sequence is unrecoverable (except in the rare case of a back mutation, for example, through gene conversion).

Types

There are several types of damage to DNA due to endogenous cellular processes:

- **Oxidation** of bases [e.g. 8-oxo-7,8-dihydroguanine (8-oxoG)] and generation of DNA strand interruptions from reactive oxygen species.
- **Alkylation** of bases (usually methylation), such as formation of 7-methylguanosine, 1-methyladenine, 6-O-Methylguanine.
- **Hydrolysis** of bases, such as deamination, depurination, and depyrimidination.
- **"Bulky adduct formation"** (e.g., benzo[a]pyrene diol epoxide-dG adduct, aristolactam I-dA adduct).
- **Mismatch** of bases, due to errors in DNA replication, in which the wrong DNA base is stitched into place in a newly forming DNA strand, or a DNA base is skipped over or mistakenly inserted.

- Monoadduct damage cause by change in single nitrogenous base of DNA.
- Di adduct damage.

Damage caused by exogenous agents comes in many forms. Some examples are:

- **UV-B light** causes crosslinking between adjacent cytosine and thymine bases creating *pyrimidine dimers*. This is called direct DNA damage.
- **UV-A light** creates mostly free radicals. The damage caused by free radicals is called indirect DNA damage.
- **Ionizing radiation** such as that created by radioactive decay or in *cosmic rays* causes breaks in DNA strands. Intermediate-level ionizing radiation may induce irreparable DNA damage (leading to replicational and transcriptional errors needed for neoplasia or may trigger viral interactions) leading to pre-mature aging and cancer.
- **Thermal disruption** at elevated temperature increases the rate of depurination (loss of purine bases from the DNA backbone) and single-strand breaks. For example, hydrolytic depurination is seen in the thermophilic bacteria, which grow in hot springs at 40–80 °C. The rate of depurination (300 purine residues per genome per generation) is too high in these species to be repaired by normal repair machinery, hence a possibility of an adaptive response cannot be ruled out.
- **Industrial chemicals** such as vinyl chloride and hydrogen peroxide, and environmental chemicals such as polycyclic aromatic hydrocarbons found in smoke, soot and tar create a huge diversity of DNA adducts-ethanoates, oxidized bases, alkylated phosphodiesters and crosslinking of DNA, just to name a few.

UV damage, alkylation/methylation, X-ray damage and oxidative damage are examples of induced damage. Spontaneous damage can include the loss of a base, deamination, sugar ring puckering and tautomeric shift. Constitutive (spontaneous) DNA damage caused by endogenous oxidants can be detected as a low level of histone H2AX phosphorylation in untreated cells.

Nuclear versus Mitochondrial

In human cells, and eukaryotic cells in general, DNA is found in two cellular locations – inside the nucleus and inside the mitochondria. Nuclear DNA (n-DNA) exists as chromatin during non-replicative stages of the cell cycle and is condensed into aggregate structures known as chromosomes during cell division. In either state the DNA is highly compacted and wound up around bead-like proteins called histones. Whenever a cell needs to express the genetic information encoded in its n-DNA the required chromosomal region is unraveled, genes located therein are expressed, and then the region is condensed back to its resting conformation. Mitochondrial DNA (mtDNA) is located inside mitochondria organelles, exists in multiple copies, and is also tightly associated with a number of proteins to form a complex known as the nucleoid. Inside mitochondria, reactive oxygen species (ROS), or free radicals, byproducts of the constant production of adenosine triphosphate (ATP) via oxidative phosphorylation, create a highly oxidative environment that is known to damage mtDNA. A critical enzyme in counteracting the toxicity of these species is superoxide dismutase, which is present in both the mitochondria and cytoplasm of eukaryotic cells.

Senescence and Apoptosis

Senescence, an irreversible process in which the cell no longer divides, is a protective response to the shortening of the chromosome ends, called telomeres. The telomeres are long regions of repetitive noncoding DNA that cap chromosomes and undergo partial degradation each time a cell undergoes division. In contrast, quiescence is a reversible state of cellular dormancy that is unrelated to genome damage. Senescence in cells may serve as a functional alternative to apoptosis in cases where the physical presence of a cell for spatial reasons is required by the organism, which serves as a "last resort" mechanism to prevent a cell with damaged DNA from replicating inappropriately in the absence of pro-growth cellular signaling. Unregulated cell division can lead to the formation of a tumor which is potentially lethal to an organism. Therefore, the induction of senescence and apoptosis is considered to be part of a strategy of protection against cancer.

Mutation

It is important to distinguish between DNA damage and mutation, the two major types of error in DNA. DNA damage and mutation are fundamentally different. Damage results in physical abnormalities in the DNA, such as single- and double-strand breaks, 8-hydroxydeoxyguanosine residues, and polycyclic aromatic hydrocarbon adducts. DNA damage can be recognized by enzymes, and thus can be correctly repaired if redundant information, such as the undamaged sequence in the complementary DNA strand or in a homologous chromosome, is available for copying. If a cell retains DNA damage, transcription of a gene can be prevented, and thus translation into a protein will also be blocked. Replication may also be blocked or the cell may die.

In contrast to DNA damage, a mutation is a change in the base sequence of the DNA. A mutation cannot be recognized by enzymes once the base change is present in both DNA strands, and thus a mutation cannot be repaired. At the cellular level, mutations can cause alterations in protein function and regulation. Mutations are replicated when the cell replicates. In a population of cells, mutant cells will increase or decrease in frequency according to the effects of the mutation on the ability of the cell to survive and reproduce.

Although distinctly different from each other, DNA damage and mutation are related because DNA damage often causes errors of DNA synthesis during replication or repair; these errors are a major source of mutation.

Given these properties of DNA damage and mutation, it can be seen that DNA damage is a special problem in non-dividing or slowly-dividing cells, where unrepaired damage will tend to accumulate over time. On the other hand, in rapidly dividing cells, unrepaired DNA damage that does not kill the cell by blocking replication will tend to cause replication errors and thus mutation. The great majority of mutations that are not neutral in their effect are deleterious to a cell's survival. Thus, in a population of cells composing a tissue with replicating cells, mutant cells will tend to be lost. However, infrequent mutations that provide a survival advantage will tend to clonally expand at the expense of neighboring cells in the tissue. This advantage to the cell is disadvantageous to the whole organism because such mutant cells can give rise to cancer. Thus, DNA damage in frequently dividing cells, because it gives rise to mutations, is a prominent cause of cancer. In contrast, DNA damage in infrequently-dividing cells is likely a prominent cause of aging.

28.2.4 Mechanisms

Cells cannot function if DNA damage corrupts the integrity and accessibility of essential information in the genome (but cells remain superficially functional when non-essential genes are missing or damaged). Depending on the type of damage inflicted on the DNA's double helical structure, a variety of repair strategies have evolved to restore lost information. If possible, cells use the unmodified complementary strand of the DNA or the sister chromatid as a template to recover the original information. Without access to a template, cells use an error-prone recovery mechanism known as translesion synthesis as a last resort.

Damage to DNA alters the spatial configuration of the helix, and such alterations can be detected by the cell. Once damage is localized, specific DNA repair molecules bind at or near the site of damage, inducing other molecules to bind and form a complex that enables the actual repair to take place.

Direct Reversal

Cells are known to eliminate three types of damage to their DNA by chemically reversing it. These mechanisms do not require a template, since the types of damage they counteract can occur in only one of the four bases. Such direct reversal mechanisms are specific to the type of damage incurred and do not involve breakage of the phosphodiester backbone. The formation of pyrimidine dimers upon irradiation with UV light results in an abnormal covalent bond between adjacent pyrimidine bases. The photoreactivation process directly reverses this damage by the action of the enzyme photolyase, whose activation is obligately dependent on energy absorbed from blue/UV light (300–500 nm wavelength) to promote catalysis. Photolyase, an old enzyme present in bacteria, fungi, and most animals no longer functions in humans, who instead use nucleotide excision repair to repair damage from UV irradiation.

Another type of damage, methylation of guanine bases, is directly reversed by the enzyme methyl guanine methyl transferase (MGMT), the bacterial equivalent of which is called ogt. This is an expensive process because each MGMT molecule can be used only once; that is, the reaction is stoichiometric rather than catalytic. A generalized response to methylating agents in bacteria is known as the adaptive response and confers a level of resistance to alkylating agents upon sustained exposure by upregulation of alkylation repair enzymes. The third type of DNA damage reversed by cells is certain methylation of the bases cytosine and adenine.

Single-Strand Damage

When only one of the two strands of a double helix has a defect, the other strand can be used as a template to guide the correction of the damaged strand. In order to repair damage to one of the two paired molecules of DNA, there exist a number of excision repair mechanisms that remove the damaged nucleotide and replace it with an undamaged nucleotide complementary to that found in the undamaged DNA strand.

- **Base Excision Repair (BER):** Damaged single bases or nucleotides are most commonly repaired by removing the base or the nucleotide involved and then inserting the correct base or nucleotide. In base excision repair, a glycosylase enzyme removes the damaged base from the DNA by cleaving the bond between the base and the deoxyribose. These enzymes remove a single base to create an apurinic or apyrimidinic site (AP site). Enzymes called AP endonucleases nick the damaged DNA backbone at the AP site. DNA polymerase then removes the damaged region using its 5' to 3' exonuclease activity and correctly synthesizes the new strand using the complementary strand as a template. The gap is then sealed by enzyme DNA ligase.
- **Nucleotide Excision Repair (NER):** Bulky, helix-distorting damage, such as pyrimidine dimerization caused by UV light is usually repaired by a three-step process. First the damage is recognized, then 12-24 nucleotide-long strands of DNA are removed both upstream and downstream of the damage site by endonucleases, and the removed DNA region is then resynthesized. NER is a highly evolutionarily conserved repair mechanism and is used in nearly all eukaryotic and prokaryotic

cells. In prokaryotes, NER is mediated by Uvr proteins. In eukaryotes, many more proteins are involved, although the general strategy is the same.

- **Mismatch Repair:** Systems are present in essentially all cells to correct errors that are not corrected by proofreading. These systems consist of at least two proteins. One detects the mismatch, and the other recruits an endonuclease that cleaves the newly synthesized DNA strand close to the region of damage. In *E. coli,* the proteins involved are the Mut class proteins: MutS, MutL, and MutH. In most Eukaryotes, the analog for MutS is MSH and the analog for MutL is MLH. MutH is only present in bacteria. This is followed by removal of damaged region by an exonuclease, resynthesis by DNA polymerase, and nick sealing by DNA ligase.

Double-Strand Breaks

Double-strand breaks, in which both strands in the double helix are severed, are particularly hazardous to the cell because they can lead to genome rearrangements. In fact, when a double-strand break is accompanied by a cross-linkage joining the two strands at the same point, neither strand can be used as a template for the repair mechanisms, so that the cell will not be able to complete mitosis when it next divides, and will either die or, in rare cases, undergo a mutation.

Three mechanisms exist to repair double-strand breaks (DSBs): non-homologous end joining (NHEJ), microhomology-mediated end joining (MMEJ), and homologous recombination (HR):

- In NHEJ, DNA Ligase IV, a specialized DNA ligase that forms a complex with the cofactor XRCC4, directly joins the two ends. To guide accurate repair, NHEJ relies on short homologous sequences called microhomologies present on the single-stranded tails of the DNA ends to be joined. If these overhangs are compatible, repair is usually accurate. NHEJ can also introduce mutations during repair. Loss of damaged nucleotides at the break site can lead to deletions, and joining of nonmatching termini forms insertions or translocations. NHEJ is especially important before the cell has replicated its DNA, since there is no template available for repair by homologous recombination. There are "backup" NHEJ pathways in higher eukaryotes. Besides its role as a genome caretaker, NHEJ is required for joining hairpin-capped double-strand breaks induced during V(D)J recombination, the process that generates diversity in B-cell and T-cell receptors in the vertebrate immune system.

- MMEJ starts with short-range end resection by MRE11 nuclease on either side of a double-strand break to reveal microhomology regions. In further steps, Poly (ADP-ribose) polymerase 1 (PARP1) is required and may be an early step in MMEJ. There is pairing of microhomology regions followed by recruitment of flap structure-specific endonuclease 1 (FEN1) to remove overhanging flaps. This is followed by recruitment of XRCC1–LIG3 to the site for ligating the DNA ends, leading to an intact DNA. MMEJ is always accompanied by a deletion, so that MMEJ is a mutagenic pathway for DNA repair.
- HR requires the presence of an identical or nearly identical sequence to be used as a template for repair of the break. The enzymatic machinery responsible for this repair process is nearly identical to the machinery responsible for chromosomal crossover during meiosis. This pathway allows a damaged chromosome to be repaired using a sister chromatid (available in G2 after DNA replication) or a homologous chromosome as a template. DSBs caused by the replication machinery attempting to synthesize across a single-strand break or unrepaired lesion cause collapse of the replication fork and are typically repaired by recombination.

In an *in vitro* system, MMEJ occurred in mammalian cells at the levels of 10–20% of HR when both HR and NHEJ mechanisms were also available.

The extremophile *Deinococcus radiodurans* has a remarkable ability to survive DNA damage from ionizing radiation and other sources. At least two copies of the genome, with random DNA breaks, can form DNA fragments through annealing. Partially overlapping fragments are then used for synthesis of homologous regions through a moving D-loop that can continue extension until complementary partner strands are found. In the final step, there is crossover by means of RecA-dependent homologous recombination.

Topoisomerases introduce both single- and double-strand breaks in the course of changing the DNA's state of supercoiling, which is especially common in regions near an open replication fork. Such breaks are not considered DNA damage because they are a natural intermediate in the topoisomerase biochemical mechanism and are immediately repaired by the enzymes that created them.

Another type of DNA double-strand breaks originates from the DNA heat-sensitive or heat-labile sites. These DNA sites are not initial DSBs. However, they convert to DSB after treating with elevated temperature. Ionizing irradiation can induces a highly complex form of DNA damage as clustered damage. It consists of different types of DNA lesions in various locations of the DNA helix. Some of these closely located lesions can probably convert to DSB by exposure to high temperatures. But the exact nature of these lesions and their interactions is not yet known

Translesion Synthesis

Translesion synthesis (TLS) is a DNA damage tolerance process that allows the DNA replication machinery to replicate past DNA lesions such as thymine dimers or AP sites. It involves switching out regular DNA polymerases for specialized translesion polymerases (i.e. DNA polymerase IV or V, from the Y Polymerase family), often with larger active sites that can facilitate the insertion of bases opposite damaged nucleotides.

The polymerase switching is thought to be mediated by, among other factors, the post-translational modification of the replication processivity factor PCNA. Translesion synthesis polymerases often have low fidelity (high propensity to insert wrong bases) on undamaged templates relative to regular polymerases. However, many are extremely efficient at inserting correct bases opposite specific types of damage. For example, Pol η mediates error-free bypass of lesions induced by UV irradiation, whereas Pol ι introduces mutations at these sites.

Pol η is known to add the first adenine across the T^T photodimer using Watson-Crick base pairing and the second adenine will be added in its syn conformation using Hoogsteen base pairing. From a cellular perspective, risking the introduction of point mutations during translesion synthesis may be preferable to resorting to more drastic mechanisms of DNA repair, which may cause gross chromosomal aberrations or cell death.

In short, the process involves specialized polymerases either bypassing or repairing lesions at locations of stalled DNA replication. For example, Human DNA polymerase eta can bypass complex DNA lesions like guanine-thymine intra-strand crosslink, G[8,5-Me]T, although it can cause targeted and semi-targeted mutations. Paromita Raychaudhury and Ashis Basu studied the toxicity and mutagenesis of the same lesion in *Escherichia coli* by replicating a G[8,5-Me]T-modified plasmid in *E. coli* with specific DNA polymerase knockouts. Viability was very low in a strain lacking pol II, pol IV, and pol V, the three SOS-inducible DNA polymerases, indicating that translesion synthesis is conducted primarily by these specialized DNA polymerases. A bypass platform is provided to these polymerases by Proliferating cell nuclear antigen (PCNA).

Under normal circumstances, PCNA bound to polymerases replicates the DNA. At a site of lesion, PCNA is ubiquitinated, or modified, by the RAD6/RAD18 proteins to provide a platform for the specialized polymerases to bypass the lesion and resume DNA replication. After translesion synthesis, extension is required. This extension can be carried out by a replicative polymerase if the TLS is error-free, as in the case of Pol η, yet if TLS results in a mismatch, a specialized polymerase is needed to extend it; Pol ζ. Pol ζ is unique in that it can extend terminal mismatches, whereas more processive polymerases cannot.

So when a lesion is encountered, the replication fork will stall, PCNA will switch from a processive polymerase to a TLS polymerase such as Pol ι to fix the lesion, then PCNA may switch to Pol ζ to extend the mismatch, and last PCNA will switch to the processive polymerase to continue replication.

28.2.5 Global Response to DNA Damage

Cells exposed to ionizing radiation, ultraviolet light or chemicals are prone to acquire multiple sites of bulky DNA lesions and double-strand breaks. Moreover, DNA damaging agents can damage other biomolecules such as proteins, carbohydrates, lipids, and RNA. The accumulation of damage, to be specific, double-strand breaks or adducts stalling the replication forks, are among known stimulation signals for a global response to DNA damage. The global response to damage is an act directed toward the cells' own preservation and triggers multiple pathways of macromolecular repair, lesion bypass, tolerance, or apoptosis. The common features of global response are induction of multiple genes, cell cycle arrest, and inhibition of cell division.

Initial Steps

The packaging of eukaryotic DNA into chromatin presents a barrier to all DNA-based processes that require recruitment of enzymes to their sites of action. To allow DNA repair, the chromatin must be remodeled. In eukaryotes, ATP dependent chromatin remodeling complexes and histone-modifying enzymes are two predominant factors employed to accomplish this remodeling process.

Chromatin relaxation occurs rapidly at the site of a DNA damage. In one of the earliest steps, the stress-activated protein kinase, c-Jun N-terminal kinase (JNK), phosphorylates SIRT6 on serine 10 in response to double-strand breaks or other DNA damage. This post-translational modification facilitates the mobilization of SIRT6 to DNA damage sites, and is required for efficient recruitment of poly (ADP-ribose) polymerase 1 (PARP1) to DNA break sites and for efficient repair of DSBs. PARP1 protein starts to appear at DNA damage sites in less than a second, with half maximum accumulation within 1.6 seconds after the damage occurs. PARP1 synthesizes polymeric adenosine diphosphate ribose (poly (ADP-ribose) or PAR) chains on itself. Next the chromatin remodeler ALC1 quickly attaches to the product of PARP1 action, a poly-ADP ribose chain, and ALC1 completes arrival at the DNA damage within 10 seconds of the occurrence of the damage. About half of the maximum chromatin relaxation, presumably due to action of ALC1, occurs by 10 seconds. This then allows recruitment of the DNA repair enzyme MRE11, to initiate DNA repair, within 13 seconds.

γH2AX, the phosphorylated form of H2AX is also involved in the early steps leading to chromatin decondensation after DNA double-strand breaks. The histone variant

H2AX constitutes about 10% of the H2A histones in human chromatin. γH2AX (H2AX phosphorylated on serine 139) can be detected as soon as 20 seconds after irradiation of cells (with DNA double-strand break formation), and half maximum accumulation of γH2AX occurs in one minute. The extent of chromatin with phosphorylated γH2AX is about two million base pairs at the site of a DNA double-strand break. γH2AX does not, itself, cause chromatin decondensation, but within 30 seconds of irradiation, RNF8 protein can be detected in association with γH2AX. RNF8 mediates extensive chromatin decondensation, through its subsequent interaction with CHD4, a component of the nucleosome remodeling and deacetylase complex NuRD.

DDB2 occurs in a heterodimeric complex with DDB1. This complex further complexes with the ubiquitin ligase protein CUL4A and with PARP1. This larger complex rapidly associates with UV-induced damage within chromatin, with half-maximum association completed in 40 seconds. The PARP1 protein, attached to both DDB1 and DDB2, then PARylates (creates a poly-ADP ribose chain) on DDB2 that attracts the DNA remodeling protein ALC1. Action of ALC1 relaxes the chromatin at the site of UV damage to DNA. This relaxation allows other proteins in the nucleotide excision repair pathway to enter the chromatin and repair UV-induced cyclobutane pyrimidine dimer damages.

After rapid chromatin remodeling, cell cycle checkpoints are activated to allow DNA repair to occur before the cell cycle progresses. First, two kinases, ATM and ATR are activated within 5 or 6 minutes after DNA is damaged. This is followed by phosphorylation of the cell cycle checkpoint protein Chk1, initiating its function, about 10 minutes after DNA is damaged.

DNA Damage Checkpoints

After DNA damage, cell cycle checkpoints are activated. Checkpoint activation pauses the cell cycle and gives the cell time to repair the damage before continuing to divide. DNA damage checkpoints occur at the G1/S and G2/M boundaries. An intra-S checkpoint also exists. Checkpoint activation is controlled by two master kinases, ATM and ATR. ATM responds to DNA double-strand breaks and disruptions in chromatin structure, whereas ATR primarily responds to stalled replication forks. These kinases phosphorylate downstream targets in a signal transduction cascade, eventually leading to cell cycle arrest. A class of checkpoint mediator proteins including BRCA1, MDC1, and 53BP1 has also been identified. These proteins seem to be required for transmitting the checkpoint activation signal to downstream proteins.

DNA damage checkpoint is a signal transduction pathway that blocks cell cycle progression in G1, G2 and metaphase and slows down the rate of S phase progression when DNA is damaged. It leads to a pause in cell cycle allowing the cell time to repair the damage before continuing to divide.

Checkpoint Proteins can be separated into four groups: phosphatidylinositol 3-kinase (PI3K)-like protein kinase, proliferating cell nuclear antigen (PCNA)-like group, two serine/threonine(S/T) kinases and their adaptors. Central to all DNA damage induced checkpoints responses is a pair of large protein kinases belonging to the first group of PI3K-like protein kinases-the ATM (Ataxia telangiectasia mutated) and ATR (Ataxia- and Rad-related) kinases, whose sequence and functions have been well conserved in evolution. All DNA damage response requires either ATM or ATR because they have the ability to bind to the chromosomes at the site of DNA damage, together with accessory proteins that are platforms on which DNA damage response components and DNA repair complexes can be assembled.

An important downstream target of ATM and ATR is p53, as it is required for inducing apoptosis following DNA damage. The cyclin-dependent kinase inhibitor p21 is induced by both p53-dependent and p53-independent mechanisms and can arrest the cell cycle at the G1/S and G2/M checkpoints by deactivating cyclin/ cyclin-dependent kinase complexes.

The Prokaryotic SOS Response

The SOS response is the changes in gene expression in *Escherichia coli* and other bacteria in response to extensive DNA damage. The prokaryotic SOS system is regulated by two key proteins: LexA and RecA. The LexA homodimer is a transcriptional repressor that binds to operator sequences commonly referred to as SOS boxes. In *Escherichia coli* it is known that LexA regulates transcription of approximately 48 genes including the lexA and recA genes. The SOS response is known to be widespread in the Bacteria domain, but it is mostly absent in some bacterial phyla, like the Spirochetes. The most common cellular signals activating the SOS response are regions of single-stranded DNA (ssDNA), arising from stalled replication forks or double-strand breaks, which are processed by DNA helicase to separate the two DNA strands. In the initiation step, RecA protein binds to ssDNA in an ATP hydrolysis driven reaction creating RecA–ssDNA filaments. RecA–ssDNA filaments activate LexA autoprotease activity, which ultimately leads to cleavage of LexA dimer and subsequent LexA degradation. The loss of LexA repressor induces transcription of the SOS genes and allows for further signal induction, inhibition of cell division and an increase in levels of proteins responsible for damage processing.

In *Escherichia coli*, SOS boxes are 20-nucleotide long sequences near promoters with palindromic structure and a high degree of sequence conservation. In other classes and phyla, the sequence of SOS boxes varies considerably, with different length and composition, but it is always highly conserved and one of the strongest short signals in the genome. The high information content of SOS boxes permits differential binding of LexA to different promoters and allows for timing of the SOS response. The lesion repair genes are induced at the beginning of SOS response. The error-prone translesion polymerases, for example, UmuCD'2 (also called DNA polymerase V), are induced later on as a last resort. Once the DNA damage is repaired or bypassed using polymerases or through recombination, the amount of single-stranded DNA in cells is decreased, lowering the amounts of RecA filaments decreases cleavage activity of LexA homodimer, which then binds to the SOS boxes near promoters and restores normal gene expression.

Eukaryotic Transcriptional Responses to DNA Damage

Eukaryotic cells exposed to DNA damaging agents also activate important defensive pathways by inducing multiple proteins involved in DNA repair, cell cycle checkpoint control, protein trafficking and degradation. Such genome wide transcriptional response is very complex and tightly regulated, thus allowing coordinated global response to damage. Exposure of yeast *Saccharomyces cerevisiae* to DNA damaging agents results in overlapping but distinct transcriptional profiles. Similarities to environmental shock response indicates that a general global stress response pathway exist at the level of transcriptional activation. In contrast, different human cell types respond to damage differently indicating an absence of a common global response. The probable explanation for this difference between yeast and human cells may be in the heterogeneity of mammalian cells. In an animal different types of cells are distributed among different organs that have evolved different sensitivities to DNA damage.

In general global response to DNA damage involves expression of multiple genes responsible for postreplication repair, homologous recombination, nucleotide excision repair, DNA damage checkpoint, global transcriptional activation, genes controlling mRNA decay, and many others. A large amount of damage to a cell leaves it with an important decision: undergo apoptosis and die, or survive at the cost of living with a modified genome. An increase in tolerance to damage can lead to an increased rate of survival that will allow a greater accumulation of mutations. Yeast Rev1 and human polymerase η are members of Y family translesion DNA polymerases present during global response to DNA damage and are responsible for enhanced mutagenesis during a global response to DNA damage in eukaryotes.

28.2.6 Aging

Pathological Effects of Poor DNA Repair

Experimental animals with genetic deficiencies in DNA repair often show decreased life span and increased cancer incidence. For example, mice deficient in the dominant NHEJ pathway and in telomere maintenance mechanisms get lymphoma and infections more often, and, as a consequence, have shorter lifespans than wild-type mice. In similar manner, mice deficient in a key repair and transcription protein that unwinds DNA helices have premature onset of aging-related diseases and consequent shortening of lifespan. However, not every DNA repair deficiency creates exactly the predicted effects; mice deficient in the NER pathway exhibited shortened life span without correspondingly higher rates of mutation.

The maximum life spans of mice, naked mole-rats and humans are respectively ~3, ~30 and ~129 years. Of these, the shortest lived species, mouse, expresses DNA repair genes, including core genes in several DNA repair pathways, at a lower level than do humans and naked mole rats. Furthermore several DNA repair pathways in humans and naked mole-rats are up-regulated compared to mouse. These observations suggest that elevated DNA repair facilitates greater longevity.

If the rate of DNA damage exceeds the capacity of the cell to repair it, the accumulation of errors can overwhelm the cell and result in early senescence, apoptosis, or cancer. Inherited diseases associated with faulty DNA repair functioning result in premature aging, increased sensitivity to carcinogens and correspondingly increased cancer risk. On the other hand, organisms with enhanced DNA repair systems, such as *Deinococcus radiodurans*, the most radiation-resistant known organism, exhibit remarkable resistance to the double-strand break-inducing effects of radioactivity, likely due to enhanced efficiency of DNA repair and especially NHEJ.

Longevity and Caloric Restriction

A number of individual genes have been identified as influencing variations in life span within a population of organisms. The effects of these genes is strongly dependent on the environment, in particular, on the organism's diet. Caloric restriction reproducibly results in extended lifespan in a variety of organisms, likely via nutrient sensing pathways and decreased metabolic rate. The molecular mechanisms by which such restriction results in lengthened lifespan are as yet unclear; however, the behavior of many genes known to be involved in DNA repair is altered under conditions of caloric restriction. Several agents reported to have anti-aging properties have been shown to attenuate constitutive level of mTOR signaling, an evidence of reduction of metabolic activity, and concurrently to reduce constitutive level of DNA damage induced by endogenously generated reactive oxygen species.

For example, increasing the gene dosage of the gene SIR-2, which regulates DNA packaging in the nematode worm *Caenorhabditis elegans*, can significantly extend lifespan. The mammalian homolog of SIR-2 is known to induce downstream DNA repair factors involved in NHEJ, an activity that is especially promoted under conditions of caloric restriction. Caloric restriction has been closely linked to the rate of base excision repair in the nuclear DNA of rodents, although similar effects have not been observed in mitochondrial DNA.

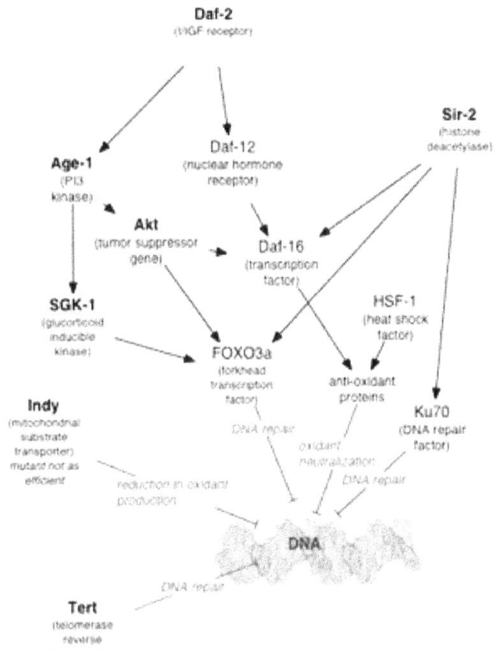

Fig 8: Most Life Span Influencing Genes Affect the Rate of DNA Damage

The *C. elegans* gene AGE-1, an upstream effector of DNA repair pathways, confers dramatically extended life span under free-feeding conditions but leads to a decrease in reproductive fitness under conditions of caloric restriction. This observation supports the pleiotropy theory of the biological origins of aging, which suggests that genes conferring a large survival advantage early in life will be selected for even if they carry a corresponding disadvantage late in life.

28.2.7 Medicine and DNA Repair Modulation

Hereditary DNA Repair Disorders

Defects in the NER mechanism are responsible for several genetic disorders, including:

- **Xeroderma Pigmentosum:** Hypersensitivity to sunlight/UV, resulting in increased skin cancer incidence and premature aging.
- **Cockayne Syndrome:** Hypersensitivity to UV and chemical agents.
- **Trichothiodystrophy:** Sensitive skin, brittle hair and nails.

Mental retardation often accompanies the latter two disorders, suggesting increased vulnerability of developmental neurons.

Other DNA repair disorders include:

- **Werner's Syndrome:** Premature aging and retarded growth.
- **Bloom's Syndrome:** Sunlight hypersensitivity, high incidence of malignancies (especially leukemias).
- **Ataxia Telangiectasia:** Sensitivity to ionizing radiation and some chemical agents.

All of the above diseases are often called "segmental progerias" ("accelerated aging diseases") because those affected appear elderly and experience aging-related diseases at an abnormally young age, while not manifesting all the symptoms of old age.

Other diseases associated with reduced DNA repair function include Fanconi anemia, hereditary breast cancer and hereditary colon cancer.

28.2.8 Cancer

Because of inherent limitations in the DNA repair mechanisms, if humans lived long enough, they would all eventually develop cancer. There are at least 34 Inherited human DNA repair gene mutations that increase cancer risk. Many of these mutations cause DNA repair to be less effective than normal. In particular, Hereditary nonpolyposis colorectal cancer (HNPCC) is strongly associated with specific mutations in the DNA mismatch repair pathway. *BRCA1* and *BRCA2*, two important genes whose mutations confer a hugely increased risk of breast cancer on carriers, are both associated with a large number of DNA repair pathways, especially NHEJ and homologous recombination.

Cancer therapy procedures such as chemotherapy and radiotherapy work by overwhelming the capacity of the cell to repair DNA damage, resulting in cell death. Cells that are most rapidly dividing – most typically cancer cells – are preferentially affected. The side-effect is that other non-cancerous but rapidly dividing cells such as progenitor cells in the gut, skin, and hematopoietic system are also affected. Modern cancer treatments attempt to localize the DNA damage to cells and tissues only associated with cancer, either by physical means (concentrating the therapeutic agent in the region of the tumor) or by biochemical means (exploiting a feature unique to cancer cells in the body). In the context of therapies targeting DNA damage response genes, the latter approach has been termed 'synthetic lethality'.

Perhaps the most well-known of these 'synthetic lethality' drugs is the poly(ADP-ribose) polymerase 1 (PARP1) inhibitor olaparib, which was approved by the Food and Drug Administration in 2015 for the treatment in women of BRCA-defective ovarian cancer. Tumor cells with partial loss of DNA damage response (specifically, homologous recombination repair) are dependent on another mechanism – single-strand break repair – which is a mechanism consisting, in part, of the PARP1 gene product. Olaparib is combined with chemotherapeutics to inhibit single-strand break repair induced by DNA damage caused by the co-administered chemotherapy. Tumor cells relying on this residual DNA repair mechanism are unable to repair the damage and hence are not able to survive and proliferate, whereas normal cells can repair the damage with the functioning homologous recombination mechanism.

Many other drugs for use against other residual DNA repair mechanisms commonly found in cancer are currently under investigation. However, synthetic lethality therapeutic approaches have been questioned due to emerging evidence of acquired resistance, achieved through rewiring of DNA damage response pathways and reversion of previously inhibited defects.

DNA Repair Defects in Cancer

It has become apparent over the past several years that the DNA damage response acts as a barrier to the malignant transformation of preneoplastic cells. Previous studies have shown an elevated DNA damage response in cell-culture models with oncogene activation and preneoplastic colon adenomas. DNA damage response mechanisms trigger cell-cycle arrest, and attempt to repair DNA lesions or promote cell death/senescence if repair is not possible. Replication stress is observed in preneoplastic cells due to increased proliferation signals from oncogenic mutations. Replication stress is characterized by: increased replication initiation/origin firing; increased transcription and collisions of transcription-replication complexes; nucleotide deficiency; increase in reactive oxygen species (ROS).

Replication stress, along with the selection for inactivating mutations in DNA damage response genes in the evolution of the tumor, leads to downregulation and/or loss of some DNA damage response mechanisms, and hence loss of DNA repair and/or senescence/programmed cell death. In experimental mouse models, loss of DNA damage response-mediated cell senescence was observed after using a short hairpin RNA (shRNA) to inhibit the double-strand break response kinase ataxia telangiectasia (ATM), leading to increased tumor size and invasiveness. Humans born with inherited defects in DNA repair mechanisms (for example, Li-Fraumeni syndrome) have a higher cancer risk.

The prevalence of DNA damage response mutations differs across cancer types; for example, 30% of breast invasive carcinomas have mutations in genes involved in homologous recombination. In cancer, downregulation is observed across all DNA damage response mechanisms (base excision repair (BER), nucleotide excision repair (NER), DNA mismatch repair (MMR), homologous recombination repair (HR), non-homologous end joining (NHEJ) and translesion DNA synthesis (TLS). As well as mutations to DNA damage repair genes, mutations also arise in the genes responsible for arresting the cell cycle to allow sufficient time for DNA repair to occur, and some genes are involved in both DNA damage repair and cell cycle checkpoint control, for example ATM and checkpoint kinase 2 (CHEK2) – a tumor suppressor that is often absent or downregulated in non-small cell lung cancer.

Genes involved in DNA damage response pathways and frequently mutated in cancer (HR = homologous recombination; NHEJ = non-homologous end joining; SSA = single-strand annealing; FA = fanconi anemia pathway; BER = base excision repair; NER = nucleotide excision repair; MMR = mismatch repair)							
	HR	NHEJ	SSA	FA	BER	NER	MMR
ATM	✓	✓	✓				
ATR	✓	✓	✓				
PAXIP	✓	✓					
RPA	✓		✓			✓	
BRCA1	✓			✓			
BRCA2	✓			✓			
RAD51	✓			✓			
RFC	✓				✓	✓	
XRCC1					✓	✓	
PCNA					✓	✓	✓
PARP1		✓			✓		
ERCC1	✓		✓	✓		✓	
MSH3	✓		✓				✓

Epigenetic DNA Repair Defects in Cancer

Classically, cancer has been viewed as a set of diseases that are driven by progressive genetic abnormalities that include mutations in tumour-suppressor genes and oncogenes, and chromosomal aberrations. However, it has become apparent that cancer is also driven by epigenetic alterations.

Epigenetic alterations refer to functionally relevant modifications to the genome that do not involve a change in the nucleotide sequence. Examples of such modifications are changes in DNA methylation (hypermethylation and hypomethylation) and histone modification, changes in chromosomal architecture (caused by inappropriate expression of proteins such as HMGA2 or HMGA1) and changes caused by microRNAs. Each of these epigenetic alterations serves to regulate gene expression without altering the underlying DNA sequence. These changes usually remain through cell divisions, last for multiple cell generations, and can be considered to be epimutations (equivalent to mutations).

While large numbers of epigenetic alterations are found in cancers, the epigenetic alterations in DNA repair genes, causing reduced expression of DNA repair proteins, appear to be particularly important. Such alterations are thought to occur early in progression to cancer and to be a likely cause of the genetic instability characteristic of cancers.

Reduced expression of DNA repair genes causes deficient DNA repair. When DNA repair is deficient DNA damages remain in cells at a higher than usual level and these excess damages cause increased frequencies of mutation or epimutation. Mutation rates increase substantially in cells defective in DNA mismatch repair or in homologous recombinational repair (HRR). Chromosomal rearrangements and aneuploidy also increase in HRR defective cells.

Higher levels of DNA damage not only cause increased mutation, but also cause increased epimutation. During repair of DNA double strand breaks, or repair of other DNA damages, incompletely cleared sites of repair can cause epigenetic gene silencing.

Deficient expression of DNA repair proteins due to an inherited mutation can cause increased risk of cancer. Individuals with an inherited impairment in any of 34 DNA repair genes have an increased risk of cancer, with some defects causing up to a 100% lifetime chance of cancer (e.g. p53 mutations). However, such germline mutations (which cause highly penetrant cancer syndromes) are the cause of only about 1 percent of cancers.

Frequencies of Epimutations in DNA Repair Genes

Deficiencies in DNA repair enzymes are occasionally caused by a newly arising somatic mutation in a DNA repair gene, but are much more frequently caused by epigenetic alterations that reduce or silence expression of DNA repair genes. For example, when 113 colorectal cancers were examined in sequence, only four had a missense mutation in the DNA repair gene MGMT, while the majority had reduced MGMT expression due to methylation of the MGMT promoter region (an epigenetic alteration).

Five different studies found that between 40% and 90% of colorectal cancers have reduced MGMT expression due to methylation of the MGMT promoter region.

Similarly, out of 119 cases of mismatch repair-deficient colorectal cancers that lacked DNA repair gene PMS2 expression, PMS2 was deficient in 6 due to mutations in the PMS2 gene, while in 103 cases PMS2 expression was deficient because its pairing partner MLH1 was repressed due to promoter methylation (PMS2 protein is unstable in the absence of MLH1).

In the other 10 cases, loss of PMS2 expression was likely due to epigenetic overexpression of the microRNA, miR-155, which down-regulates MLH1.

In a further example, epigenetic defects were found in various cancers (e.g. breast, ovarian, colorectal and head and neck). Two or three deficiencies in the expression of ERCC1, XPF or PMS2 occur simultaneously in the majority of 49 colon cancers evaluated by Facista et al.

Red-highlighted genes are frequently reduced or silenced by epigenetic mechanisms in various cancers. When these genes have low or absent expression, DNA damages can accumulate. Replication errors past these damages can lead to increased mutations and, ultimately, cancer. Epigenetic repression of DNA repair genes in accurate DNA repair pathways appear to be central to carcinogenesis.

The two gray-highlighted genes *RAD51* and *BRCA2*, are required for homologous recombinational repair. They are sometimes epigenetically over-expressed and sometimes under-expressed in certain cancers. These repair deficiencies would likely cause increased unrepaired DNA damages. The over-expression of *RAD51* and *BRCA2* seen in these cancers may reflect selective pressures for compensatory *RAD51* or *BRCA2* over-expression and increased homologous recombinational repair to at least partially deal with such excess DNA damages. In those cases where *RAD51* or *BRCA2* are under-expressed, this would itself lead to increased unrepaired DNA damages. Replication errors past these damages could cause increased mutations and cancer, so that under-expression of *RAD51* or *BRCA2* would be carcinogenic in itself.

Cyan-highlighted genes are in the microhomology-mediated end joining (MMEJ) pathway and are up-regulated in cancer. MMEJ is an additional error-prone *inaccurate* repair pathway for double-strand breaks. In MMEJ repair of a double-strand break, an homology of 5–25 complementary base pairs between both paired strands is sufficient to align the strands, but mismatched ends (flaps) are usually present. MMEJ removes the extra nucleotides (flaps) where strands are joined, and then ligates the strands to create an intact DNA double helix. MMEJ almost always involves at least a small deletion, so that it is a mutagenic pathway. FEN1, the flap endonuclease in MMEJ, is epigenetically increased by promoter hypomethylation and is over-expressed in the majority of cancers of the breast, prostate, stomach, neuroblastomas, pancreas, and lung. PARP1 is also over-expressed when its promoter region ETS site is epigenetically hypomethylated, and this contributes to progression to endometrial cancer and BRCA-mutated serous ovarian cancer. Other genes in the MMEJ pathway are also over-expressed in a number of cancers and are also shown in cyan.

Genome-Wide Distribution of DNA Repair in Human Somatic Cells

Differential activity of DNA repair pathways across various regions of the human genome causes mutations to be very unevenly distributed within tumor genomes. In particular, the gene-rich, early-replicating regions of the human genome exhibit lower mutation frequencies than the gene-poor, late-replicating heterochromatin. One mechanism underlying this involves the histone modification H3K36me3, which can recruit mismatch repair proteins, thereby lowering mutation rates in H3K36me3-marked regions. Another important mechanism concerns nucleotide excision repair, which can be recruited by the transcription machinery, lowering somatic mutation rates in active genes and other open chromatin regions.

28.2.9 Epigenetic Alterations due to DNA Repair

Damage to DNA is very common and is constantly being repaired. Epigenetic alterations can accompany DNA repair of oxidative damage or double-strand breaks. In human cells, oxidative DNA damage occurs about 10,000 times a day and DNA double-strand breaks occur about 10 to 50 times a cell cycle in somatic replicating cells. The selective advantage of DNA repair is to allow the cell to survive in the face of DNA damage. The selective advantage of epigenetic alterations that occur with DNA repair is not clear.

Repair of Oxidative DNA Damage can Alter Epigenetic Markers

In the steady state (with endogenous damages occurring and being repaired), there are about 2,400 oxidatively damaged guanines that form 8-oxo-2'-deoxyguanosine (8-OHdG) in the average mammalian cell DNA. 8-OHdG constitutes about 5% of the oxidative damages commonly present in DNA. The oxidized guanines do not occur randomly among all guanines in DNA. There is a sequence preference for the guanine at a methylated CpG site (a cytosine followed by guanine along its 5' $\rightarrow$ 3' direction and where the cytosine is methylated (5-mCpG)). A 5-mCpG site has the lowest ionization potential for guanine oxidation.

Oxidized guanine has mispairing potential and is mutagenic. Oxoguanine glycosylase (OGG1) is the primary enzyme responsible for the excision of the oxidized guanine during DNA repair. OGG1 finds and binds to an 8-OHdG within a few seconds. However, OGG1 does not immediately excise 8-OHdG. In HeLa cells half maximum removal of 8-OHdG occurs in 30 minutes, and in irradiated mice, the 8-OHdGs induced in the mouse liver are removed with a half-life of 11 minutes.

When OGG1 is present at an oxidized guanine within a methylated CpG site it recruits TET1 to the 8-OHdG lesion. This allows TET1 to demethylate an adjacent methylated cytosine. Demethylation of cytosine is an epigenetic alteration.

As an example, when human mammary epithelial cells were treated with H_2O_2 for six hours, 8-OHdG increased about 3.5-fold in DNA and this caused about 80% demethylation of the 5-methylcytosines in the genome. Demethylation of CpGs in a gene promoter by TET enzyme activity increases transcription of the gene into messenger RNA. In cells treated with H_2O_2, one particular gene was examined, *BACE1*. The methylation level of the *BACE1* CpG island was reduced (an epigenetic alteration) and this allowed about 6.5 fold increase of expression of *BACE1* messenger RNA.

While six-hour incubation with H_2O_2 causes considerable demethylation of 5-mCpG sites, shorter times of H_2O_2 incubation appear to promote other epigenetic alterations. Treatment of cells with H_2O_2 for 30 minutes causes the mismatch repair protein heterodimer MSH2-MSH6 to recruit DNA methyltransferase 1 (DNMT1) to sites of some kinds of oxidative DNA damage. This could cause increased methylation of cytosines (epigenetic alterations) at these locations.

Jiang et al. treated HEK 293 cells with agents causing oxidative DNA damage, (potassium bromate (KBrO3) or potassium chromate (K2CrO4)). Base excision repair (BER) of oxidative damage occurred with the DNA repair enzyme polymerase beta localizing to oxidized guanines. Polymerase beta is the main human polymerase in short-patch BER of oxidative DNA damage. Jiang et al. also found that polymerase beta recruited the DNA methyltransferase protein DNMT3b to BER repair sites. They then evaluated the methylation pattern at the single nucleotide level in a small region of DNA including the promoter region and the early transcription region of the BRCA1 gene. Oxidative DNA damage from bromate modulated the DNA methylation pattern (caused epigenetic alterations) at CpG sites within the region of DNA studied. In untreated cells, CpGs located at -189, -134, -29, -19, +16, and +19 of the BRCA1 gene had methylated cytosines (where numbering is from the messenger RNA transcription start site, and negative numbers indicate nucleotides in the upstream promoter region). Bromate treatment-induced oxidation resulted in the loss of cytosine methylation at -189, -134, +16 and +19 while also leading to the formation of new methylation at the CpGs located at -80, -55, -21 and +8 after DNA repair was allowed.

Homologous Recombinational Repair Alters Epigenetic Markers

During homologous recombinational repair (HR) of the double-strand break, the involvement of DNMT1 causes the two repaired strands of DNA to have different levels of methylated cytosines. One strand becomes frequently methylated at about 21 CpG sites downstream of the repaired double-strand break. The other DNA strand loses methylation at about six CpG sites that were previously methylated downstream of the double-strand break, as well as losing methylation at about five CpG sites that were previously methylated upstream of the double-strand break. When the chromosome is replicated, this gives rise to one daughter chromosome that is heavily methylated downstream of the previous break site and one that is unmethylated in the region both upstream and downstream of the previous break site. With respect to the gene that was broken by the double-strand break, half of the progeny cells express that gene at a high level and in the other half of the progeny cells expression of that gene is repressed. When clones of these cells were maintained for three years, the new methylation patterns were maintained over that time period.

In mice with a CRISPR-mediated homology-directed recombination insertion in their genome there were a large number of increased methylations of CpG sites within the double-strand break-associated insertion.

Non-Homologous End Joining can Cause Some Epigenetic Marker Alterations

Non-homologous end joining (NHEJ) repair of a double-strand break can cause a small number of demethylations of pre-existing cytosine DNA methylations downstream of the repaired double-strand break. Further work by Allen et al. showed that NHEJ of a DNA double-strand break in a cell could give rise to some progeny cells having repressed expression of the gene harboring the initial double-strand break and some progeny having high expression of that gene due to epigenetic alterations associated with NHEJ repair.

The frequency of epigenetic alterations causing repression of a gene after an NHEJ repair of a DNA double-strand break in that gene may be about 0.9%.

28.3 DNA RECOMBINATION

Recombinant DNA (rDNA) molecules are DNA molecules formed by laboratory methods of genetic recombination (such as molecular cloning) that bring together genetic material from multiple sources, creating sequences that would not otherwise be found in the genome.

Recombinant DNA is the general name for a piece of DNA that has been created by combining two or more fragments from different sources. Recombinant DNA is possible because DNA molecules from all organisms share the same chemical structure, differing only in the nucleotide sequence. Recombinant DNA molecules are sometimes called chimeric DNA because they can be made of material from two different species like the mythical chimera. rDNA technology uses palindromic sequences and leads to the production of sticky and blunt ends.

The DNA sequences used in the construction of recombinant DNA molecules can originate from any species.

For example, plant DNA can be joined to bacterial DNA, or human DNA can be joined with fungal DNA. In addition, DNA sequences that do not occur anywhere in nature can be created by the chemical synthesis of DNA and incorporated into recombinant DNA molecules. Using recombinant DNA technology and synthetic DNA, any DNA sequence can be created and introduced into living organisms.

Proteins that can result from the expression of recombinant DNA within living cells are termed *recombinant proteins.* When recombinant DNA encoding a protein is introduced into a host organism, the recombinant protein is not necessarily produced. Expression of foreign proteins requires the use of specialized expression vectors and often necessitates significant restructuring by foreign coding sequences.

Recombinant DNA differs from genetic recombination in that the former results from artificial methods while the latter is a normal biological process that results in the remixing of existing DNA sequences in essentially all organisms.

28.3.1 History

The idea of recombinant DNA was first proposed by Peter Lobban, a graduate student of Prof. Dale Kaiser in the Biochemistry Department at Stanford University Medical School. The first publications describing the successful production and intracellular replication of recombinant DNA appeared in 1972 and 1973, from Stanford and UCSF. In 1980 Paul Berg, a professor in the Biochemistry Department at Stanford and an author on one of the first papers was awarded the Nobel Prize in Chemistry for his work on nucleic acids "with particular regard to recombinant DNA". Werner Arber, Hamilton Smith, and Daniel Nathans shared the 1978 Nobel Prize in Physiology or Medicine for the discovery of restriction endonucleases which enhanced the techniques of rDNA technology. Stanford University applied for a U.S. patent on recombinant DNA on November 4, 1974, listing the inventors as Herbert W. Boyer (professor at the University of California, San Francisco) and Stanley N. Cohen (professor at Stanford University); this patent, U.S. 4,237,224A, was awarded on December 2, 1980. The first licensed drug generated using recombinant DNA technology was human insulin, developed by Genentech and licensed by Eli Lilly and Company.

28.3.2 Production

Molecular cloning is the laboratory process used to produce recombinant DNA. It is one of two most widely used methods, along with polymerase chain reaction (PCR), used to direct the replication of any specific DNA sequence chosen by the experimentalist. There are two fundamental differences between the methods. One is that molecular cloning involves replication of the DNA within a living cell, while PCR replicates DNA in the test tube, free of living cells. The other difference is that cloning involves cutting and pasting DNA sequences, while PCR amplifies by copying an existing sequence.

Formation of recombinant DNA requires a cloning vector, a DNA molecule that replicates within a living cell. Vectors are generally derived from plasmids or viruses, and represent relatively small segments of DNA that contain necessary genetic signals for replication, as well as additional elements for convenience in inserting foreign DNA, identifying cells that contain recombinant DNA, and, where appropriate, expressing the foreign DNA. The choice of vector for molecular cloning depends on the choice of host organism, the size of the DNA to be cloned, and whether and how the foreign DNA is to be expressed. The DNA segments can be combined by using a variety of methods, such as restriction enzyme/ligase cloning or Gibson assembly.

In standard cloning protocols, the cloning of any DNA fragment essentially involves seven steps:

- Choice of host organism and cloning vector,
- Preparation of vector DNA,
- Preparation of DNA to be cloned,
- Creation of recombinant DNA,
- Introduction of recombinant DNA into the host organism,
- Selection of organisms containing recombinant DNA, and
- Screening for clones with desired DNA inserts and biological properties.

28.3.3 DNA Expression

DNA expression requires the transfection of suitable host cells. Typically, either bacterial, yeast, insect, or mammalian cells (such as Human Embryonic Kidney cells or CHO cells) are used as host cells. Following transplantation into the host organism, the foreign DNA contained within the recombinant DNA construct may or may not be expressed. That is, the DNA may simply be replicated without expression, or it may be transcribed and translated and a recombinant protein is produced. Generally speaking, expression of a foreign gene requires restructuring the gene to include sequences that are required for producing an mRNA molecule that can be used by the host's translational apparatus (e.g. promoter, translational initiation signal, and transcriptional terminator). Specific changes to the host organism may be made to improve expression of the ectopic gene. In addition, changes may be needed to the coding sequences as well, to optimize translation, make the protein soluble, direct the recombinant protein to the proper cellular or extracellular location, and stabilize the protein from degradation.

28.3.4 Properties of Organisms Containing Recombinant DNA

In most cases, organisms containing recombinant DNA have apparently normal phenotypes. That is, their appearance, behavior and metabolism are usually unchanged, and the only way to demonstrate the presence of recombinant sequences is to examine the DNA itself, typically using a polymerase chain reaction (PCR) test. Significant exceptions exist, and are discussed below.

If the rDNA sequences encode a gene that is expressed, then the presence of RNA and/or protein products of the recombinant gene can be detected, typically using RT-PCR or western hybridization methods. Gross phenotypic changes are not the norm, unless the recombinant gene has been chosen and modified so as to generate biological activity in the host organism. Additional phenotypes that are encountered include toxicity to the host organism induced by the recombinant gene product, especially if it is over-expressed or expressed within inappropriate cells or tissues.

In some cases, recombinant DNA can have deleterious effects even if it is not expressed. One mechanism by which this happens is insertional inactivation, in which the rDNA becomes inserted into a host cell's gene. In some cases, researchers use this phenomenon to "knock out" genes to determine their biological function and importance. Another mechanism by which rDNA insertion into chromosomal DNA can affect gene expression is by inappropriate activation of previously unexpressed host cell genes. This can happen, for example, when a recombinant DNA fragment containing an active promoter becomes located next to a previously silent host cell gene, or when a host cell gene that functions to restrain gene expression undergoes insertional inactivation by recombinant DNA.

28.3.5 Applications of Recombinant DNA

Recombinant DNA is widely used in biotechnology, medicine and research. Today, recombinant proteins and other products that result from the use of DNA technology are found in essentially every western pharmacy, physician or veterinarian office, medical testing laboratory, and biological research laboratory. In addition, organisms that have been manipulated using recombinant DNA technology, as well as products derived from those organisms, have found their way into many farms, supermarkets, home medicine cabinets, and even pet shops, such as those that sell GloFish and other genetically modified animals.

The most common application of recombinant DNA is in basic research, in which the technology is important to most current work in the biological and biomedical sciences. Recombinant DNA is used to identify, map and sequence genes, and to determine their function. rDNA probes are employed in analyzing gene expression within individual cells, and throughout the tissues of whole organisms. Recombinant proteins are widely used as reagents in laboratory experiments and to generate antibody probes for examining protein synthesis within cells and organisms.

Many additional practical applications of recombinant DNA are found in industry, food production, human and veterinary medicine, agriculture, and bioengineering. Some specific examples are identified below.

Recombinant Chymosin

Found in rennet, chymosin is the enzyme responsible for hydrolysis of κ-casein to produce para-κ-casein and glycomacropeptide, which is the first step in formation of cheese, and subsequently curd, and whey. It was the first genetically engineered food additive used commercially. Traditionally, processors obtained chymosin from rennet, a preparation derived from the fourth stomach of milk-fed calves. Scientists engineered a non-pathogenic strain (K-12) of *E. coli* bacteria for large-scale laboratory production of the enzyme. This microbiologically produced recombinant enzyme, identical structurally to the calf derived enzyme, costs less and is produced in abundant quantities. Today about 60% of U.S. hard cheese is made with genetically engineered chymosin. In 1990, FDA granted chymosin "generally recognized as safe" (GRAS) status based on data showing that the enzyme was safe.

Recombinant Human Insulin

Recombinant human insulin has almost completely replaced insulin obtained from animal sources (e.g. pigs and cattle) for the treatment of type 1 diabetes. A variety of different recombinant insulin preparations are in widespread use. Recombinant insulin is synthesized by inserting the human insulin gene into *E. coli*, or yeast (Saccharomyces cerevisiae) which then produces insulin for human use. Insulin produced by E. coli requires further post translational modifications (e.g. glycosylation) whereas yeasts are able to perform these modifications themselves by virtue of being more complex host organisms. The advantage of recombinant human insulin is after chronic use patients don't develop an immune defence against it the way animal sourced insulin stimulates the human immune system.

Recombinant Human Growth Hormone (HGH, Somatotropin)

Administered to patients whose pituitary glands generate insufficient quantities to support normal growth and development. Before recombinant HGH became available, HGH for therapeutic use was obtained from pituitary glands of cadavers. This unsafe practice led to some patients developing Creutzfeldt–Jakob disease. Recombinant HGH eliminated this problem, and is now used therapeutically. It has also been misused as a performance-enhancing drug by athletes and others.

Recombinant Blood Clotting Factor VIII

It is the recombinant form of factor VIII, a blood-clotting protein that is administered to patients with the bleeding disorder hemophilia, who are unable to produce factor VIII in quantities sufficient to support normal blood coagulation. Before the development of recombinant factor VIII, the protein was obtained by processing large quantities of human blood from multiple donors, which carried a very high risk of transmission of blood borne infectious diseases, for example HIV and hepatitis B.

Recombinant Hepatitis B Vaccine

Hepatitis B infection can be successfully controlled through the use of a recombinant subunit hepatitis B vaccine, which contains a form of the hepatitis B virus surface antigen that is produced in yeast cells. The development of the recombinant subunit vaccine was an important and necessary development because hepatitis B virus, unlike other common viruses such as polio virus, cannot be grown in vitro.

Recombinant Antibodies

Recombinant antibodies (rAbs) are produced in vitro by the means of expression systems based on mammalian cells. Their monospecific binding to a specific epitope makes rAbs eligible not only for research purposes, but also as therapy options against certain cancer types, infections and autoimmune diseases.

Diagnosis of HIV Infection

Each of the three widely used methods for diagnosing HIV infection has been developed using recombinant DNA. The antibody test (ELISA or western blot) uses a recombinant HIV protein to test for the presence of antibodies that the body has produced in response to an HIV infection. The DNA test looks for the presence of HIV genetic material using reverse transcription polymerase chain reaction (RT-PCR).

Development of the RT-PCR test was made possible by the molecular cloning and sequence analysis of HIV genomes. HIV testing page from US Centers for Disease Control (CDC)

Golden Rice

Golden rice is a recombinant variety of rice that has been engineered to express the enzymes responsible for β-carotene biosynthesis. This variety of rice holds substantial promise for reducing the incidence of vitamin A deficiency in the world's population. Golden rice is not currently in use, pending the resolution of regulatory and intellectual property issues.

Herbicide-Resistant Crops

Commercial varieties of important agricultural crops (including soy, maize/corn, sorghum, canola, alfalfa and cotton) have been developed that incorporate a recombinant gene that results in resistance to the herbicide glyphosate (trade name *Roundup*), and simplifies weed control by glyphosate application. These crops are in common commercial use in several countries.

Insect-Resistant Crops

Bacillus thuringiensis is a bacterium that naturally produces a protein (Bt toxin) with insecticidal properties. The bacterium has been applied to crops as an insect-control strategy for many years, and this practice has been widely adopted in agriculture and gardening. Recently, plants have been developed that express a recombinant form of the bacterial protein, which may effectively control some insect predators. Environmental issues associated with the use of these transgenic crops have not been fully resolved.

28.3.6 Controversy

Scientists associated with the initial development of recombinant DNA methods recognized that the potential existed for organisms containing recombinant DNA to have undesirable or dangerous properties. At the 1975 Asilomar Conference on Recombinant DNA, these concerns were discussed and a voluntary moratorium on recombinant DNA research was initiated for experiments that were considered particularly risky.

This moratorium was widely observed until the National Institutes of Health (USA) developed and issued formal guidelines for rDNA work. Today, recombinant DNA molecules and recombinant proteins are usually not regarded as dangerous. However, concerns remain about some organisms that express recombinant DNA, particularly when they leave the laboratory and are introduced into the environment or food chain.

Furthermore, there are concerns about the by-products in biopharmaceutical production, where recombinant DNA result in specific protein products. The major by-product, termed host cell protein, comes from the host expression system and poses a threat to the patient's health and the overall environment.

28.4 INTERCONNECTIONS BETWEEN REPLICATION, REPAIR, AND RECOMBINATION

The processes of DNA replication, repair, and recombination are intricately linked. DNA replication is a continuous process that requires the integrity of the DNA. Any errors during replication can lead to mutations, which are subsequently addressed by DNA repair mechanisms. Conversely, the recombination processes that occur during meiosis can help correct replication errors by introducing genetic diversity and ensuring that any deleterious mutations are not passed on.

Moreover, the efficiency and fidelity of these processes are tightly regulated. Proteins involved in replication often have dual roles in repair and recombination, underscoring the interconnected nature of these processes. For example, proteins like Rad51 and BRCA1 are crucial for homologous recombination and also play roles in maintaining replication fork stability. Defects in these proteins can lead to genomic instability and are associated with various cancers.

28.5 CONCLUSION

In summary, DNA replication, repair, and recombination are essential processes that uphold the integrity of genetic material in all living organisms. These processes ensure accurate cell division, correct DNA damage, and promote genetic diversity, playing vital roles in evolution and adaptation. Understanding these mechanisms is crucial for advancements in medicine and biotechnology, particularly in addressing genetic disorders and developing targeted therapies for cancer.

Continued research in this field promises to unveil further complexities and intricacies that govern DNA behavior, ultimately enhancing our understanding of life itself.

CHAPTER 29

RNA FUNCTIONS, BIOSYNTHESIS AND PROCESSING

Ribonucleic acid (RNA) plays a crucial role in the cellular processes of all living organisms, functioning not only as a messenger between DNA and proteins but also participating in various regulatory and catalytic activities. RNA is a versatile molecule that comes in several forms, each with distinct functions in the cell.

29.1 FUNCTIONS OF RNA

RNA is essential for various biological functions that extend beyond serving merely as a template for protein synthesis. The most well-known function of RNA is its role in protein synthesis, where messenger RNA (mRNA) acts as the intermediary between DNA and ribosomes, the cellular machinery that assembles proteins. During translation, ribosomes read the sequence of codons in mRNA and synthesize polypeptides by linking the appropriate amino acids in a sequence determined by the mRNA template.

In addition to mRNA, several other types of RNA have critical functions in the cell. Ribosomal RNA (rRNA) is a fundamental component of ribosomes, providing structural support and catalyzing peptide bond formation during protein synthesis. Transfer RNA (tRNA) serves as an adapter molecule that brings specific amino acids to the ribosome, recognizing the codons on the mRNA through its anticodon sequence.

Moreover, non-coding RNAs (ncRNAs) play vital roles in gene regulation and other cellular processes. For example, microRNAs (miRNAs) and small interfering RNAs (siRNAs) are involved in RNA interference (RNAi), a regulatory mechanism that silences gene expression by degrading target mRNAs or inhibiting their translation. Long non-coding RNAs (lncRNAs) also contribute to gene regulation by interacting with chromatin, transcription factors, and other components of the transcriptional machinery.

RNA also has roles in catalysis. Ribozymes are RNA molecules that can catalyze specific biochemical reactions, challenging the traditional view that only proteins can serve as enzymes. This discovery has implications for our understanding of the origin of life, suggesting that RNA may have played a central role in early biological systems.

29.2 RNA BIOSYNTHESIS

The biosynthesis of RNA, known as transcription, is a fundamental process that involves the synthesis of RNA from a DNA template. This process is initiated by the binding of RNA polymerase to a specific region of the DNA called the promoter. The promoter region contains regulatory sequences that determine the efficiency and timing of transcription. Once RNA polymerase is bound, it unwinds a portion of the DNA double helix, exposing the template strand.

During transcription, RNA polymerase synthesizes a single-stranded RNA molecule by adding ribonucleotides complementary to the DNA template strand. Unlike DNA replication, RNA synthesis does not require a primer; RNA polymerase can initiate synthesis de novo. The synthesis proceeds in the 5' to 3' direction, with RNA polymerase catalyzing the formation of phosphodiester bonds between adjacent ribonucleotides. Transcription can be divided into three main stages: initiation, elongation, and termination. Initiation involves the assembly of the transcription machinery at the promoter, followed by the unwinding of the DNA. During elongation, RNA polymerase moves along the DNA template, synthesizing the RNA transcript. Termination occurs when RNA polymerase encounters a specific sequence in the DNA that signals the end of transcription, resulting in the release of the newly synthesized RNA molecule.

In eukaryotes, transcription takes place in the nucleus, where the primary RNA transcript (pre-mRNA) undergoes several modifications before it can function in the cytoplasm. In prokaryotes, transcription and translation occur simultaneously in the cytoplasm, allowing for a more rapid response to environmental changes.

29.3 RNA PROCESSING

After transcription, eukaryotic pre-mRNA undergoes a series of processing steps that are crucial for its maturation and functionality. These modifications include capping, polyadenylation, and splicing, all of which enhance the stability and translational efficiency of the RNA molecule.

One of the first modifications is the addition of a 5' cap, which is a modified guanine nucleotide. This cap protects the RNA from degradation by exonucleases, facilitates ribosome binding during translation, and plays a role in nuclear export. The 5' cap is added shortly after the transcription begins, ensuring that the RNA is adequately protected from degradation.

Another significant processing step is polyadenylation, where a stretch of adenine nucleotides (the poly-A tail) is added to the 3' end of the RNA transcript. This modification enhances RNA stability and influences its transport from the nucleus to the cytoplasm. The poly-A tail also aids in the initiation of translation, as it interacts with proteins that facilitate ribosome assembly.

Splicing is the process by which introns, non-coding regions of the RNA, are removed, and exons, the coding regions, are joined together. This process is carried out by a complex known as the spliceosome, composed of small nuclear RNAs (snRNAs) and protein factors. Splicing can be classified into two types: constitutive splicing, where all introns are removed, and alternative splicing, where different combinations of exons are joined together. Alternative splicing allows a single gene to produce multiple protein isoforms, significantly increasing the diversity of the proteome.

In addition to these modifications, RNA molecules may undergo editing, a process that alters specific nucleotide sequences after transcription. RNA editing can involve the insertion, deletion, or substitution of nucleotides, resulting in a final RNA product that differs from the original transcript. This process can have significant functional implications, such as altering the coding potential of the RNA or affecting its stability and translational efficiency.

29.4 RNA STABILITY AND DEGRADATION

The stability of RNA molecules is vital for their function and regulation. Various factors influence RNA stability, including the presence of protective modifications (such as the 5' cap and poly-A tail) and the sequences of the RNA itself. Unprotected RNA molecules are susceptible to degradation by ribonucleases, enzymes that hydrolyze RNA.

The cellular degradation of RNA is a critical aspect of gene regulation, allowing for the timely removal of RNA that is no longer needed. This process can be regulated by specific sequences in the RNA, such as elements in the 3' untranslated region (3' UTR) that interact with RNA-binding proteins. These proteins can either promote RNA stability or facilitate degradation, thereby influencing the overall levels of gene expression in the cell.

29.5 CONCLUSION

In conclusion, RNA plays multifaceted roles in cellular biology, from serving as a template for protein synthesis to participating in gene regulation and catalysis. The processes of RNA biosynthesis and processing are intricately regulated and essential for producing functional RNA molecules. Understanding RNA functions, biosynthesis, and processing has significant implications for molecular biology, genetics, and biotechnology.

As research continues to unveil the complexities of RNA biology, new therapeutic strategies targeting RNA processes are being developed, promising advancements in the treatment of various diseases, including cancer and genetic disorders. The versatility and dynamism of RNA underscore its importance in the fundamental mechanisms of life.

CHAPTER 30
PROTEIN BIOSYNTHESIS

Protein biosynthesis is a fundamental biological process through which cells produce proteins, which are essential for a wide variety of cellular functions. Proteins serve as structural components, enzymes, signaling molecules, and play critical roles in immune response and metabolism. The process of protein biosynthesis involves two main stages: transcription and translation.

30.1 OVERVIEW OF PROTEIN BIOSYNTHESIS

The biosynthesis of proteins begins with the genetic instructions encoded in DNA. The information contained in genes is transcribed into messenger RNA (mRNA) in a process that takes place in the cell nucleus. This mRNA serves as a template for translating the genetic code into a specific protein sequence in the cytoplasm. Protein biosynthesis is a highly regulated process, ensuring that proteins are synthesized at the right time, in the right amounts, and in response to the cell's needs.

30.2 TRANSCRIPTION: FROM DNA TO MRNA

The first stage of protein biosynthesis, transcription, involves the synthesis of mRNA from a DNA template. This process begins with the binding of RNA polymerase, an enzyme responsible for synthesizing RNA, to a specific region of the gene known as the promoter. The promoter contains sequences that regulate the initiation of transcription, allowing the cell to control when and how much mRNA is produced.

Once RNA polymerase is bound to the promoter, it unwinds the DNA double helix, exposing the template strand. As the enzyme moves along the DNA, it synthesizes a single-stranded RNA molecule by adding ribonucleotides complementary to the DNA template strand.

This process occurs in the 5' to 3' direction, meaning that RNA polymerase adds nucleotides to the growing RNA strand at the 3' end. Importantly, RNA polymerase does not require a primer to initiate RNA synthesis, a significant difference from DNA polymerase.

Transcription proceeds through three main phases: initiation, elongation, and termination. During initiation, RNA polymerase assembles the transcription machinery and begins RNA synthesis. In the elongation phase, the enzyme moves along the DNA template, elongating the RNA transcript. Termination occurs when RNA polymerase reaches a termination signal in the DNA sequence, signaling the end of transcription. The result is a primary RNA transcript, which may undergo further processing before it is translated into a protein.

30.3 RNA PROCESSING

In eukaryotic cells, the primary mRNA transcript (pre-mRNA) undergoes several processing steps before it can be translated into a protein. These modifications are crucial for ensuring the stability, transport, and functionality of the mRNA. The primary processing steps include capping, polyadenylation, and splicing.

Capping involves the addition of a modified guanine nucleotide to the 5' end of the mRNA. This cap protects the RNA from degradation, facilitates ribosome binding during translation, and assists in nuclear export. The addition of the 5' cap occurs shortly after transcription begins.

Polyadenylation is the addition of a poly-A tail, a stretch of adenine nucleotides, to the 3' end of the mRNA. This tail enhances mRNA stability and influences its transport from the nucleus to the cytoplasm. The poly-A tail also plays a role in the initiation of translation, as it interacts with proteins that help assemble the ribosome.

Splicing is the process of removing non-coding sequences (introns) from the pre-mRNA and joining the coding sequences (exons) together. This process is carried out by a complex known as the spliceosome, which is composed of small nuclear RNAs (snRNAs) and protein factors. Splicing can be classified into two types: constitutive splicing, where all introns are removed, and alternative splicing, which allows for different combinations of exons to be joined together. This mechanism significantly increases the diversity of proteins that can be produced from a single gene.

30.4 TRANSLATION: THE SYNTHESIS OF PROTEINS

Once mRNA is processed and transported to the cytoplasm, it undergoes translation, the second major stage of protein biosynthesis. Translation is the process by which the sequence of codons in the mRNA is read by ribosomes to synthesize a polypeptide chain, ultimately folding into a functional protein. Translation occurs in three stages: initiation, elongation, and termination.

The initiation of translation begins with the assembly of the ribosomal subunits at the start codon on the mRNA. The ribosome is a complex composed of ribosomal RNA (rRNA) and proteins, and it serves as the site of protein synthesis. The start codon, typically AUG, codes for the amino acid methionine and signals the beginning of the polypeptide chain. Transfer RNA (tRNA) molecules play a crucial role during this stage, as they bring the appropriate amino acids to the ribosome. Each tRNA molecule carries a specific amino acid and has an anticodon that is complementary to the corresponding codon on the mRNA.

During elongation, the ribosome moves along the mRNA, and tRNA molecules continue to deliver amino acids based on the codon sequence. The ribosome catalyzes the formation of peptide bonds between the amino acids, linking them together to form a growing polypeptide chain. This process continues until the ribosome reaches a stop codon (UAA, UAG, or UGA) on the mRNA.

Termination of translation occurs when the ribosome encounters a stop codon. This signals the release factors to bind to the ribosome, prompting the release of the completed polypeptide chain from the ribosome. The ribosomal subunits then disassemble, and the mRNA is released, ready to be translated again or degraded.

30.5 POST-TRANSLATIONAL MODIFICATIONS

Once a polypeptide is synthesized, it often undergoes post-translational modifications (PTMs) that are critical for its functionality. PTMs can include phosphorylation, glycosylation, methylation, and ubiquitination, among others. These modifications can affect the protein's activity, stability, localization, and interaction with other molecules.

For example, phosphorylation involves the addition of phosphate groups to specific amino acids, which can alter a protein's conformation and activity, often serving as a regulatory mechanism. Glycosylation, the addition of carbohydrate moieties, can influence protein folding, stability, and cell signaling. Ubiquitination tags proteins for degradation by the proteasome, regulating the levels of proteins within the cell.

30.6 REGULATION OF PROTEIN BIOSYNTHESIS

The biosynthesis of proteins is a tightly regulated process that ensures cells respond appropriately to internal and external signals. Regulation occurs at multiple levels, including transcriptional, post-transcriptional, translational, and post-translational stages.

At the transcriptional level, various transcription factors can enhance or inhibit the binding of RNA polymerase to promoters, influencing the rate of mRNA synthesis. Additionally, chromatin remodeling and epigenetic modifications can affect gene accessibility, thereby impacting transcription.

Post-transcriptional regulation involves mechanisms such as RNA interference (RNAi), where small non-coding RNAs can bind to mRNA and prevent its translation or promote its degradation. This process allows for rapid and specific responses to cellular conditions. Translational regulation can occur through the availability of tRNA, ribosomal subunits, and initiation factors. Certain mRNAs may also contain regulatory elements that influence their translation efficiency. At the post-translational level, proteins can be regulated through modifications that affect their stability, localization, and activity. The interplay of these regulatory mechanisms ensures that proteins are synthesized in response to the cell's needs, maintaining homeostasis and enabling adaptation to changing environments.

30.7 CONCLUSION

Protein biosynthesis is a vital and complex process that involves the transcription of DNA into mRNA, the processing of mRNA, and the translation of that mRNA into functional proteins. Each stage of this process is intricately regulated to ensure that proteins are synthesized accurately and efficiently, allowing cells to perform their diverse functions.

Understanding protein biosynthesis is essential for advancing our knowledge of cellular biology and has significant implications for fields such as medicine, biotechnology, and genetic engineering. As research continues to uncover the complexities of protein synthesis and regulation, new therapeutic strategies are likely to emerge, harnessing the power of proteins in the treatment of diseases and disorders.

CHAPTER 31
CONTROL OF GENE EXPRESSION

Gene expression is the process by which information from a gene is used to synthesize functional gene products, primarily proteins. This intricate process is tightly regulated to ensure that the right genes are expressed at the right times, in the right cells, and in the right amounts. The control of gene expression is vital for cellular differentiation, adaptation to environmental changes, and maintaining homeostasis within an organism.

31.1 IMPORTANCE OF GENE EXPRESSION CONTROL

The control of gene expression is essential for the development and functioning of multicellular organisms. It allows cells to respond to internal and external stimuli, differentiate into specialized cell types, and coordinate complex processes such as growth, metabolism, and immune responses. Misregulation of gene expression can lead to a variety of diseases, including cancer, genetic disorders, and autoimmune diseases. Understanding the mechanisms that control gene expression is crucial for developing therapeutic strategies to address these conditions.

31.2 TRANSCRIPTIONAL REGULATION

Transcriptional regulation is the first and most crucial level of control in gene expression. This process involves various mechanisms that determine whether a gene is transcribed into mRNA. The regulation of transcription can occur through several factors, including transcription factors, enhancers, silencers, and chromatin structure.

31.2.1 Transcription Factors

Transcription factors are proteins that bind to specific DNA sequences in the promoter region of a gene to either promote or inhibit transcription. These factors can act as activators or repressors. Activators enhance the recruitment of RNA polymerase to the promoter, facilitating the initiation of transcription, while repressors inhibit transcription by blocking the binding of RNA polymerase or by recruiting co-repressors that modify chromatin structure.

The binding of transcription factors is often influenced by signaling pathways that respond to various stimuli, such as hormones, nutrients, and environmental changes. For example, steroid hormones can enter the cell and bind to specific nuclear receptors, which then act as transcription factors to regulate the expression of target genes involved in processes such as metabolism and growth.

31.2.2 Enhancers and Silencers

Enhancers are regulatory DNA elements that can be located far from the promoter of the gene they regulate. They contain binding sites for transcription factors and can significantly increase the likelihood of transcription initiation when activated. Enhancers achieve this by looping the DNA to bring transcription factors and the transcriptional machinery into close proximity to the promoter.

Conversely, silencers are DNA elements that inhibit gene expression. They can function in a similar manner to enhancers but work to reduce the recruitment of RNA polymerase or the assembly of the transcription machinery. The interplay between enhancers and silencers allows for fine-tuning of gene expression, enabling cells to respond dynamically to changing conditions.

31.2.3 Chromatin Structure

The structure of chromatin, the material that makes up chromosomes, also plays a crucial role in gene regulation. DNA is wrapped around histone proteins, forming nucleosomes that create a compact structure. The accessibility of DNA for transcription depends on the chromatin's state—whether it is in a relaxed (euchromatin) or condensed (heterochromatin) form.

Chemical modifications to histones, such as acetylation, methylation, and phosphorylation, can alter chromatin structure and influence gene expression. For instance, histone acetylation generally promotes a more open chromatin configuration, facilitating transcription, while methylation can either enhance or repress transcription, depending on the specific context.

31.3 POST-TRANSCRIPTIONAL REGULATION

Once mRNA is synthesized, it undergoes several processing steps before it can be translated into protein. Post-transcriptional regulation refers to the control mechanisms that affect mRNA stability, splicing, transport, and translation efficiency.

31.3.1 RNA Splicing

In eukaryotic cells, the primary mRNA transcript (pre-mRNA) contains both coding regions (exons) and non-coding regions (introns). Splicing is the process by which introns are removed, and exons are joined together to form a mature mRNA molecule. This process is carried out by a complex known as the spliceosome, which consists of small nuclear RNAs (snRNAs) and protein factors.

Alternative splicing allows a single gene to produce multiple mRNA isoforms by including or excluding specific exons. This mechanism greatly increases the diversity of proteins that can be generated from a single gene, enabling cells to adapt to different functional requirements.

31.3.2 mRNA Stability and Degradation

The stability of mRNA molecules is a key factor in determining gene expression levels. mRNA molecules with longer half-lives can be translated more times, leading to increased protein production. Various sequences within the mRNA, particularly in the 3' untranslated region (3' UTR), can influence its stability. These regions often contain binding sites for RNA-binding proteins that can either stabilize the mRNA or promote its degradation.

mRNA degradation is an essential regulatory mechanism that allows cells to rapidly respond to changes in gene expression. Processes such as RNA interference (RNAi) can target specific mRNAs for degradation, effectively silencing gene expression and modulating protein levels.

31.3.3 Nuclear Export

After processing, mature mRNA must be exported from the nucleus to the cytoplasm for translation. This export process is highly regulated and involves interactions between mRNA and transport proteins. Proper nuclear export is essential for ensuring that only fully processed and functional mRNAs are translated into proteins.

31.4 TRANSLATIONAL REGULATION

Translational regulation controls the efficiency and rate at which mRNA is translated into protein. This regulation can occur at multiple levels, including initiation, elongation, and termination of translation.

31.4.1 Initiation of Translation

The initiation phase of translation is a critical regulatory step. Various initiation factors bind to the mRNA's 5' cap and facilitate the assembly of the ribosome at the start codon. The availability of initiation factors, as well as the presence of specific sequences in the mRNA, can influence the efficiency of translation initiation.

For instance, upstream open reading frames (uORFs) in the mRNA can regulate the translation of downstream coding sequences by affecting ribosome scanning and initiation. Additionally, mRNA secondary structures can impact ribosome access to the start codon, further influencing translation efficiency.

31.4.2 Regulatory Peptides and Small RNAs

Certain peptides can also regulate translation by interacting with the ribosome or translation factors. Moreover, small non-coding RNAs, such as microRNAs (miRNAs) and small interfering RNAs (siRNAs), play essential roles in post-transcriptional regulation by binding to complementary sequences in target mRNAs, leading to translational repression or degradation.

31.5 POST-TRANSLATIONAL MODIFICATIONS

After translation, proteins may undergo various post-translational modifications (PTMs) that can affect their stability, localization, and activity. These modifications include phosphorylation, glycosylation, acetylation, and ubiquitination, among others.

31.5.1 Phosphorylation

Phosphorylation is one of the most common PTMs and can regulate protein function by inducing conformational changes. The addition of phosphate groups can activate or deactivate enzymes, alter protein-protein interactions, and impact cellular localization. This regulatory mechanism plays a crucial role in signaling pathways, cell cycle progression, and responses to stress.

31.5.2 Ubiquitination

Ubiquitination involves the attachment of ubiquitin molecules to a target protein, marking it for degradation by the proteasome. This process is essential for regulating protein turnover and maintaining cellular homeostasis. Ubiquitination can also affect protein activity and localization, highlighting its importance in controlling gene expression.

31.6 CONCLUSION

The control of gene expression is a complex and dynamic process that involves multiple layers of regulation, from transcription to post-translational modifications. This regulation ensures that genes are expressed at the appropriate times and levels, allowing cells to respond to internal and external signals effectively. The intricate interplay between transcription factors, enhancers, silencers, and chromatin structure highlights the sophistication of gene regulatory networks. Furthermore, post-transcriptional and translational regulation adds additional layers of control, ensuring that proteins are synthesized accurately and efficiently.

Understanding the mechanisms underlying gene expression control is crucial for advancements in fields such as medicine, biotechnology, and genetics. As research continues to unveil the complexities of gene regulation, new therapeutic strategies may emerge, harnessing the power of gene expression to address various diseases and disorders. Ultimately, the regulation of gene expression is fundamental to the functioning of life, underpinning the processes that allow organisms to grow, develop, and adapt to their environments.

CHAPTER 32

PRINCIPLES OF DRUG DISCOVERY AND DEVELOPMENT

The journey from a novel idea for a drug to an approved pharmaceutical product is a complex and lengthy process, encompassing multiple stages of research, development, and testing. This process, known as drug discovery and development, involves a multidisciplinary approach integrating biology, chemistry, pharmacology, and regulatory science. Understanding the principles of drug discovery and development is essential for researchers, healthcare professionals, and policymakers involved in the pharmaceutical industry.

32.1 TARGET IDENTIFICATION AND VALIDATION

The first step in drug discovery is identifying a biological target associated with a disease. Targets can be proteins, nucleic acids, or other biomolecules that play crucial roles in the disease's pathology. The process begins with extensive research to understand the molecular mechanisms of the disease and to identify potential targets that can be modulated by a drug.

Once a target is identified, it must undergo validation to establish its relevance in the disease process. This can involve various experimental techniques, such as genetic studies, biochemical assays, and animal models, to demonstrate that modulating the target has a therapeutic effect. Validation is critical, as targeting irrelevant or non-essential proteins may lead to unsuccessful drug development.

32.2 LEAD DISCOVERY

After target validation, the next phase involves discovering lead compounds that can interact with the identified target. Lead discovery typically employs high-throughput screening (HTS), where large libraries of compounds are tested for their ability to bind to or modulate the target's activity. This process can generate thousands of potential leads, but not all compounds are suitable for further development.

The selection of promising lead compounds often requires additional studies to assess their potency, selectivity, and pharmacokinetic properties—how a drug is absorbed, distributed, metabolized, and excreted in the body (ADME). Structure-activity relationship (SAR) studies are also conducted to understand how modifications to the chemical structure of lead compounds affect their biological activity, guiding the optimization of drug candidates.

32.3 PRECLINICAL DEVELOPMENT

Once lead compounds are identified and optimized, they progress to preclinical development, which involves testing in vitro (in test tubes) and in vivo (in living organisms) to evaluate their safety and efficacy. Preclinical studies aim to gather data on pharmacodynamics (the effects of the drug on the body) and toxicology (the potential harmful effects of the drug).

In vitro studies involve assessing the lead compounds' effects on cultured cells, measuring their ability to inhibit or activate the target, and determining their cytotoxicity. These studies help to establish initial safety profiles and determine the therapeutic index—the ratio of the drug's toxic dose to its effective dose.

In vivo studies are conducted using animal models to assess the drug's efficacy in a living system and to evaluate its pharmacokinetic properties. These studies also provide crucial information on dosage, administration routes, and potential side effects. The data obtained from preclinical studies are essential for justifying the transition to clinical trials.

32.4 CLINICAL DEVELOPMENT

Clinical development involves testing the drug in human subjects and is conducted in several phases, each with specific objectives and increasing numbers of participants.

32.4.1 Phase I Trials

Phase I trials are the first studies conducted in humans, typically involving a small group of healthy volunteers. The primary objectives of Phase I trials are to assess the safety, tolerability, and pharmacokinetics of the drug. Researchers carefully monitor participants for adverse effects and determine the maximum tolerated dose (MTD) to inform dosing for subsequent trials.

32.4.2 Phase II Trials

Phase II trials involve a larger group of patients who have the disease targeted by the drug. The primary goal of this phase is to evaluate the drug's efficacy and further assess its safety. These trials often include different dose levels to identify the optimal dosing regimen. Phase II trials provide crucial data on the drug's therapeutic potential and help refine the patient population for Phase III studies.

32.4.3 Phase III Trials

Phase III trials are large-scale studies conducted in diverse patient populations. The primary objective of Phase III trials is to confirm the drug's efficacy and safety compared to standard treatments or placebos. These trials are often randomized and controlled, providing robust evidence to support the drug's therapeutic benefits.

Successful completion of Phase III trials leads to the submission of a New Drug Application (NDA) to regulatory authorities, such as the U.S. Food and Drug Administration (FDA) or the European Medicines Agency (EMA). The NDA includes comprehensive data from all clinical trials, manufacturing information, and proposed labeling.

32.5 REGULATORY CONSIDERATIONS

Regulatory agencies play a critical role in drug development, ensuring that new drugs are safe and effective for public use. The drug development process is heavily regulated, and companies must adhere to strict guidelines and protocols established by these agencies.

Before conducting clinical trials, sponsors must obtain Investigational New Drug (IND) status, which allows them to test the drug in humans. Regulatory authorities review the data from preclinical studies and the proposed clinical trial design to ensure patient safety and scientific validity.

Once the clinical trials are completed, the NDA submission is thoroughly reviewed by regulatory agencies. The review process can take several months to years, during which agencies evaluate the safety and efficacy data, assess manufacturing practices, and consider the proposed labeling and marketing plans. If the NDA is approved, the drug can be marketed and made available to patients.

32.6 POST-MARKETING SURVEILLANCE

After a drug is approved, post-marketing surveillance, also known as Phase IV trials, is conducted to monitor the drug's long-term safety and effectiveness in the general population. This phase is crucial for identifying rare adverse events or long-term effects that may not have been apparent during clinical trials. Regulatory agencies require ongoing reporting from manufacturers regarding any adverse effects observed in patients using the drug.

Pharmacovigilance programs are established to collect and analyze data on drug safety post-approval. These programs aim to detect, assess, understand, and prevent adverse effects or any other drug-related problems. If significant safety concerns arise, regulatory agencies may issue warnings, recommend changes to labeling, or, in some cases, withdraw the drug from the market.

32.7 COMMERCIALIZATION OF DRUGS

The final stage in drug development involves commercialization, where the approved drug is marketed and sold to healthcare providers and patients. This process requires strategic planning and execution, including pricing, marketing strategies, and distribution channels.

Pharmaceutical companies must navigate a competitive landscape, differentiating their product from existing therapies and demonstrating its value to healthcare providers and patients. Effective communication of clinical benefits, safety profiles, and cost-effectiveness is essential for gaining acceptance in the market.

Additionally, companies must engage with healthcare professionals, payers, and regulatory bodies to ensure reimbursement and access to their products. Understanding the healthcare ecosystem and addressing the needs of various stakeholders is critical for the successful commercialization of a new drug.

32.8 CONCLUSION

The principles of drug discovery and development encompass a multifaceted approach that involves target identification, lead discovery, preclinical and clinical development, regulatory considerations, and commercialization. Each stage of this process requires rigorous scientific investigation and adherence to regulatory guidelines to ensure the safety and efficacy of new drugs.

The complexities of drug discovery and development underscore the importance of collaboration among scientists, clinicians, and regulatory agencies. As the field continues to evolve with advances in technology, personalized medicine, and novel therapeutic modalities, understanding these principles becomes increasingly vital for addressing the healthcare challenges of the future. Effective drug discovery and development processes not only lead to the creation of life-saving medications but also contribute to the overall advancement of medical science and public health.

BIBLIOGRAPHY

Advanced Biochemistry: A Comprehensive Guide Author: Alfred L. B. Marz et al. Publisher: CRC Press Year: 2016

Biochemical Analysis: A Practical Guide to Techniques Author: M. L. G. de Oliveira et al. Publisher: Springer Year: 2017

Biochemical Pathways: An Atlas of Biochemistry and Molecular Biology Author: Gerhard Michal et al. Publisher: Wiley-VCH Year: 2011

Biochemistry and Molecular Biology Author: David T. Plummer Publisher: McGraw-Hill Year: 2012

Biochemistry Author: Donald Voet and Judith G. Voet Publisher: Wiley Year: 2016

Biochemistry Author: Gary G. Hammes Publisher: Wiley Year: 2014

Biochemistry Author: Jeremy M. Berg, John L. Tymoczko, and Lubert Stryer Publisher: W. H. Freeman Year: 2015

Biochemistry Author: Mary K. Campbell and Shawn O. Farrell Publisher: Cengage Learning Year: 2016

Biochemistry Author: Robert K. Scopes Publisher: Springer Year: 2013

Biochemistry for Dummies Author: John T. Moore and Richard H. Langley Publisher: Wiley Year: 2014

Biochemistry of Endocrinology Author: Allen L. McMillan Publisher: Elsevier Year: 2017

Biochemistry of Lipids: Metabolism of Lipids Author: J. de Jong et al. Publisher: Springer Year: 2018

Biochemistry of Neurotransmission Author: David. Hughes Publisher: Academic Press Year: 2018

Biochemistry of the Nucleic Acids Author: Roberts Publisher: Wiley Year: 2015

Biochemistry: A Short Course Author: John L. Tymoczko et al. Publisher: W. H. Freeman Year: 2018

Biochemistry: A Student Companion Author: Judith A. T. McGarry Publisher: Wiley Year: 2015

Biochemistry: A Very Short Introduction Author: Mark R. P. Wright Publisher: Oxford University Press Year: 2016

Biochemistry: An Illustrated History Author: John D. Robson Publisher: Wiley Year: 2014

Biochemistry: Concepts and Connections Author: Kim S. F. Lee et al. Publisher: Pearson Year: 2017

Biochemistry: The Interface between Chemistry and Biology Author: Phillips Publisher: Cambridge University Press Year: 2018

Biochemistry: The Molecular Basis of Life Author: Trudy McKee and James R. McKee Publisher: Oxford University Press Year: 2018

Bioenergetics Author: David G. Nicholls and Stuart J. Ferguson Publisher: Academic Press Year: 2013

Clinical Biochemistry Author: Nader Rifai et al. Publisher: Elsevier Year: 2018

Clinical Biochemistry: An Illustrated Colour Text Author: Kelly Publisher: Elsevier Year: 2016

Enzyme Kinetics: Principles and Methods Author: Robert A. Copeland Publisher: Wiley Year: 2013

Essentials of Biochemistry Author: Arthur M. Lesk Publisher: Oxford University Press Year: 2014

Fundamentals of Biochemistry Author: J. W. P. William et al. Publisher: Cengage Learning Year: 2015

Fundamentals of Biochemistry: Life at the Molecular Level Author: Donald Voet et al. Publisher: Wiley Year: 2016

Introduction to Biochemistry Author: Robert G. Mortimer Publisher: Oxford University Press Year: 2016

Molecular Biochemistry: Principles and Practice Author: Brian W. Mattson Publisher: Academic Press Year: 2015

Molecular Biology of the Cell Author: Bruce Alberts et al. Publisher: Garland Science Year: 2014

Nucleic Acids in Chemistry and Biology Author: AK. Sewa Tanaka Publisher: Royal Society of Chemistry Year: 2015

Principles of Biochemistry Author: A M. Frizz. Scopes Publisher: Springer Year: 2014

Principles of Biochemistry Author: David L. Nelson and Michael M. Cox Publisher: W. H. Freeman Year: 2017

Principles of Biochemistry Author: H. Robert Horton et al. Publisher: Pearson Year: 2018

Principles of Biochemistry with a Human Focus Author: David C. H. S. Leong Publisher: Wiley Year: 2016

Protein Biochemistry Author: Gregory M. S. K. Davis Publisher: Wiley Year: 2015

INDEX

A

Affinity 10, 13

Allosteric Regulation 27, 175, 252

Alpha Helices 6

Amino Acids 5, 119, 240, 241, 246, 249

B

Beta Sheets 6

Binding 9, 10, 11, 12, 100, 101, 132

Biochemical Gradients 1

Building Blocks 5

C

Catalysis 10, 17, 18, 23, 24, 25

Cellular Timers 1

Chromatography 13

Circadian Rhythms 1

Circular Dichroism 14

Covalent Modification 27

Cryo-Electron 15

crystallography 15, 16, 284

D

Distortion 18

E

Electrostatic 25

Enzymes 1, 10, 17, 18, 20, 23, 26, 27, 79, 94, 117, 134, 149, 167, 185, 191, 205, 211, 262, 264, 285, 297, 312

Extreme Environments 2

F

Fine-Tuning 27

Fluorescence 14

G

Gluconeogenesis 125, 128, 142, 148, 149, 150, 155, 156, 157, 158, 159, 160, 161, 162, 163, 177, 179, 202, 216, 219, 224, 234, 240, 241, 243, 244, 245, 246

Glycoproteins 49, 50, 57, 58, 59, 60, 61, 62, 63, 64, 79

Gonadotropins 60

Gradient 63, 83, 84, 85, 125, 126, 154, 168, 181, 183, 185, 186, 187, 189

Greenhouse 199

H

Human Focus 355

Hydrogen Bonds 9

Hydrophobic Interactions 10

I

Inhibitors 20, 195
Intermediate Stabilization 24
Ionic Interactions 9

J

Junowicz 140

K

Kinetics 17, 18, 354

L

Large Protein 15

M

Mass Spectrometry 16
Metal Ion catalysis 25
Michaelis-Menten 19, 21, 27
Microenvironments 1
Molecular 9, 10, 11, 194, 330, 353, 354

N

Nuclear Magnetic Resonance 15
Nucleic Acids 36, 353, 355
Nucleophilic Attack 24

O

Orientation Effects 18, 23

P

Peptide Bonds 5
Protein misfolding 7
Proximity 18, 23

Q

Quaternary Structure 6

R

Receptors 11, 79, 94, 97, 111
Reversible Chemical Changes 27

S

Secondary Structure 6
Signal Transduction 11, 90, 91, 115, 116
Size-Exclusion 13
Space-Time 3
Spatial Dimension 1
Spectroscopy 14, 15
Stabilization 24, 25
Strain 18
Substrate 10, 19

T

Tertiary Structure 6, 7, 99, 102, 308
Therapeutic Development 11
Time Scales 2

U

UV-Visible 14

V

Vertebrate Cells 303
Viral Interactions 309

W

Water Molecules 25

X

X-ray Crystallography 15

Y

Yeast 64, 93, 95, 118, 135, 139, 152, 255, 293, 300, 302, 303, 319, 331, 333, 334

Z

Zinc Ion 24, 25

Zona Pellucida 60

Zymogen 28

www.ingramcontent.com/pod-product-compliance
Lightning Source LLC
LaVergne TN
LVHW062112200926
848484LV00007B/1648